Plant Genetics and Genomics: Crops and Models

Volume 18

Series Editor
Richard A. Jorgensen

More information about this series at http://www.springer.com/series/7397

John P. Vogel
Editor

Genetics and Genomics of Brachypodium

Springer

Editor
John P. Vogel
DOE Joint Genome Institute
Walnut Creek, CA, USA

ISSN 2363-9601 ISSN 2363-961X (electronic)
Plant Genetics and Genomics: Crops and Models
ISBN 978-3-319-26942-9 ISBN 978-3-319-26944-3 (eBook)
DOI 10.1007/978-3-319-26944-3

Library of Congress Control Number: 2016930291

Springer Cham Heidelberg New York Dordrecht London

Printed on acid-free paper

Springer International Publishing AG Switzerland is part of Springer Science+Business Media (www.springer.com)

Preface

Grasses dominate many natural ecosystems and produce most of the calories consumed by humans either directly in the form of grains or indirectly through forage/grain fed animals. In addition, grasses grown as biomass crops are poised to become a significant source of renewable energy. Thus, the enormous economic and environmental importance of grasses is undeniable. Despite this, research into grass biology has been hampered by the lack of a truly tractable experimental model system. While much valuable research has been conducted in crops like maize, rice, wheat, and barley, none of these grasses possess the full suite of traits necessary for a truly tractable model system (e.g., small size, rapid generation time, small diploid genome, self-fertile, simple growth requirements, and easily transformed). The power of applying a model system approach to plant biology has been amply demonstrated by the tremendous advances achieved using *Arabidopsis thaliana* as a model organism. Unfortunately, *A. thaliana* is a eudicot and about 150–200 million years of evolution separates it from the grasses. Thus, there are many biological traits unique to the grasses for which *A. thaliana* is an unsuitable model. Furthermore, even for shared traits the specific genes and regulatory networks often differ considerable between *A. thaliana* and the grasses. Thus, there is a pressing need for a grass model system. Over the past decade, *Brachypodium distachyon* has emerged to fill this void.

This book describes the current state of *B. distachyon* research tools and how they have been applied to a wide range of topics. Specific chapters describe the development of key resources and techniques including a high-quality reference genome sequence, the development of high-efficiency transformation methods, the creation of a large collection of insertional mutants, the assembly of extensive germplasm collections, genome sequencing of natural accessions, the optimization of efficient crossing techniques, and the creation of recombinant inbred lines. Chapters focused on the application of these tools cover a wide range of topics including cell wall biosynthesis, seed development including starch and storage protein biosynthesis, microRNA and small RNA biology, cytogenetics, adaptation to local environments, abiotic stress, plant–pathogen interactions, root biology, and flowering time. In addition, recent advances using closely related *Brachypodium* species as models for perenniality and polyploidy are described.

That the current state of *Brachypodium* research can fill a book is a testament to the maturity of the system. Additional evidence in this regard includes the enormous number of seed orders distributed and the rapidly rising number of publications using *B. distachyon* as an experimental model. In comparison to other model systems, this maturity has come at an exceptional pace. In a little over a decade *B. distachyon* has gone from a curiosity to a full-blown model organism used by over 350 laboratories around the world. This explosive growth followed a new paradigm in which a genome sequencing project was initiated early on in the development of a model system. The genome project catalyzed both the rapid development of experimental tools and the adoption of the system by many researchers. While this "sequence it and they will come" strategy helps explain the rapid rise of *B. distachyon* as a model for the grasses, another crucial factor was the conscious effort by a handful of early *B. distachyon* researchers to freely share material and information long before publication. This fostered the development of a collaborative and open community ethos that is the fertile ground necessary to grow a model system. As *B. distachyon* emerges from adolescence, its future is indeed bright.

Walnut Creek, CA, USA John P. Vogel

Contents

Contributors

Richard M. Amasino Department of Biochemistry, University of Wisconsin, Madison, WI, USA

United States Department of Energy Great Lakes Bioenergy Research Center, University of Wisconsin-Madison, Madison, WI, USA

Sébastien Antelme Institut Jean-Pierre Bourgin, INRA, AgroParisTech, CNRS, Université Paris-Saclay, Versailles Cedex, France

Alexander Betekhtin Faculty of Biology and Environmental Protection, Department of Plant Anatomy and Cytology, University of Silesia in Katowice, Katowice, Poland

Justin Borevitz Centre of Excellence in Plant Energy Biology, Australian National University, Canberra, ACT, Australia

Natalia Borowska-Zuchowska Faculty of Biology and Environmental Protection, Department of Plant Anatomy and Cytology, University of Silesia in Katowice, Katowice, Poland

Agnieszka Braszewska-Zalewska Faculty of Biology and Environmental Protection, Department of Plant Anatomy and Cytology, University of Silesia in Katowice, Katowice, Poland

Pilar Catalan Department of Agricultural and Environmental Sciences, High Polytechnic School of Huesca, University of Zaragoza, Huesca, Spain

Department of Botany, Institute of Biology, Tomsk State University, Tomsk, Russia

Joshua H. Coomey Plant Biology Graduate Program, Biology Department, University of Massachusetts, Amherst, MA, USA

Antonio Díaz-Pérez Department of Agricultural and Environmental Sciences, High Polytechnic School of Huesca, University of Zaragoza, Huesca, Spain

Department of Genetics, School of Agronomy, Universidad Central de Venezuela, Maracay, Venezuela

Karl R. Franke Delaware Biotechnology Institute, University of Delaware, Newark, DE, USA

Department of Biological Sciences, University of Delaware, Newark, DE, USA

David F. Garvin USDA-ARS Plant Science Research Unit, University of Minnesota, St. Paul, MN, USA

Sean P. Gordon DOE, Joint Genome Institute, Walnut Creek, CA, USA

Fabienne Granier Institut Jean-Pierre Bourgin, INRA, AgroParisTech, CNRS, Université Paris-Saclay, Versailles Cedex, France

Pamela J. Green Delaware Biotechnology Institute, University of Delaware, Newark, DE, USA

Department of Plant and Soil Sciences, University of Delaware, Newark, DE, USA

Yong Gu USDA-ARS, Crop Improvement and Genetics Research Unit, Western Regional Research Center, Albany, CA, USA

Christian S. Hardtke Department of Plant Molecular Biology, University of Lausanne, Lausanne, Switzerland

Robert Hasterok Faculty of Biology and Environmental Protection, Department of Plant Anatomy and Cytology, University of Silesia in Katowice, Katowice, Poland

Samuel P. Hazen Plant Biology Graduate Program, Biology Department, University of Massachusetts, Amherst, MA, USA

Mon Mandy Hsia USDA ARS Western Regional Research Center, Albany, CA, USA

Dominika Idziak-Helmcke Faculty of Biology and Environmental Protection, Department of Plant Anatomy and Cytology, University of Silesia in Katowice, Katowice, Poland

Thomas E. Juenger Department of Integrative Biology and Institute for Cellular and Molecular Biology, The University of Texas at Austin, Austin, TX, USA

Debbie Laudencia-Chingcuanco USDA-ARS, Western Regional Research Center, Albany, CA, USA

Philippe LeBris Institut Jean-Pierre Bourgin, INRA, AgroParisTech, CNRS, Université Paris-Saclay, Versailles Cedex, France

Aurélie Lemaire Institut Jean-Pierre Bourgin, INRA, AgroParisTech, CNRS, Université Paris-Saclay, Versailles Cedex, France

Lifeng Liu DOE, Joint Genome Institute, Walnut Creek, CA, USA

USDA-Agriculture Research Service, Western Regional Research Center, Albany, CA, USA

Diana López-Álvarez Department of Agricultural and Environmental Sciences, High Polytechnic School of Huesca, University of Zaragoza, Huesca, Spain

María Luisa López-Herránz Department of Agricultural and Environmental Sciences, High Polytechnic School of Huesca, University of Zaragoza, Huesca, Spain

Christopher W.P. Lyons Department of Plant Pathology and Microbiology, Texas A&M University, College Station, TX, USA

David L. Des Marais Department of Organismic and Evolutionary Biology and The Arnold Arboretum, Harvard University, Cambridge, MA, USA

Valentina Mazzamurro Department of Life Sciences, University of Modena and Reggio Emilia, Reggio Emilia, Italy

Keiichi Mochida Cellulose Production Research Team, Biomass Engineering Research Division, RIKEN Center for Sustainable Resource Science, Yokohama, Kanagawa, Japan

Kihara Institute for Biological Research, Yokohama City University, Yokohama, Kanagawa, Japan

Todd C. Mockler Donald Danforth Plant Science Center, St. Louis, MO, USA

Luis A.J. Mur Institute of Environmental and Rural Science, Aberystwyth University, Aberystwyth, UK

David Pacheco-Villalobos Department of Plant Molecular Biology, University of Lausanne, Lausanne, Switzerland

Nicola Pecchioni CRA-CER Cereal Research Centre, CRA—Council for Agricultural Research and Economic Analysis, Foggia, Italy

Ewa Robaszkiewicz Faculty of Biology and Environmental Protection, Department, of Plant Anatomy and Cytology, University of Silesia in Katowice, Katowice, Poland

Jérôme Salse INRA/UBP UMR 1095 GDEC 'Génétique, Diversité et Ecophysiologie des Céréales', Institut National de la Recherche Agronomique (INRA), Clermont Ferrand, France

Rubén Sancho Department of Agricultural and Environmental Sciences, High Polytechnic School of Huesca, University of Zaragoza, Huesca, Spain

Karen-Beth G. Scholthof Department of Plant Pathology and Microbiology, Texas A&M University, College Station, TX, USA

Alan H. Schulman Institute of Biotechnology, University of Helsinki, Helsinki, Finland

Green Technology, Luke Natural Resources Institute, Helsinki, Finland

Kazuo Shinozaki Gene Discovery Research Group, RIKEN Center for Sustainable Resource Science, Yokohama, Kanagawa, Japan

Biomass Research Platform Team, Biomass Engineering Research Division, RIKEN Center for Sustainable Resource Science, Tsukuba, Ibaraki, Japan

Richard Sibout Institut Jean-Pierre Bourgin, INRA, AgroParisTech, CNRS, Université Paris-Saclay, Versailles Cedex, France

Jared Streich Centre of Excellence in Plant Energy Biology, Australian National University, Canberra, ACT, Australia

Roger Thilmony USDA-ARS, Crop Improvement and Genetics Research Unit, Western Regional Research Center, Albany, CA, USA

Robert VanBuren Donald Danforth Plant Science Center, St. Louis, MO, USA

John P. Vogel DOE Joint Genome Institute, Walnut Creek, CA, USA

USDA-Agriculture Research Service, Western Regional Research Center, Albany, CA, USA

Yin Wang Institut Jean-Pierre Bourgin, INRA, AgroParisTech, CNRS, Université Paris-Saclay, Versailles Cedex, France

Pip Wilson Centre of Excellence in Plant Energy Biology, Australian National University, Canberra, ACT, Australia

Elzbieta Wolny Faculty of Biology and Environmental Protection, Department of Plant Anatomy and Cytology, University of Silesia in Katowice, Katowice, Poland

Daniel P. Woods Laboratory of Genetics, University of Wisconsin, Madison, WI, USA

United States Department of Energy Great Lakes Bioenergy Research Center, University of Wisconsin-Madison, Madison, WI, USA

Jiajie Wu State Key Laboratory of Crop Biology, Shandong Key Laboratory of Crop Biology, Shandong Agricultural University, Taían, China

The Rise of *Brachypodium* as a Model System

John P. Vogel

Abstract Grasses have played a central role in the formation of human civilization. Indeed, the inception of agriculture and the production of grains that could be easily stored and traded is one of the major factors that led to the creation of cities and the specialization of labor. Today, grasses provide the bulk of calories consumed by humans either directly through the consumption of grain or indirectly through grain and grass fed animals. Furthermore, due to their high productivity, grasses are increasingly utilized as a source of renewable biomass for the sustainable production of bioenergy and liquid biofuels. Grasses also play a fundamental role in many terrestrial ecosystems that benefit humans in numerous ways. Given the importance of grasses to humanity, there is considerable value in understanding their biology in great detail. Model biological systems greatly facilitate scientific research and many of the rapid advances in molecular biology and genetics would have been difficult to achieve without them. The model plant *Arabidopsis thaliana* has been used to make tremendous gains in our understanding of plant biology. However, as a eudicot, *A. thaliana* is unsuitable to study the unique aspects of grass biology. Several crop grasses (e.g. maize and rice) have been used as model systems and while each has certain strengths, they have some disadvantages when compared to a model like *A. thaliana*. *Brachypodium distachyon* has emerged to fill the need for a truly tractable model grass that is compatible with modern high-throughput molecular-genetic experiments. An overview of the development and widespread adoption of *B. distachyon* as a model grass is presented.

Keywords Grass • Bioenergy • Genome sequence • *Brachypodium* • Model system

Importance of Grasses and the Need for a Model System

It is difficult to overstate the importance of grasses to humanity. They were a source of nutrition even before the advent of agriculture. The first known grinding of grass seed to produce flour occurred 30,000 years ago, long before the dawn of agriculture ~10,000 years ago (Aranguren et al. 2007; Revedin et al. 2010). Curiously, while no

J.P. Vogel (✉)
DOE Joint Genome Institute, Walnut Creek, CA
e-mail: jpvogel@lbl.gov

J. Vogel (ed.), *Genetics and Genomics of Brachypodium*,
Plant Genetics and Genomics: Crops Models, DOI 10.1007/7397_2015_14
Published online: 15 October 2015

Brachypodium species are widely used as human food today, it was a species of *Brachypodium* that was ground for food ~30,000 years ago (Aranguren et al. 2007; Revedin et al. 2010). The advent of agriculture and the domestication of wheat and other grains ~10,000 years ago had certain advantages over hunting and gathering. The predictable production, high yields and storability of grains allowed larger numbers of humans to live in a single location. This ultimately led to the formation of cities and modern civilization. Grasses are still essential for modern civilization. Indeed, the majority of calories consumed by humans come directly or indirectly from grasses. In 2013, the four largest agricultural crops in terms of tonnage produced were all grasses (sugar cane, maize, rice, wheat) and these crops together had a higher production than the next 40 crops combined (http://www.fao.org).

Due to increasing human population and a rising standard of living in the developing world it is estimated that grain yields must increase by 70–100 % by 2050 to keep pace with demand (Editorial 2010; Godfray et al. 2011). Further complicating matters, these increases must occur despite the increasingly volatile and uncertain weather patterns associated with global climate change (Godfray et al. 2011; Wheeler and Von Braun 2013). Unfortunately, the current rate of yield increase for the major grains are not sufficient to meet projected needs (Godfray et al. 2010). As an example, in the 1990s, despite enormous investments, corn yield only increased 0.78 % annually (Kucharik and Ramankutty 2005). Thus, it will be necessary to use multiple strategies to accelerate yield increases if farmers are to meet demand (Fedoroff et al. 2010). Knowledge gained from model plants can be used to design rationale strategies for crop improvement and accelerate yield gains.

In addition to driving demand for food, increasing population and rising living standards are also driving an increase in demand for energy and transportation fuel. It is estimated that by 2024 global energy needs will increase 37 % (Development 2014). To minimize the effects of climate change, it is essential that sustainable energy sources be increasingly utilized and it is projected that biofuel production will increase 300 % by 2024 (Development 2014). The conversion of biomass derived from crop residues and dedicated energy crops into liquid transportation fuels may be a particularly important source of sustainable energy because electricity from wind or solar cannot fully substitute for gasoline and diesel with current battery technology. Due to their high productivity and ability to grow on marginal lands perennial grasses are projected to be extensively used as feedstocks for the production of biofuels (DOE 2006; Carroll and Somerville 2009). Unfortunately, the leading candidates for biomass crops such as switchgrass and *Miscanthus* are difficult to breed due to their perenniality, complex genetics, long generation time and polyploid genomes. Not surprisingly, these grasses are not fully domesticated. Indeed, current cultivars are essentially wild selections. Thus, knowledge gained from model systems may be especially useful for accelerating the domestication of these emerging crops (DOE 2006). In light of the need for fundamental knowledge to improve both food and biomass crops, it would be highly desirable to have a truly tractable model grass.

Limitations of Other Models

Several cereal crops have served as model grasses over the years, but none possess the full suite of model organism traits: small size, rapid generation time, easily grown, self-fertile, diploid, small genome size, easily transformed (Brutnell et al. 2015). Maize and rice in particular have been widely utilized for a broad array of experiments. In some respects, maize is an ideal genetic model due to the ease with which crosses can be made, the huge amount of natural diversity and highly active endogenous transposons. However, stature and generation time present difficulties for researchers outside of large institutions with significant field programs and winter nurseries. Large size also makes growing large numbers of maize plants under controlled conditions prohibitively expensive. In addition, the complex maize genome and difficulties with transformation are impediments to many molecular genetic studies. Like maize, there is a large rice research community and significant experimental resources. Though much smaller than maize, its size and demanding growth conditions remain significant obstacles for researchers outside of large institutions in rice growing regions. In addition, the economic importance of rice has resulted in restrictive intellectual property rights and import quarantine restrictions that limit the free sharing of germplasm and other resources. Finally, as a semi-aquatic tropical grass, rice is not an ideal model for temperate grains, forage, and biomass crops.

In contrast to maize and rice, *A. thaliana* is an ideal model plant. Its small size, simple growth requirements and rapid generation time facilitate the growth of large numbers of plants under controlled conditions. Further increasing its values as a model system, it has one of the smallest genomes of any plant and is extremely easy to transform (Clough and Bent 1998; Kaul et al. 2000). Thus, it is no surprise that a large, vibrant, and open community of *A. thaliana* researchers has emerged (Meyerowitz 2001). Together, this community created a wealth of easily accessible experimental resources that have further increased the power of *A. thaliana* as a model system (*https://www.arabidopsis.org/*). Unfortunately, as a eudicot *A. thaliana* is not an ideal model for the grasses. The aspects of grass biology not shared with *A. thaliana* (e.g. cell wall composition, development, grain production, mycorrhizal associations) simply cannot be studies in *A. thaliana*. Also, given the evolutionary separation of the eudicots and grasses, even conserved areas of plant biology are expected to have significant differences at the gene level that may limit the utility of *A. thaliana* as a model for the grasses (Hayama and Coupland 2004). Thus, it would be extremely useful to have a truly tractable model grass with the attributes that have made *A. thaliana* such a powerful model.

Early *Brachypodium* Research

Prior to 1995, most publications that mentioned *Brachypodium* species focused on ecology (e.g. Davis et al. 1985) while a few studies described the phylogenetic relationships in the genus, which was particularly interesting because of the high degree of variation in chromosome number and ploidy (e.g. Khan 1992; Shi et al. 1993). In 1995, Bablak et al. published a paper that explicitly proposed *B. distachyon* as model grass (Bablak et al. 1995). In this paper they described tissue culture conditions for *B. distachyon* as a first step toward developing a transformation system, a prerequisite for any modern model system. It was not until 2001 that a second paper from the same group presented a much more extensive case for using *B. distachyon* as a model system (Draper et al. 2001). This paper described biological traits that were compatible with its use as a model system including genome size, generation time, and mature size. In addition, they challenged several accessions with various pathogens and noted a range of responses indicating the potential for *B. distachyon* to serve as a model for plant-pathogen interactions. They also used biolistics to successfully transform a polyploid accession (now classified as a distinct species, *B. hybridum* (Catalán et al. 2012)). Taken together these results made a powerful case for the utility of *B. distachyon* as a model grass.

The first significant genomic resource created for the genus *Brachypodium* was a *B. sylvaticum* BAC library (Foote et al. 2004). In this case, the ultimate goal was to use the *B. sylvaticum* genome as a structural model for the much more complex wheat genome. Significantly, they noted that *B. sylvaticum* probes were more likely to work as markers for wheat than were rice probes, presumably due to greater sequence conservation between wheat and *B. sylvaticum* than between wheat and rice. The next publication developing *B. distachyon* as a model system came in 2006. This publication presented two key advances (Vogel et al. 2006). The first was a set of freely available inbred lines. These lines have since been distributed to hundreds of laboratories around the world. It is noteworthy that the lines used in the study by Draper et al. in 2001 were only available through an MTA that was too restrictive for many institutions (for an excellent review of the early days of *B. distachyon* research see Lyons and Scholthof 2015). The second advance in this paper was an *Agrobacterium*-mediated transformation protocol for the true *B. distachyon*. While the transformation efficiency was not high, it demonstrated that *B. distachyon* could be transformed using *Agrobacterium tumefaciens*, an absolute requirement for a truly tractable model system (see the chapter Phylogeny and evolution of the genus *Brachypodium*).

A New Paradigm for Model Systems

Up to this point, the development of *B. distachyon* as a model followed a traditional trajectory of slowly increasing use and gradual development of experimental resources. However, a confluence of factors led to an accelerated developmental path in which a genome sequencing project greatly stimulated the adoption of *B. distachyon* as a model system (reviewed in Lyons and Scholthof 2015). In 2005 The U.S. Department of Energy held a workshop to develop a research plan to accelerate the development of cellulosic biofuels. During this workshop it was decided that *B. distachyon* could serve as a tractable model for the large perennial

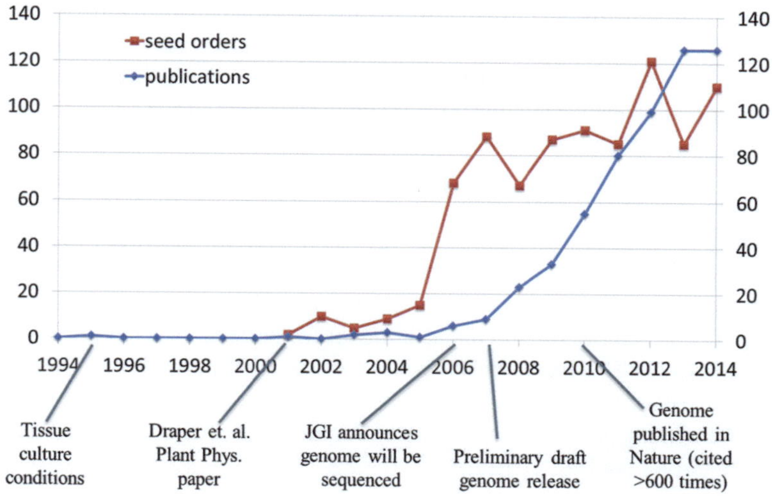

Fig. 1 A model takes hold. Seed distribution and publications are indicators of the size of the *B. distachyon* research community. Total seed orders distributed by the Vogel lab, the Garvin lab (prior to 2010 only) and the National Plant Germplasm System (prior to 2014 only) are plotted. The total number of seed packets shipped is 15,186. In addition, many secondary distributions are presumed to have occurred. Publications using *B. distachyon* as a molecular, genetic or physiological model system are plotted. The total number of publications is 564. Publications focused on ecology were not included. Key developments are noted below the graph

grasses being developed as biomass crops (e.g. switchgrass and *Miscanthus*). The report from this workshop outlined a plan for developing resources necessary to use *B. distachyon* as a model system (DOE 2006). Subsequent funding from the DOE led to the development of many resources for this nascent model system (Vogel and Hill 2008; IBI 2010; Bragg et al. 2012; Jeong et al. 2013; Gordon et al. 2014; Tyler et al. 2014). In addition to standard research grants, the DOE invested in *B. distachyon* research by sequencing the genome through the DOE Joint Genome Institute (IBI 2010). This represented a significant investment since it was based on Sanger sequencing alone. As soon as the genome sequencing project was announced, the demand for *B. distachyon* seed spiked upward and has remained high ever since (Fig. 1). Similarly, shortly after the first draft genome sequence was released the number of publications rose dramatically (Fig. 1).

Widespread Adoption and Outlook

Concurrent with the genome sequencing project, other essential resources were developed (reviewed in Brkljacic et al. 2011). In 2008, two methods for high-efficiency transformation were published (Vain et al. 2008; Vogel and Hill 2008). These methods and further refinements, made *B. distachyon* one of, if not the, easiest grass to transform (Bragg et al. 2015; Chapter Transformation and T-DNA mutagenesis). The next logical step was the creation of insertional mutant collections (Thole et al. 2010; Bragg et al. 2012; http://jgi.doe.gov/our-science/science-programs/plant-genomics/brachypodium/brachypodium-t-dna-collection/). These key resources further accelerated the adoption of *B. distachyon* as a model grass. The growth of the *B. distachyon* research community can be estimated by the number of seeds distributed and the number of publications. By both measures, the community is continuing to grow rapidly (Fig. 1). The resources available for *B. distachyon* are now being leveraged to use other species in the genus as models for polyploid genome evolution and regulation and for the study of perenniality (Chapter The genus *Brachypodium* as a model for perenniality and polyploidy).

The outlook for *B. distachyon* and sister species as model systems is very bright. A critical mass of researchers and resources has clearly been established. These resources are being used for an ever-expanding range of projects as described in the remainder of this book.

Acknowledgement I would like to thank Samuel Hazen for critically reading the manuscript. The work conducted by the U.S. Department of Energy Joint Genome Institute, a DOE Office of Science User Facility, is supported under Contract No. DE-AC02-05CH11231.

References

Aranguren B, Becattini R, Lippi MM, Revedin A. Grinding flour in Upper Palaeolithic Europe (25 000 years bp). Antiquity. 2007;81(314):845–55.

Bablak P, Draper J, Davey MR, Lynch PT. Plant regeneration and micropropagation of *Brachypodium distachyon*. Plant Cell Tiss Org Cult. 1995;42(1):97–107.

Bragg JN, Anderton A, Nieu R, Vogel JP. *Brachypodium distachyon*. Methods Mol Biol. 2015;1223:17–33.

Bragg JN, Wu J, Gordon SP, Guttman ME, Thilmony R, Lazo GR, et al. Generation and characterization of the Western Regional Research Center *Brachypodium* T-DNA insertional mutant collection. PLoS One. 2012;7(9):e41916.

Brkljacic J, Grotewold E, Scholl S, Mockler T, Garvin D, Vain P, et al. *Brachypodium* as a model for the grasses: today and the future. Plant Physiol. 2011;157:3–13.

Brutnell TP, Bennetzen JL, Vogel JP. *Brachypodium distachyon* and *Setaria viridis*: model genetic systems for the grasses. Annu Rev Plant Biol. 2015;66:465–85.

Carroll A, Somerville C. Cellulosic biofuels. Annu Rev Plant Biol. 2009;60:165–82.

Catalán P, Müller J, Hasterok R, Jenkins G, Mur LAJ, Langdon T, et al. Evolution and taxonomic split of the model grass *Brachypodium distachyon*. Ann Bot. 2012;109(2):385–405.

Clough SJ, Bent AF. Floral dip: a simplified method for *Agrobacterium*-mediated transformation of *Arabidopsis thaliana*. Plant J. 1998;16(6):735–43.

Davis BNK, Lakhani KH, Brown MC, Park DG. Early seral communities in a limestone quarry: an experimental study of treatment effects on cover and richness of vegetation. J Appl Ecol. 1985;22(2):473–90.

DOE, editor. Breaking the biological barriers to cellulosic ethanol: a joint research agenda. U.S. Department of Energy, Office of Science and Office of Energy Efficiency; 2006. http://genomicsgtl.energy.gov/biofuels/b2bworkshop.shtml.

Draper J, Mur LAJ, Jenkins G, Ghosh-Biswas GC, Bablak P, Hasterok R, et al. *Brachypodium distachyon*. A new model system for functional genomics in grasses. Plant Physiol. 2001;127(4):1539–55.

Editorial. How to feed a hungry world. Nature. 2010;466(7306):531–2.

Fedoroff NV, Battisti DS, Beachy RN, Cooper PJM, Fischhoff DA, Hodges CN, et al. Radically rethinking agriculture for the 21st century. Science. 2010;327(5967):833–4.

Foote TN, Griffiths S, Allouis S, Moore G. Construction and analysis of a BAC library in the grass *Brachypodium sylvaticum*: its use as a tool to bridge the gap between rice and wheat in elucidating gene content. Funct Integr Genomics. 2004;4(1):26–33.

Godfray HCJ, Beddington JR, Crute IR, Haddad L, Lawrence D, Muir JF, et al. Food security: the challenge of feeding 9 billion people. Science. 2010;327(5967):812–8.

Godfray HCJ, Pretty J, Thomas SM, Warham EJ, Beddington JR. Linking policy on climate and food. Science. 2011;331(6020):1013–4.

Gordon SP, Priest H, Des Marais DL, Schackwitz W, Figueroa M, Martin J, et al. Genome diversity in *Brachypodium distachyon*: deep sequencing of highly diverse inbred lines. Plant J. 2014;79(3):361–74.

Hayama R, Coupland G. The molecular basis of diversity in the photoperiodic flowering responses of Arabidopsis and rice. Plant Physiol. 2004;135(2):677–84.

IBI. Genome sequencing and analysis of the model grass *Brachypodium distachyon*. Nature. 2010;463:763–8.

Jeong DH, Schmidt SA, Rymarquis LA, Park S, Ganssmann M, German MA, et al. Parallel analysis of RNA ends enhances global investigation of microRNAs and target RNAs of Brachypodium distachyon. Genome Biol. 2013;14(12):R145.

Kaul S, Koo HL, Jenkins J, Rizzo M, Rooney T, Tallon LJ, et al. Analysis of the genome sequence of the flowering plant *Arabidopsis thaliana*. Nature. 2000;408(6814):796–815.

Khan MA. Seed-protein electrophoretic pattern in *Brachypodium* P. Beauv. species. Ann Bot. 1992;70(1):61–8.

Kucharik CJ, Ramankutty N. Trends and variability in U.S. Corn yields over the twentieth century. Earth Interact. 2005;9(1):1–29.

Lyons C, Scholthof K. Watching grass grow: the emergence of *Brachypodium distachyon* as a model for the Poaceae. In: Phillips D, Kingsland S, editors. New perspectives on the history of life sciences and agriculture. Archimedes 40; 2015:479–501. http://link.springer.com/chapter/10.1007%2F978-3-319-12185-7_23

Meyerowitz EM. Prehistory and history of Arabidopsis research. Plant Physiol. 2001;125(1):15–9.

Organisation for Economic Co-Operation and Development. World Energy Outlook 4014. Organisation for Economic Co-Operation and Development; 2014. ISBN: 978-92-64-20804-9.

Revedin A, Aranguren B, Becattini R, Longo L, Marconi E, Lippi MM, et al. Thirty thousand-year-old evidence of plant food processing. Proc Natl Acad Sci U S A. 2010;107(44):18815–9.

Shi Y, Draper J, Stace C. Ribosomal DNA variation and its phylogenetic implication in the genus *Brachypodium* (Poaceae). Plant Syst Evol. 1993;188(3–4):125–38.

Thole V, Worland B, Wright J, Bevan MW, Vain P. Distribution and characterization of more than 1000 T-DNA tags in the genome of *Brachypodium distachyon* community standard line Bd21. Plant Biotechnol J. 2010;8(6):734–47.

Tyler L, Fangel JU, FagerstrÃm AD, Steinwand MA, Raab TK, Willats WGT, et al. Selection and phenotypic characterization of a core collection of *Brachypodium distachyon* inbred lines. BMC Plant Biol. 2014;14(1):25.

Vain P, Worland B, Thole V, McKenzie N, Alves SC, Opanowicz M, et al. *Agrobacterium*-mediated transformation of the temperate grass *Brachypodium distachyon* (genotype Bd21) for T-DNA insertional mutagenesis. Plant Biotechnol J. 2008;6:236–45.

Vogel J, Hill T. High-efficiency *Agrobacterium*-mediated transformation of *Brachypodium distachyon* inbred line Bd21-3. Plant Cell Rep. 2008;27(3):471–8.

Vogel JP, Garvin DF, Leong OM, Hayden DM. *Agrobacterium*-mediated transformation and inbred line development in the model grass *Brachypodium distachyon*. Plant Cell Tiss Org Cult. 2006;85:199–211.

Wheeler T, Von Braun J. Climate change impacts on global food security. Science. 2013;341(6145):508–13.

Phylogeny and Evolution of the Genus *Brachypodium*

Pilar Catalan, Diana López-Álvarez, Antonio Díaz-Pérez, Rubén Sancho, and María Luisa López-Herránz

Abstract We present an updated review of the phylogenetic and evolutionary studies conducted on the model genus *Brachypodium*. The genus, which contains approximately 20 globally distributed taxa (17 species, 1 variety, and 2 undescribed cytotypes) shows an intermediate evolutionary placement within the grass temperate pooid clade, being closer to the basal than to the recent Pooideae lineages. Our comprehensive molecular phylogenetic survey of all the currently known *Brachypodium* lineages illustrates a complex reticulate scenario of recently evolved diploid and allopolyploid lineages. Haplotypic statistical parsimony networks, multilabelled (multigenic) Minimum Evolution gene tree discordances, and Bayesian dating analysis have provided a testable hypothesis for the reconstruction of the *Brachypodium* species tree and for the estimation of its nodal divergence times. Our results support the early splits of the annual and short-rhizomatose lineages (*B. stacei*, *B. mexicanum*, *B. distachyon*) in the Holarctic region during the early-Middle Miocene (and *B. hybridum* in the Pleistocene), and a profusion of rapid splits for the perennial lineages since the late Miocene to the Pleistocene in the Mediterranean and Eurasian regions, with sporadic colonizations of more remote areas. Several perennial allopolyploid species (*B. boissieri*, *B. retusum*, *B. phoenicoides*, *B. rupestre* 4x, *B. pinnatum* 4x) showed homeologous copies from both ancestral and recent genome donors. More in-depth studies of the species of the *B. distachyon* complex have demonstrated the polyphyletic origin of the allotetraploid *B. hybridum* from bidirectional crosses of its diploid *B. stacei* and

P. Catalan (✉)
Department of Agricultural and Environmental Sciences, High Polytechnic School of Huesca, University of Zaragoza, Ctra. Cuarte km 1, 22071 Huesca, Spain

Department of Botany, Institute of Biology, Tomsk State University, Tomsk, Russia
e-mail: pcatalan@unizar.es

D. López-Álvarez • R. Sancho • M.L. López-Herránz
Department of Agricultural and Environmental Sciences, High Polytechnic School of Huesca, University of Zaragoza, Ctra. Cuarte km 1, 22071 Huesca, Spain

A. Díaz-Pérez
Department of Agricultural and Environmental Sciences, High Polytechnic School of Huesca, University of Zaragoza, Ctra. Cuarte km 1, 22071 Huesca, Spain

Department of Genetics, School of Agronomy, Universidad Central de Venezuela, Maracay, Venezuela

J. Vogel (ed.), *Genetics and Genomics of Brachypodium*,
Plant Genetics and Genomics: Crops Models, DOI 10.1007/7397_2015_17
Published online: 17 December 2015

B. distachyon parents. Our niche modeling analysis has also detected distinct adaptations to different ecological tolerances in the diploids and evidence of niche conservatism for *B. hybridum* and each of its parents in their native Mediterranean region. Future perspectives include ongoing comparative genomics, phylogenomic and genotype-based phylogeographic studies of *Brachypodium*.

Keywords: Annual and perennial *Brachypodium* species • Dated phylogeny • Environmental niche modeling (*B. distachyon* group) • Haplotypic networks • Multigenic Minimum Evolution species tree • (allo)polyploid complexes

Evolutionary Placement of *Brachypodium* within the Poaceae Tree

The genus *Brachypodium* has received considerable attention since the selection of the annual species *B. distachyon* as model functional plant for temperate cereals and biofuel grasses (IBI 2010; Catalán et al. 2014; Mur et al. 2011). Recently, the three segregated annual species of the *B. distachyon* complex (*B. distachyon, B. stacei, B. hybridum*; Catalán et al. 2012) have been proposed as a model system for grass polyploid speciation (Catalán et al. 2014) and the whole genus, containing taxa characterized by their small-size and compact genomes (Mur et al. 2011; Betekhtin et al. 2014), is also seen as an ideal candidate for comparative genomics of monocots.

Decades of systematic and phylogenetic studies were necessary, however, to frame its evolutionary position within the grasses. *Brachypodium* is considered today the single representative genus of the monotypic tribe Brachypodieae, which constitutes one of the intermediate diverging lineages of the temperate Pooideae grasses (Catalán et al. 1997; Bouchenak-Khelladi et al. 2008; Schneider et al. 2011) (Fig. 1). Its controversial position was caused by its shared or similar morphological and anatomical traits with distinct pooid groups (Catalán et al. 1995, and references therein). Consequently, it was classified in different tribes, based on the possession of embryo with mesocotyl (Poeae), hairy terminal ovary appendage and long narrow caryopsis and hilum (Bromeae), or spicate to racemose inflorescence and hairy lodicles (Triticeae), until its definitive adscription to its own tribe Brachypodieae (Jacques-Félix 1962; Schippmann 1991; Watson and Dallwitz 1992). Subsequently, its separate tribal treatment was confirmed by a number of private biological (embryo development), biochemical (exclusive seed storage proteins, seed globulins, seed storage polysaccharides and stem and leaf fructosans) (Schippmann 1991), and karyotype (large disploidy) (Robertson 1981; Khan 1984) characters.

The most recent phylogenetic works have consistently resolved Brachypodieae as the sister lineage of the recently evolved core pooid clade of temperate cereals and forages [Triticodae (Triticeae + Bromeae)/Poodae (Poeae + Aveneae)] (Fig. 1a). Its intermediate placement between the basal (Brachyelytreae, Lygeae-Nardeae, Phaenospermatae, Meliceae, Stipeae) and the recently evolved (Triticodae/Poodae) Pooideae lineages has been recovered from both plastid and nuclear based topologies (Catalán et al. 1997; Davis and Soreng 2007; Bouchenak-

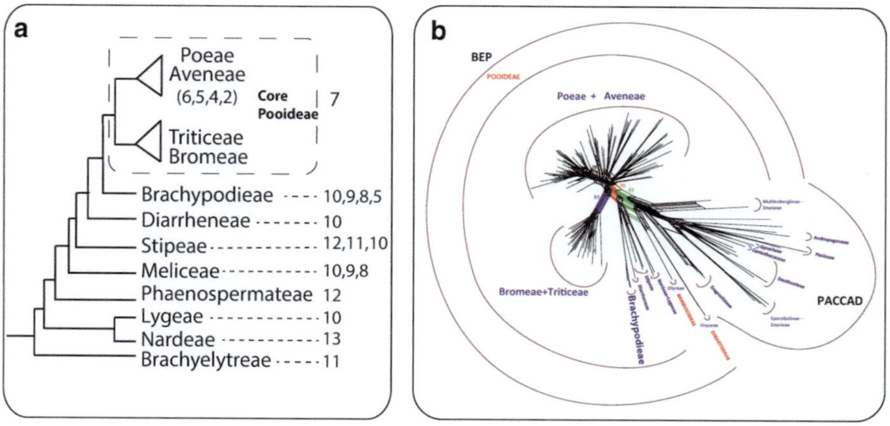

Fig. 1 (**a**) Summarized plastid phylogeny of the temperate grasses showing the evolutionary placement of *Brachypodium* (Brachypodieae) between the early diverging and the recently evolved Pooideae tribes, and the intermediacy of its chromosome base numbers. (**b**) NeighborNet partition network tree based on nuclear *ß-amylase* sequences showing the phylogenetic relationships of major tribal and subtribal grass lineages; *Brachypodium* is resolved close to the basal pooids. Pooideae (*green*), core pooids (*red*) and Triticeae + Bromeae (*purple*) splits showing bootstrap support values. Subfigure (**a**) partially adapted and updated from Catalán et al. (1997; Fig. 4); subfigure (**b**) adapted from Minaya et al. (2015; Fig. 3)

Khelladi et al. 2008; Schneider et al. 2011) and from combined analysis of molecular and morphological data (GPWG 2001). An intermediate position in the Pooideae tree is also reconstructed for the isolated *Diarrhena* (Diarrheneae) lineage, which apparently split earlier than *Brachypodium* (Catalán et al. 1997; Davis and Soreng 2007; Schneider et al. 2011). Recent phylogenetic studies based on a low copy nuclear gene (*ß-amylase*) showed, however, that *Brachypodium* and *Diarrhena* could be closer to the basal pooids than to the recently evolved core pooid clade (Minaya et al. 2015; Fig. 1b). The two independent and small monogeneric Brachypodieae and Diarrheneae tribes present remarkable embryo features (bambusoid-like in *Diarrhena*, first lateral stem developing from coleoptile in *Brachypodium*), with *Brachypodium* also showing intermediate chromosome base numbers when mapped into the pooid tree (Catalán et al. 1997). A karyotype evolutionary trend of increasing chromosome sizes and decreasing chromosome base numbers is observed in the Pooideae, ranging from basal tribes with small chromosomes and high chromosome base numbers (Brachyelytreae = 11; Lygeae = 10; Nardeae = 13; Phaenospermatae = 12; Meliceae = 10, 9, 8; Stipeae = 12, 11, 10; Diarrheneae = 10), through the intermediate ones of Brachypodieae (10, 9, 8, but also 5), to the large chromosomes and almost constant chromosome base number of $x = 7$ present in the more recently evolved Triticodae + Poodae although $x = 6, 5, 4, 2$ occasionally occur in Aveneae (Poodae) (Fig. 1a).

The isolated monophyly of *Brachypodium*, close but divergent from the core pooid clade (Fig. 1a, b), corroborates other unique genomic features reported for this genus, like the possession of small genomes with low amounts of repetitive DNA (Shi et al. 1993) and of private repetitive DNA and ribosomal DNA families and nuclear RFLP markers (Catalán et al. 1995). Recent studies have confirmed that *Brachypodium* combines both genus-specific and core-pooids-type or basal-pooids-BEP-type genomic traits. *Brachypodium* exhibits EST (expressed sequence tag)-contig chromosomal orthology, and similar globulin gene duplication and loci controlling phenotypic traits [e.g., spiking Eps-A (m)1, earliness Mot1 and FtsH4] and pathogen resistance (e.g., stem rust resistance, Rpg1 and Rpg4) responses with the Triticeae; however, it lacks colinearity for several STS (sequence tagged sites) and other stress controlling genes with this tribe (Mur et al 2011, and references therein). Also, the *Brachypodium* genome shows greater synteny with the more ancestral *Oryza* (Ehrartoideae, early BEP lineage) genome than with the more recently evolved Triticeae genomes, probably due to accelerated genomic rearrangements in the Triticeae (Mur et al. 2011). Despite these findings, the *Brachypodium* genome is more closely related to the core pooid genomes than the rice genome, and, together with its intermediate evolutionary position within the BEP clade (Fig. 1b), is well placed to serve as model plant not only for the temperate cereals and forages but also for tropical PACCMAD grasses including species proposed as biofuel crops (e.g. *Miscanthus*, *Panicum* (switchgrass), *Paspalum*) (Mur et al. 2011; Catalán et al. 2014).

Systematics of *Brachypodium*

Brachypodium is a relatively small genus that contains ca. 18 species distributed worldwide (Schippmann 1991; Catalán and Olmstead 2000; Catalán et al. 2012) (Table 1; Fig. 2). According to the most recent taxonomic updating (Catalán et al. 2012; Diaz-Pérez et al. unpub. data), 3 of them are annual species and 15 are perennial taxa. It has been recently demonstrated that the three annuals have a large distribution in their native circumMediterranean region (*B. distachyon*, *B. stacei*, *B. hybridum*) (Catalán et al. 2012; López-Alvarez et al. 2012, 2015). Among the perennials, few species show a large native Eurasian (*B. sylvaticum*, *B. pinnatum*, *B. rupestre*) or Mediterranean (*B. retusum*) distribution, whereas the rest have a restrict disjunct distributions in their respective native ranges [W Mediterranean (*B. phoenicoides*), C Mediterranean (*B. genuense*), E Mediterranean—SW Asia (*B. glaucovirens*), S Spain (*B. boissieri*), Canary isles (*B. arbuscula*), South Africa (*B. bolusii*), tropical and South Africa (*B. flexum*), Madagascar (*B. madagascariense*), Taiwan (*B. kawakamii*), SE Asia—New Guinea (*B. sylvaticum* var. *pseudodistachyon*), and America (*B. mexicanum*)] (Schippmann 1991; Diaz-Pérez et al. unpub. data; Fig. 2). Since 1812 two segregated genera were erected, *Trachynia* Link, to cover the annual species, and *Brevipodium* Lovë & Lovë, to accommodate *B. sylvaticum*; however, in almost all modern works neither of these two segregates were recognized (Catalán et al. 1995), and all the newly described species have been subsumed within *Brachypodium* (Schippmann 1991; Catalán and Olmstead 2000; Catalán et al. 2012).

The annual species are characterized by their short life-cycle, ephemeral habit and self-fertility (Catalán and Olmstead 2000; Catalán et al. 2012). Recent analysis of cryptic phenotypic, cytogenetic and molecular traits allowed us to separate the three species (Catalán et al. 2012). By contrast, most of the perennial taxa show long-rhizomes and self-incompatibility (Catalán et al. 1995; Khan and Stace 1999),

Table 1 List of world *Brachypodium* taxa, cytotypes and ecotypes used in the phylogenetic analysis

Taxon	Code	Geographical distribution (native range)	2n	x	Ploidy	Genome size (pg/2C)
Annuals						
B. distachyon (L.) P. Beauv.	Bdis	circumMediterranean (Mediterranean, SW Asia)	2n = 10	5	2x	0.63
B. stacei Catalán, Joch. Müll., Mur & Langdon	Bsta	circumMediterranean (Mediterranean, Macaronesia, SW Asia)	2n = 20	10	2x	0.56
B. hybridum Catalán, Joch. Müll., Hasterok & Jenkins	Bhyb	circumMediterranean (Mediterranean, Macaronesia, SW Asia)	2n = 30	5 + 10	4x	1.26
Short-rhizomatose perennial						
B. mexicanum (Roem. & Schult.) Link	Bmex	America (from Mexico to N Bolivia)	2n = 40	10?	4x?	?
Long-rhizomatose perennials						
B. arbuscula Gay ex Knoche	Barb	Macaronesia: Canary isles (Spain)	2n = 18	9	2x	0.70
B. boissieri Nym.	Bboi	Spain: Betic mountain ranges (southern Spain)	2n = 42, 46	?	6x–8x?	?
B. bolusii Stapf	Bbol	South Africa	?	?	?	?
B. flexum Nees	Bflex	Tropical Africa and South Africa	?	?	?	?
B. genuense (D.C.) Roem. & Schult.	Bgen	Italy	2n = 18	9	2x	?
B. glaucovirens (Murb.) Sagorski	Bgla	East Mediterranean and SW Asia	2n = 16	8	2x	0.88
B. kawakamii Hayata	Bkaw	Taiwan	?	?	?	?
B. madagascariense Camus & Perrier	Bmad	Madagascar	?	?	?	?
B. phoenicoides (L.) P. Beauv. ex Roem. & Schultes	Bpho	West Mediterranean	2n = 28	5 + 9	4x	1.49
B. pinnatum (L.) P. Beauv. (diploid A)	Bpin2xA	Eurasia	2n = 18	9	2x	0.88
B. pinnatum (L.) P. Beauv. (tetraploid)	Bpin4x	Eurasia	2n = 28	5 + 9	4x	1.57
B. pinnatum (L.) P. Beauv. (diploid B)	Bpin2xB	SW Asia	2n = 16	8	2x	?

(continued)

Table 1 (continued)

Taxon	Code	Geographical distribution (native range)	2n	x	Ploidy	Genome size (pg/2C)
B. retusum (Pers.) P. Beauv.	Bret	Mediterranean	2n = 36	?	6x?	2.57
B. rupestre (Host) Roem. & Schult. (diploid)	Brup2x	West Eurasia	2n = 18	9	2x	0.84
B. rupestre (Host) Roem. & Schult. (tetraploid)	Brup4x	West Eurasia	2n = 28	5 + 9	4x	?
B. sylvaticum (Huds.) P. Beauv.	Bsyl	panEurasia (Eurasia, Macaronesia)	2n = 18	9	2x	0.87
B. sylvaticum (Huds.) P. Beauv. (Eastern lineage)	BsylEAs	East Asia (China)	?	?	?	?
B. sylvaticum var. *pseudodistachyon* J. D. Hook.	Bsylpse	Malesia and New Guinea	?	?	?	?

Geographical distribution taken from Schippmann (1991), Catalán et al. (1995, 2012), Catalán and Olmstead (2000), López-Alvarez et al. (2012, 2015); 2n, x, ploidy level and genome size taken from Robertson (1981), Khan (1984), Schippmann (1991), Wolny and Hasterok (2009), Catalán et al. (2012), Betekhtin et al. (2014)

Information on geographical distribution in the native range, code, chromosome number (2n), chromosome base number (x), ploidy level and genome size is provided for each taxon. ? = unknown or unclear

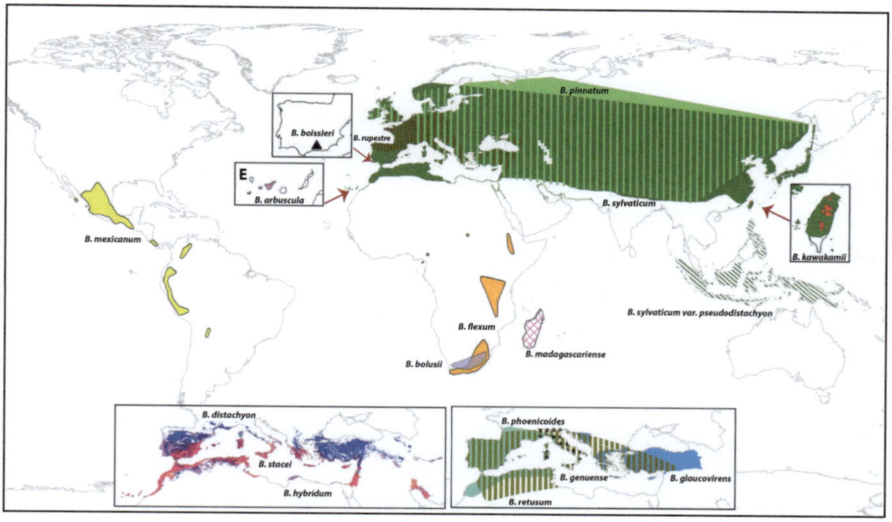

Fig. 2 Geographic distributions of the 18 world *Brachypodium* taxa (*B. arbuscula*, pink; *B. boissieri*, black triangle; *B. bolusii*, violet; *B. distachyon*, dark blue; *B. flexum*, orange; *B. genuense*, black star; *B. glaucovirens*, pale blue; *B. hybridum*, purple; *B. kawakamii*, red dot; *B. madagascariense*, open red square; *B. mexicanum*, yellow; *B. phoenicoides*, aquamarine; *B. pinnatum*, bright green; *B. retusum*, pale brown; *B. rupestre*, dark brown; *B. stacei*, red; *B. sylvaticum*, dark green; *B. sylvaticum* var. pseudodistachyon, diagonal dark green line)

except *B. mexicanum* and *B. sylvaticum* that are self-compatible (Khan and Stace 1999; Steinwand et al. 2013). *B. mexicanum* differs from them in its short-rhizomatous habit and self-compatibility (Khan and Stace 1999), taxonomically resembling more the annual than the perennial taxa in those traits and in seed protein contents and RFLP patterns (Khan 1992; Shi et al. 1993). The rhizomatous perennials are separated by their morphoanatomical and karyotypic traits. *B. arbuscula*, *B. retusum* and *B. boissieri* bear branched woody stems and long-lasting innovations. The Canarian *B. arbuscula* possesses top branched buds and dispersed root xylem and phloem, and grows in more humid places, whereas *B. retusum* and *B. boissieri* are adapted to xeric Mediterranean habitats and show strongly inrolled leaves. The narrow endemic *B. boissieri*, previously circumscribed within the broadly spread *B. retusum*, differs from it in its single-spikelet inflorescence, short habit and leaf blade morphology and anatomy (Schippmann 1990); the species is confined to dolomitic mountain ranges of southern Spain. The remaining taxa of the core perennial clade do not bear branched stems. The endemic alpines *B. kawakamii* and *B. bolusii* have a relatively short stature and dense, erect, and glabrous leaves; the inflorescences of *B. bolusii* present more spikelets but with less fertile florets than those of *B. kawakamii*. *B. pinnatum*, *B. rupestre* and *B. phoenicoides* show erect panicles. *B. phoenicoides*, adapted to dry places, is glabrous and presents partially inrolled leaves, semi-patent twisted spikelets and awnless lemmas, whereas the mesic *B. pinnatum* and *B. rupestre* have short awns and bright green colored leaves.

B. rupestre, considered until recently a subspecies of *B. pinnatum*, differs from it in its glabrous leaves and spikelets and in leaf epidermal traits (Schippmann 1991). The central Mediterranean endemic *B. genuense*, classified within *B. pinnatum* by some authors (Clayton et al. 2015), departs from it based on its particular karyotype, showing co-localized 5S and 25S rDNA loci in the same chromosome (Betekhtin pers. com.) and minor morphological differences (Valdés and Scholz 2009).

Brachypodium sylvaticum is the most distinct and widespread species of the genus. Its native Palearctic area ranges from Macaronesia in the west to New Guinea in the east (Fig. 2). It is characterized by its nodding panicle, densely hairy habit and long-awned lemma. Most of these features are also shared by the tropical and South African *B. flexum* and the Malagasy *B. madagascariense*, though they differ from the former in their shorter panicles, spikelets and awns, and from each other in the overall smaller habit of the mountain endemic island species. The '*B. sylvaticum*' complex also includes the eastern Mediterranean—SW Asian endemic *B. glaucovirens*. This taxon, formerly synonymized to *B. sylvaticum*, or even considered a hybrid between this species and *B. pinnatum* (Schippmann 1991), has been recently recognized as a separate species (Scholz 2007). Morphologically it shows intermediate features, resembling *B. sylvaticum* in its short rhizome and long awn, and *B. pinnatum* in its bright green leaf color, broad leaf ribs and erect panicle. Furthermore, some of the six infraspecific *B. sylvaticum* taxa described in eastern Asia and Malesia—New Guinea (Schippmann 1991), like *B. sylvaticum* subsp. *pseudodistachyon*, which is characterized by its mountain dwarf habit and stiff leaves, could correspond to independent species.

Taxonomic uncertainty still persists among some poorly known extra-European taxa and within some Eurasian cryptic complex taxa (Schippmann 1991; Catalán and Olmstead 2000). Among the less known extra-European taxa, up to 5 different species have been described in America, 11 in Africa and 15 in Asia; however, most of them could probably be synonymized to currently recognized species from those regions (cf. Schippmann 1991). Regarding the Eurasian cryptic taxa, they correspond to ploidy complexes of putative diploid parents and their derived allopolyploids, involving different cytotypes of *B. pinnatum* (2x, 4x) and *B. rupestre* (2x, 4x) (Khan and Stace 1999; Wolny and Hasterok 2009; Betekhtin et al. 2014, and references therein). The intraspecific cytotypes could hardly be differentiated based on morphological traits; however, cytogenetic studies using Comparative Chromosome Painting (CCP) approaches suggest that the allopolyploids derive from interspecific crosses of distinct diploid progenitors, involving their respective diploid perennial counterparts (Wolny and Hasterok 2009; Idziak et al. 2014), or even those perennials and the annual *B. distachyon* (Wolny and Hasterok 2009; Betekhtin et al. 2014). The genus shows a remarkable disploidy, with chromosome base numbers of diploids ranging from the presumably more ancestral $x = 10$ (*B. stacei*), through $x = 9$ (*B. arbuscula*, *B. sylvaticum*, *B. pinnatum*, *B. rupestre*) and $x = 8$ (*B. glaucovirens*), to $x = 5$ (*B. distachyon*) (Robertson 1981; Betekhtin et al. 2014). Noticeably, the 'recently evolved' chromosome base number $x = 7$ (Robertson 1981), which is almost fixed in most species of the large and young core pooid clade (Catalán et al. 1997), is apparently absent in *Brachypodium*, where tetraploid species with 2n = 28 chromosomes have been found to be hybrid allopolyploids, potentially derived from diploid 2n = 18 (x = 9) and 2n = 10 (x = 5) progenitors (Khan and Stace 1999; Wolny and Hasterok 2009; Betekhtin et al. 2014). Betekhtin et al. (2014) proposed two alternative hypotheses for karyotype evolution in *Brachypodium*, continuous descendant disploidy (x = 10 to x = 9, 8 to x = 5) vs. descendant + ascendant disploidy (x = 10 to x = 5 to x = 9, 8), with allotetraploid 2n = 28 species originating always in a later stage.

The taxonomic identity of these allotetraploid cytotypes is still unclear, though they might constitute separate species, paralleling the case of the segregated annual species of the diploid-allopolyploid *B. distachyon* complex (Catalán et al. 2012). Overall, *Brachypodium* constitutes a small isolated genus of approximately 20 species, with native ranges distributed in five continents. Two of its species, the annual *B. hybridum* and the perennial *B. sylvaticum*, are invasive plants. *B. sylvaticum* has been introduced and is spread in western N America and in Australia, and *B. hybridum* has successfully colonized C Europe, western N America (California), S America (Uruguay, Argentina), South Africa and Oceania (Australia, New Zealand) (Jenkins et al. 2003; Garvin et al. 2008; Bakker et al. 2009; Catalán et al. 2012).

Phylogeny of *Brachypodium*

All phylogenetic studies conducted on *Brachypodium* support a rapid and relatively recent radiation of its crown ancestor, after a long time span from the earlier split of the stem ancestor and the recent split of the crown clade (Catalán and Olmstead 2000; Catalán et al. 1995, 2012; Diaz-Perez et al. unpub. data). This long isolation, followed by recent divergence, is corroborated by its exclusive nuclear genomic families (Catalán et al. 1995; Mur et al. 2011) and by its confounding assorted lineages (Catalán and Olmstead 2000; Wolny et al. 2011; Catalán et al. 2012). Successive phylogenetic works based mostly on analysis of plastid and nuclear rDNA sequences and on nuclear (RAPD) markers, including approximately half of the species of *Brachypodium*, recovered a congruent evolutionary framework for the genus (Catalán et al. 1995, 2012; Catalán and Olmstead 2000). Dated phylogenies based on combined analysis of nuclear ribosomal genes and plastid genes have estimated the origin of the common ancestor of *Brachypodium* in the mid Miocene, showing the early successive divergences of *B. boissieri*, *B. stacei*, and *B. mexicanum*, a later Pliocene split of *B. distachyon* and the recent Pliocene-Pleistocene radiation of the core perennial clade (Catalán et al. 2012). Within the latter group, a congruent trend was observed in the early divergence of *B. arbuscula*, followed by that of *B. retusum*, though uncertainty affected the rapid splits of the most recent nodes, ending in an unresolved scenario for the divergence of the *B. pinnatum*, *B. rupestre*, *B. phoenicoides*, *B. glaucovirens* and *B. sylvaticum* lineages (Catalán et al. 1995, 2012; Catalán and Olmstead 2000; Wolny et al. 2011). Phylogenetic trees reconstructed from low copy nuclear genes concurred with this hypothesis, but also showed basal homeologous copies in one allopolyploid member of the core perennial clade (*B. retusum*; Wolny et al. 2011; Catalán et al. 2012).

Deep evolutionary analysis of the perennial *Brachypodium* genomes has been hampered, however, by the intrincate reticulate nature of the species in this core clade, which shows a prevalence of allopolyploid taxa, and by their explosive radiation, manifested in the mostly unresolved or weakly supported topologies (Catalán et al. 2012). Recently, a thorough taxonomic and geographic sampling of all the currently recognized species of the genus allowed us to conduct the largest and most comprehensive phylogenetic study of *Brachypodium* to date (Diaz-Pérez et al. unpub. data described below). A total of 110 samples representing the 17 recognized species plus one geographically isolated infraspecific taxon (*B. sylvaticum* var. *pseudodistachyon*) were included in the study (Table 1; Fig. 2). Six taxa (35.3 % of the total taxonomic diversity) were studied molecularly for the first time (*B. bolusii*, *B. flexum*, *B. genuense*, *B. kawakamii*, *B. madagascariense*, *B. sylvaticum* var. *pseudodistachyon*). Our study also included representatives of both diploid and allotetraploid cytotypes of the perennial *B. pinnatum* and *B. rupestre* species. Chromosomal, genome size and ploidy data information was collected for all samples except for some poorly known taxa which have not been karyotyped yet (Table 1). One thousand one hundred fifty-four DNA sequences from three nuclear (ETS, ITS, GI) and two plastid (*ndh*F, *trn*LF) loci were used to reconstruct the phylogeny of *Brachypodium*. The non-recombinant plastid *ndh*F + *trn*LF sequences were concatenated into a combined (cpDNA) data set and provided information about the maternal genomic inheritance in the hybrids, and the cloned sequences of the nuclear loci retrieved homeologous copies in the allopolyploids.

Exploratory phylogenetic and haplotypic network analyses were conducted with the respective sets of sequences (Fig. 3a–e). Phylogenetic analyses based on Maximum Likelihood (ML; RAxML) and Bayesian Inference (BI; MrBAYES) methods recovered the evolutionary relationships among the *Brachypodium* lineages, using other pooid representatives and *Oryza* (Ehrartoideae) as outgroups (Fig. 3e). Haplotypic networks were constructed to infer the genealogical relationships of the *Brachypodium* haplotypes (species and samples) obtained from each separate data set using statistical parsimony approaches (NETWORK) (Fig. 3a–d). The maternally inherited plastid haplotypic network consisted of 43 haplotypes (Fig. 3a) and was relatively well resolved for the early divergences of the monophyletic *B. boissieri*, *B. stacei*, *B. mexicanum* and *B. distachyon* clusters, each separated by a number of mutational steps in a star-like net (with highly supported divergences in the phylogenetic tree; Fig. 3e). The *B. hybridum* haplotypes were shared with its *B. stacei* and *B. distachyon* parents, though more frequently with the former. However, the cluster of the recently evolved core perennial species showed a lack of genealogical and taxonomic structure (Fig. 3a), denoted by the high number of interspecific shared haplotypes (with some haplotypes shared by up to three species; e.g., h. 28: *B. pinnatum*, *B. retusum*, *B. rupestre*), and an ambiguous resolution, manifested in the high number of internal loops and few internal mutational steps. The high number of interspecific shared maternal haplotypes reflects a history of repeated introgressions among lineages of the core perennial

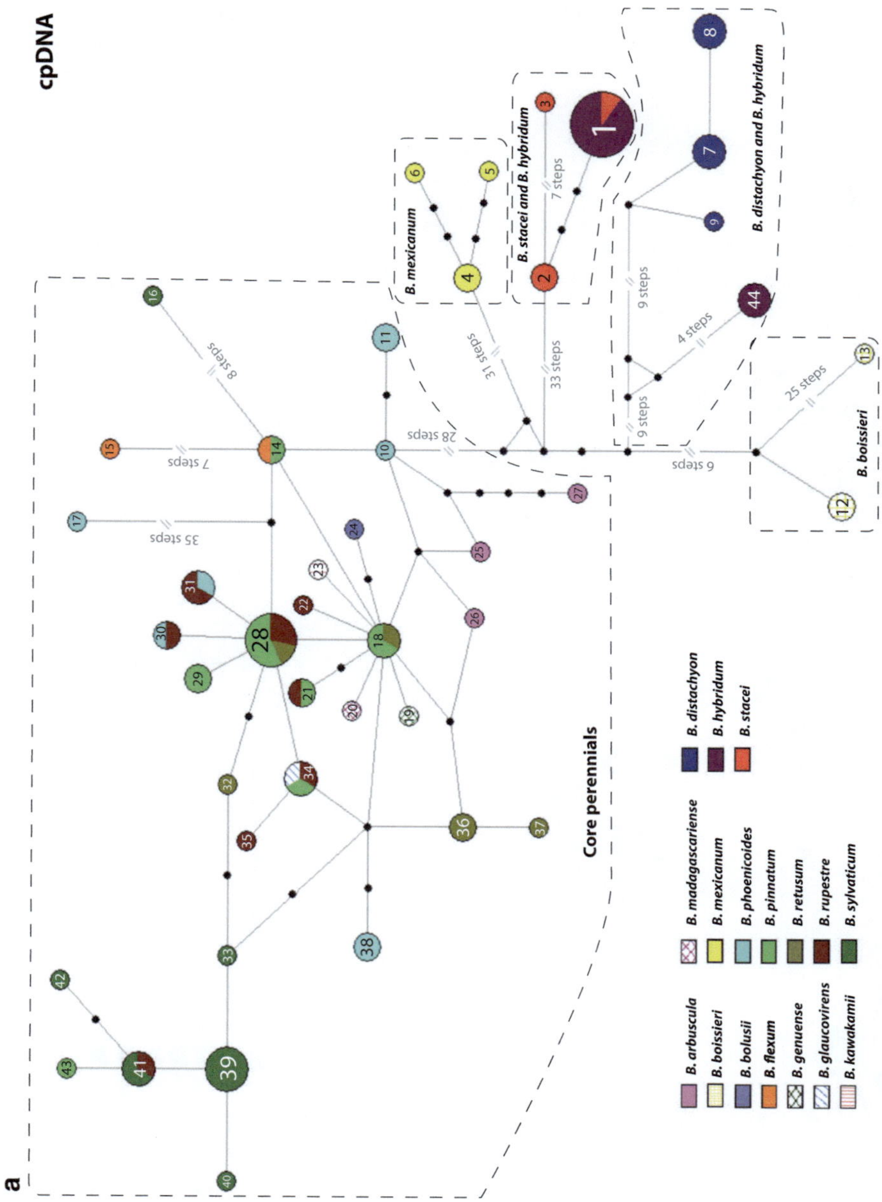

Fig. 3 Haplotypic statistical parsimony networks (**a–d**) and summarized Bayesian phylogenetic trees (**e**) of *Brachypodium* reconstructed from the four independent studied loci. (**a**) cpDNA; (**b**) ITS; (**c**) ETS, (**d**) GI. Species colors are indicated in the charts. *Numbers inside the circles* indicate haplotype number. *Size* of the *circles* is correlated with number of samples showing the haplotype

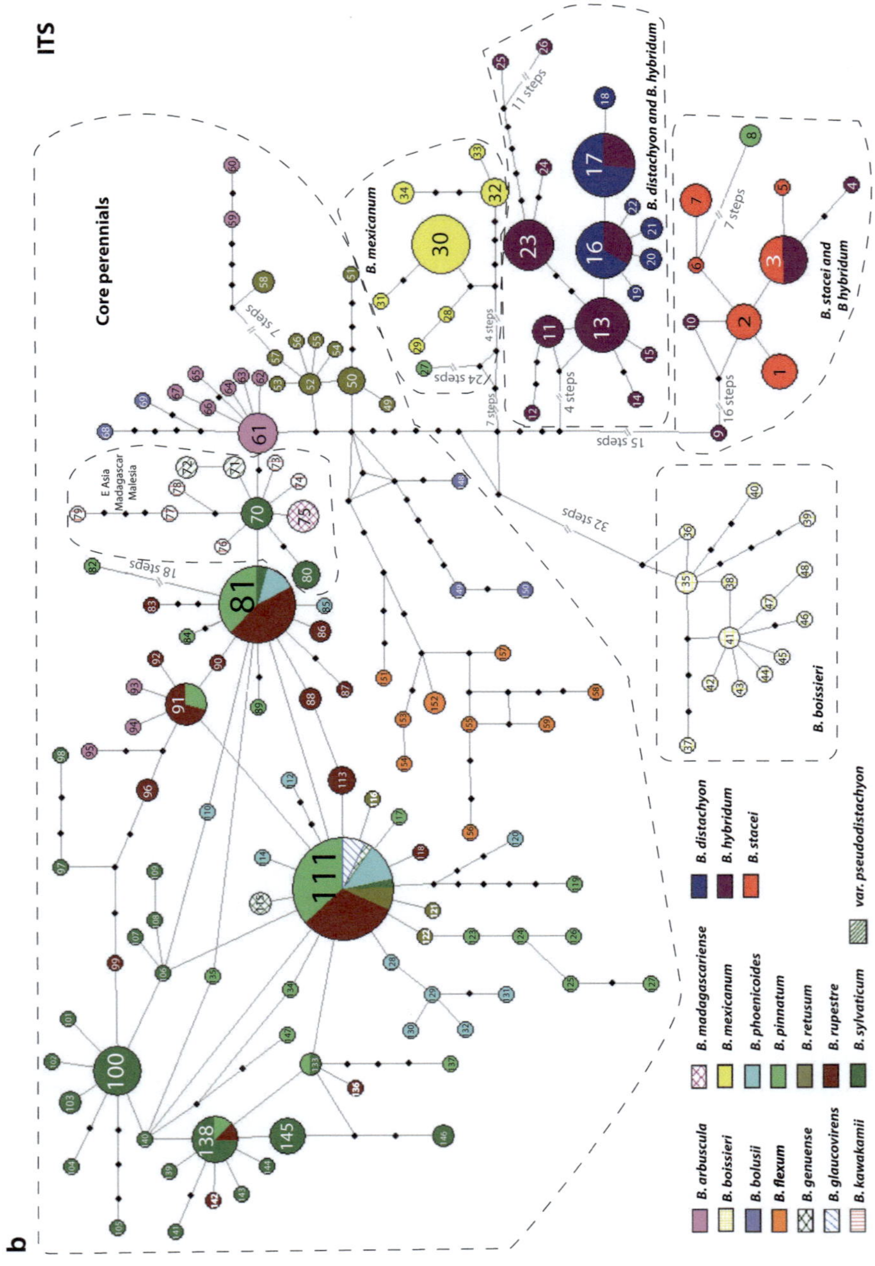

Fig. 3 (continued)

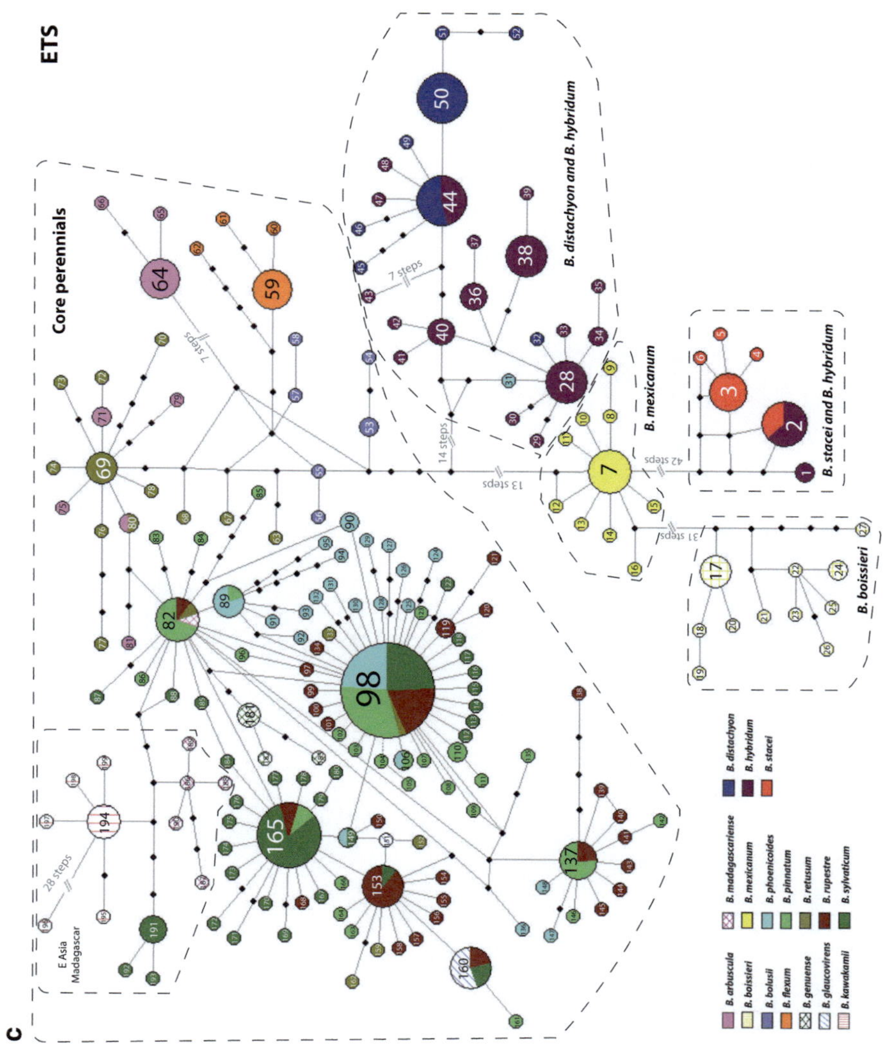

Fig. 3 (continued)

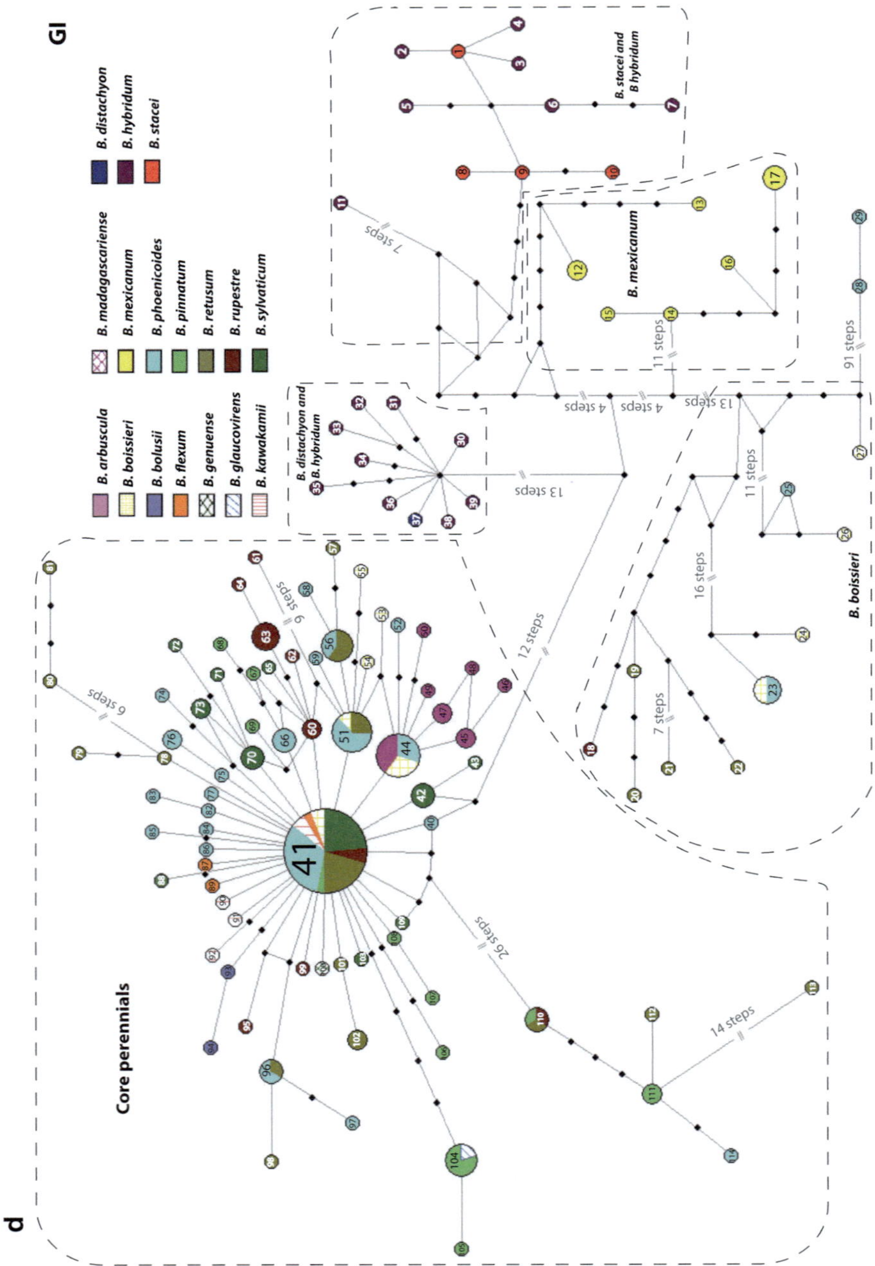

Fig. 3 (continued)

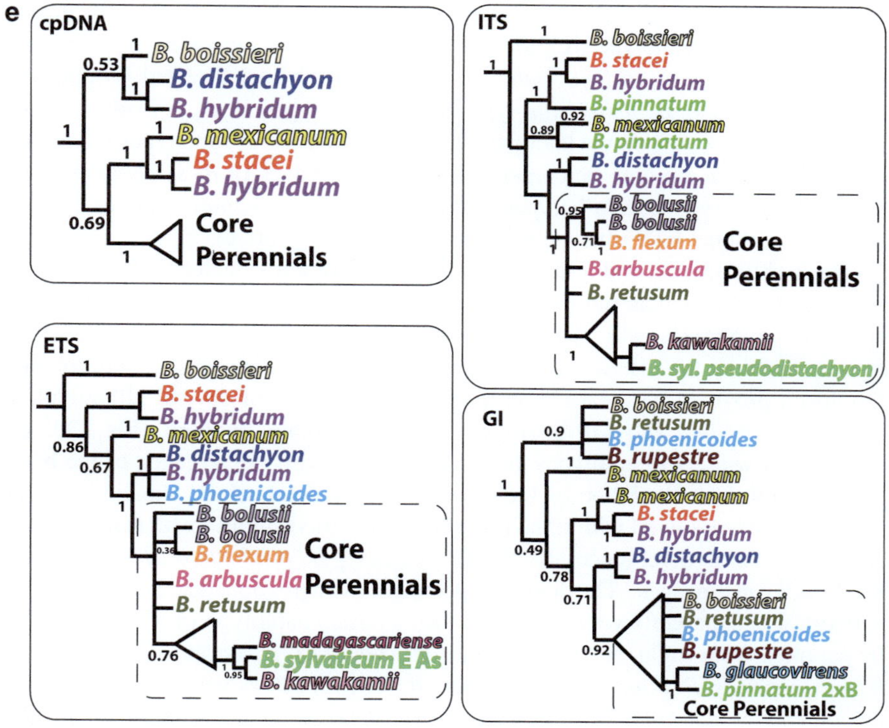

Fig. 3 (continued)

group, whereas the lack of resolution of the network (and the phylogenetic tree; Fig. 3e) indicates a high level of homoplasy in the data.

The biparentally-inherited and more variable ITS and ETS loci were overall congruent with the plastid data but reconstructed more resolved trees and networks for the earlier splits within the core perennials clade and for some geographical groupings. These loci also detected 'ancestral' homeologous ribotypic copies in some core perennials allopolyploids. The ITS and ETS phylogenies and haplotypic networks (Fig. 3b, c, e) constructed, respectively, with 159 and 199 haplotypes, were congruent in the separate basal divergences of the *B. boissieri*, *B. stacei*, *B. mexicanum* and *B. distachyon* lineages and in the complex reticulate structure of the core perennials group. They further detected the early divergences of the *B. bolusii*/*B. flexum*, *B. arbuscula* and *B. retusum* lineages within the core perennials clade, and the clustering of endemic East Asia (*B. sylvaticum* (China)/ *B. kawakamii*)—Madagascar (*B. madagascariense*)—New Guinea (*B. sylvaticum* var. *pseudodistachyon*) haplotypes in their respective regional subnetworks (Fig. 3b, c, e). The introgression and homoplasy levels detected by these loci were much higher than those detected by the plastid data within the core perennial cluster, and mostly affected the Eurasian and Mediterranean species. Thus, the commonest ITS haplotype (h. 111) was shared by seven species (*B. genuense*, *B. glaucovirens*,

B. phoenicoides, *B. pinnatum*, *B. retusum*, *B. rupestre*, *B. sylvaticum*; Fig. 3b) and the commonest ETS haplotype (h. 98) by five species (*B. phoenicoides*, *B. pinnatum*, *B. retusum*, *B. rupestre*, *B. sylvaticum*; Fig. 3c). Both loci detected co-inherited *B. stacei*-type and *B. distachyon*-type parental ribotypes in *B. hybridum*, being more frequent those from the later parent, and a few ancestral haplotypes in individual samples of some core perennials allopolyploid species (a *B. stacei*-type and a *B. mexicanum*-type ITS ribotypes in tetraploid *B. pinnatum* samples, and a *B. distachyon*-type ETS ribotype in *B. phoenicoides*; Fig. 3b, c).

The highly variable and bi-parentally inherited low copy nuclear gene GI which, in contrast to the nuclear multicopy ribosomal ITS and ETS loci, is not subjected to concerted evolution, was also congruent with the main evolutionary patterns reconstructed by the plastid and ribosomal genes for *Brachypodium*. Additionally, this locus provided new data about the putative origins of several allopolyploid perennial species. The GI phylogeny and haplotypic network (Fig. 3d, e), constructed with 114 haplotypes, also supported the early divergence of the basal lineages and the reticulation of the recent core perennials clade, though relationships were less resolved within the last group and varied slightly with respect to the successive basal divergences of the *B. boissieri*, *B. mexicanum*, *B. stacei* and *B. distachyon* lineages. The level of potential introgression detected by the GI network was apparently very high, the most common haplotype (h. 41) was shared by samples from eight perennial species (*B. boissieri*, *B. flexum*, *B. kawakamii*, *B. phoenicoides*, *B. pinnatum*, *B. retusum*, *B. rupestre*, *B. sylvaticum*). In contrast, the GI clones detected the highest number of homeologous copies among the perennial *Brachypodium* allopolyploid species. Most interestingly, highly divergent GI sequences of *B. boissieri*, *B. retusum*, *B. phoenicoides* and *B. rupestre* 4x were nested within both the basal '*B. boissieri*' cluster and the recent core perennial cluster. The *B. hybridum* individuals showed homeologous copies from each *B. stacei* and *B. distachyon* parent. The analyses also recovered two close but separate homeologous lineages within *B. mexicanum* (Fig. 3d, e).

Our new findings provide new insights into the evolutionary history of *Brachypodium* (Diaz-Pérez et al. unpub. data, Fig. 3). All the analyzed plastid and nuclear loci agree with previous studies (Catalán et al. 2012; Catalán et al. 2014) in the more ancestral divergences of the annual *B. stacei* and the shortly-rhizomatose *B. mexicanum*, and in the sister relationship of the annual *B. distachyon* to the recentmost core perennial clade (Fig. 3a–e). However, the homeologous 'ancestral' and 'recently evolved' copies detected in several perennial species at the GI locus and, to a lesser extent, at the ITS and ETS loci, provide a new scenario for the likely allopolyploid origins of these perennial plants. Our results indicate that *B. boissieri* is not an early divergent lineage within the *Brachypodium* clade but, most probably, an allopolyploid species originated from the cross of at least one ancestral genome donor and one recent perennial-core genome donor (Fig. 3d, e). Similarly, *B. retusum* and *B. rupestre* 4x (GI network, Fig. 3d), *B. phoenicoides* (ETS and GI networks, Fig. 3c, d) and *B. pinnatum* 4x (ITS network, Fig. 3b) would have also resulted from different crosses of at least one ancestral genome donor and one recent perennial-core genome donor. Wolny et al. (2011) and Catalán et al. (2012) also detected both ancestral and recent homeologous copies in the purported allohexaploid *B. retusum* for the low copy nuclear genes GI, CAL and STT3A, and CAL, DGAT and GI, respectively; and Catalán et al. (2012) detected two close homeologous genomes in *B. mexicanum* also for CAL. Betekhtin et al. (2014) hypothetized about the potential origins of the allotetraploids *B. pinnatum*, *B. rupestre* and *B. phoenicoides* from a cross of a *B. distachyon*-type diploid genome (x = 5) and a perennial-core-type diploid genome (x = 9). Their CCP study used BAC libraries from the *B. distachyon* genome, which is more closely related to the core perennials' genomes (Fig. 3); however, no other CCP analyses have been conducted yet with BAC clones derived from other *Brachypodium* genomes. Noticeably, this cytogenetic analysis revealed the absence of CCP hybridization signals of the *B. distachyon* BAC probes in approximately half of the chromosomes of the basal *B. stacei* and *B. mexicanum*, and in some chromosomes of *B. retusum* (*B. boissieri* was not included in the analysis; Betekhtin et al. 2014, pers. com.). These phylogenetic and cytogenetic evidences support the existence of an ancestral-type *Brachypodium* genome, divergent from the recently evolved *B. distachyon* and core perennial genomes, in the nuclear genomes of those species.

Our results point towards to alternative sources of ancestral genome donors for the perennial allopolyploids (e.g., *B. stacei*-type or *B. mexicanum*-type for *B. pinnatum* 4x, *B. distachyon*-type for *B. phoenicoides*, '*B. boissieri*' ancestral-type for *B. boissieri*, *B. retusum* and *B. phoenicoides*; Fig. 3b, c, d). Conversely, the low and ambiguous resolutions obtained for the core perennial group in all assayed loci preclude the identification of the recent core perennial genome donors. Regarding the morphologically close high allopolyploid Mediterranean *B. boissieri* (cf. 6x–8x) and *B. retusum* (6x) species, it could be hypothetized that convergent evolution has apparently converted the *B. boissieri* ITS and ETS copies into the ancestral '*B. boissieri*'ancestral-type ribotypes and the *B. retusum* copies into the recent core perennial-type ribotypes (Fig. 3b, c). This might be also the reason for the low number of 'ancestral' ITS and ETS ribotypes found in the *B. pinnatum* 4x and *B. phoenicoides* samples (convergence towards the recent core perennial types). These plausible hypotheses on alternative origins of the allopolyploids could not be confirmed, however, with the current data. Low copy nuclear genes are prone to several phylogenetic distorting events, such as paralogy (duplication), losses, recombination and pseudogenization (Diaz-Pérez et al. 2014; Minaya et al. 2015), and multicopy nuclear ribosomal genes are constrained by convergent evolution. These recently diverged species could also have experienced a number of additional confounding evolutionary events, apart from hybridization, such as incomplete lineage sorting and vector-mediated or other horizontal gene transfers (Minaya et al. 2013). A larger genomic coverage would be necessary to dissect the evolutionary history of the *Brachypodium* allopolyploids. Ongoing phylogenomic studies are under way to target it.

Phylogenetic studies of highly reticulate polyploid groups confront the difficulty or impossibility of reconstructing bifurcate tree-like topologies from genome-mergers and genome-doubled species, which render network-like phylogenies (Jones et al. 2013; Marcussen et al. 2015). In our attempt to construct a robust explicit phylogenetic framework for *Brachypodium*, we used comparative statistical analysis of diploid/polyploid multiple gene tree discordances (Cai et al. 2012) to build the species tree of this highly reticulate and allopolyploid genus. In order to discard spurious variation generated from PCR or cloning artifacts, intraindividual consensus sequences were generated collapsing closely related sequences that showed a p-distance value lower than 0.01 (cf. Diaz-Pérez et al. 2014). Multilabelled gene phylogenies were constructed for each diploid and diploid +polyploid data set through ML and BI methods. The resulting topologies were used as reference trees to select the best consensus sequences and to build the respective species trees. We followed the method of Cai et al. (2012), based on Minimum Evolution (FASTME) analysis, to calculate the internodal distances (NJst) among tips in each multilabelled gene tree for every polyploid consensus allele, using averaged pairwise distances for all the diploid and diploid-polyploid combinations. Integrated data matrices of averaged distances from the four loci were constructed for all diploid species and for one polyploid allele each time, generating as many integrated data matrices as polyploid alleles. The unrooted species tree of all diploid *Brachypodium* taxa had 15 branches. To estimate the optimal placement(s) of the polyploid alleles in the diploid skeleton tree (Fig. 4),

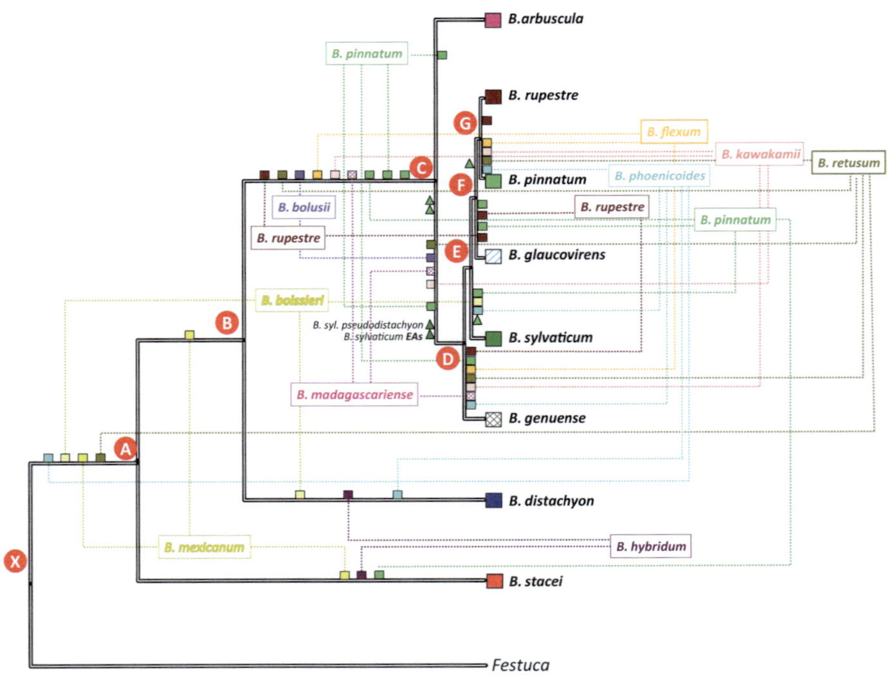

Fig. 4 Multilabelled Nst-FastME species tree of *Brachypodium* obtained from a four gene-tree-discordance Minimum Evolution approach. Insertion of the allelic copies of the allopolyploids (discontinuous lines) on the branches of the diploid skeleton tree (*solid lines*) was done according to their positions in the selected trees (minimum tree values) for each polyploidy allele per locus. Species colors correspond to those indicated in Fig. 3

each polyploid allele was inserted in any potential branch, rendering 15 species trees per allele. The lengths of the trees were calculated according to the minimum evolution method, using the integrated distance matrices and the 15 species trees per polyploid allele, and selecting the minimum tree length value each time as indicator of the optimal placement of each polyploid allele in a particular branch of the diploid tree.

Our diploid skeleton tree included only *Brachypodium* species of confirmed diploid nature (Fig. 4). According to this multigene tree, the lineage of the annual *B. stacei* diverged first from the common ancestor, followed by that of the annual *B. distachyon* and then by the clade of core perennial taxa, which showed the successive divergences of the *B. arbuscula*, *B. genuense*, *B. sylvaticum*, *B. glaucovirens*, and the *B. pinnatum* 2x/*B. rupestre* 2x lineages. The FASTME reconstruction placed the multigene alleles of *B. mexicanum*, and therefore the potential origins of its genome donors, in three basal and subbasal positions of the tree (stem branch of *Brachypodium*, stem branch of *B. distachyon*/core perennials clade, terminal branch of *B. stacei*), those of *B. hybridum* along each terminal branch of its *B. stacei* and *B. distachyon* diploid parents, and those of *B. boissieri*, *B. retusum*, *B. phoenicoides* and *B. pinnatum* 4x in basal and subbasal branches of the tree but also in more recent branches of the core perennial subtree and in its stem branch (Fig. 4). The topological placement of the allelic copies of the remaining polyploid or unknown ploidy *Brachypodium* species was restricted to the recent stem branch and internal branches of the core perennial clade. Among them, *B. bolusii* showed a more ancestral putative origin, whereas a highly reticulate scenario of multiple potential crosses was reconstructed for the respective origins of the other species (Fig. 4). Though some of these reconstructions might have been affected by homoplasy, the evolutionary trends depicted in this ME species tree explain the reproductive biology and hybridization capability observed in the *Brachypodium* species. Artificial crosses involving five perennial species (*B. glaucovirens, B. phoenicoides, B. pinnatum, B. retusum, B. sylvaticum*) demonstrated the ability of these taxa to interbreed and produce viable F_1 hybrids in all directional crosses (Khan 1984; Khan and Stace 1999). Furthermore, all the obtained F_1 hybrids exhibited some degree of fertility. Successful hybridisations were also performed between those perennials and the annual *B. hybridum* ("*B. distachyon*" $2n = 30$); however, despite the fact that hybrids were vigorous and long-lived, they were sterile. By contrast, no hybrids could be raised from crosses of these perennials and *B. hybridum* with *B. mexicanum* (Khan and Stace 1999). Recent attempts to cross current *B. stacei* and *B. distachyon* individuals and to produce a synthetic artificial *B. hybridum* plant from bidirectional crosses of these parental-genomes species have also failed (López-Alvarez et al. 2012) though Boulos Chalhoub and Vinh-Ha Dinh-Thi made a viable but sterile F1 (Chalhoub pers. com.). These experimental evidences support the reproductive isolation of the ancestral *B. stacei* and *B. mexicanum* lineages (and genomes), the reproductive compatibility of the more recently diverged *B. hybridum* (*B. distachyon* genome) with the perennials, and the high reproductive success of the recentmost perennial lineages in all interspecific crosses. All these events fit well the reticulate scenario recovered in our evolutionary study (Fig. 4).

Divergence time estimations for the nodal splits of the *Brachypodium* species tree were obtained from a multigene diploid + polyploid data set using a Bayesian dating approach (BEAST). A consensus topology derived from the diploid skeleton tree (Fig. 4) was enforced for the placement of the unknown-ploidy and allopolyploid species, considering subtree resolutions supported by one or more loci (Fig. 3b–e) and the topology of the multigene ME species tree (Fig. 4). It was based on the premise that the allopolyploid hybrid lineages could never have originated before its genome donor lineages (Fig. 4). Nonetheless, due to the absence of a current diploid 'ancestral' lineage (e.g. 'ancestral'-type genome), a basal '*B. boissieri*' ancestral-type allele was also included in the analysis in order to estimate the divergence time of the potential earliest split of the *Brachypodium* crown node. We calibrated the crown node of the BEP clade using a secondary calibration (mean age 54.9 ± 5.7 Ma), according to Bouchenak-Khelladi et al. (2010), and the crown node of the Pooideae clade using a fossil calibration (48.4 ± 4.0 Ma), based on a pooid-type phytolit from the Middle-Eocene (Zucol et al. 2010). Our dating analysis (Diaz-Perez et al. unpub. data) indicated that the *Brachypodium* lineage branched off from its stem node in the Late Eocene (38.8 Ma) and that the split of the crown node could have occurred earlier (21.6 Ma) or later (16.2 Ma) in the early Miocene, considering or not the basal divergence of the 'ancestral'-type *B. boissieri* genome. The long time elapsed between the stem and crown splits (17.2–22.6 Ma) would explain the evolutionary and genomic isolation of *Brachypodium* from its closest pooid relatives. Our analysis also showed successive early and Mid-Miocene divergences for the basalmost currently extant *Brachypodium* lineages (*B. stacei*, 16.2 Ma; *B. mexicanum*, 14.0 Ma; *B. distachyon*, 10.6 Ma), followed by a rapid radiation of the core perennial lineages since the end of the Miocene (6.1 Ma) and through the Pliocene and the Pleistocene, with the annual allotetraploid *B. hybridum* also arisen in the Pleistocene. Our nodal datings were older than those proposed by Catalán et al. (2012) though both studies showed overlapping values in their respective 95 % highest posterior density (HPD) intervals for the divergence time estimates.

Evolutionary History and Environmental Niche Variation of the Model *B. distachyon* Complex Species (*B. distachyon*, *B. stacei*, *B. hybridum*)

The most exhaustively studied species of *Brachypodium* are those of the annual *B. distachyon* complex, which include the model grass plant *B. distachyon* and its close allies *B. stacei* and *B. hybridum* (which show, respectively, 2n = 10, 20 and 30 chromosomes; Catalán et al. 2012; Catalán et al. 2014). The three cytotypes were previously attributed to different ploidy levels of the same taxon *B. distachyon* s. l.

(Robertson 1981); however, phylogenetic, cytogenetic and phenotypic analyses demonstrated that they should be treated as three different species. They consist of two diploids, each with a different chromosome base number [*B. distachyon* (x = 5, 2n = 10); *B. stacei* (x = 10, 2n = 20)], and their derived allotetraploid *B. hybridum* (x = 5 + 10, 2n = 30). Phylogenetic analyses of two plastid (*ndh*F, *trn*LF) and five nuclear (ITS, ETS, CAL, GI, DGAT) genes indicated that the more basally-diverging *B. stacei* and the more recently evolved *B. distachyon* emerged from two independent lineages, confirming their contribution as genome donors of *B. hybridum* (Catalán et al. 2012). Further evidence from different molecular sources, like seed protein data (Hammami et al. 2011), nuclear SSRs (Giraldo et al. 2012), DNA barcoding (López-Alvarez et al. 2012), isozymes (Jaaska 2014), and metabolite fingerprinting (López-Alvarez et al. unpub. data) have also confirmed the co-occurrence of progenitor *B. distachyon* and *B. stacei* markers in the *B. hybridum* background.

Statistical analysis of morphometric traits showed that five characters (stomata leaf guard cell length, pollen grain length, upper glume length, lemma length, and awn length) significantly discriminated among the three species when they were grown under controlled greenhouse conditions (Catalán et al. 2012). An enlarged morphological study of both inbred and wild individuals from widespread Mediterranean populations has corroborated the phenotypic differentiation of the three species, with the number of species-specific discriminant traits increasing to 13 when only wild populations are considered (Catalán et al. unpub. data; López-Alvarez et al. unpub. data). Although the three species can be differentiated through several phenotypic and cytogenetic traits, their direct identification is not always straightforward as wild populations show overlapping phenotypic variation for some characters and a similar diploid genome size. This has led to taxonomic uncertainty, or even to taxonomic misclassifications of the model species and its close allies when using currently employed identification methods such as morphology or flow cytometry (López-Alvarez et al. 2012). This is particularly problematic in natural admixed populations, where *B. hybridum* lives in sympatry with one or the other parental species (Fig. 2), and in germplasm collections, where *B. stacei* and *B. hybridum* samples have been found within the *B. distachyon* stocks. Recently, López-Alvarez et al. (2012) provided a reliable method to differentiate the individuals of the three species using a DNA barcoding system that offered a suitable approach to this problem. Three genomic barcoding loci, the plastid *trn*LF region and the nuclear multicopy ribosomal ITS spacer and low copy GI gene successfully discriminated among the three species. Any one of the three assayed loci could unambiguously differentiate the two monophyletic diploid species from direct sequencing of PCR amplicons, whereas the identity of the allotetraploid required combined analysis of direct *trn*LF and direct or cloned ITS sequences or through analysis of cloned GI sequences.

The study also revealed other particularities of the *B. distachyon*, *B. stacei* and *B. hybridum* genomes. Pairwise substitution rates showed high interspecific sequence divergence values and low intraspecific values between and among the diploids *B. distachyon* and *B. stacei*. Regarding *B. hybridum*, the differences between the intra-parental and inter-parental (*B. distachyon*-like vs. *B. stacei*-like) mean values were equivalent to those found between and within the sequences of the two diploids for the three loci. These results suggested that the original diploid progenitor genomes have remained largely intact in the allotetraploid *B. hybridum* genome and that the time elapsed since the hybridization took part was a brief one. Haplotype networks constructed for each of the separate loci using statistical parsimony methods showed a clear-cut separation between the *B. distachyon*-type and *B. stacei*-type classes of sequences in all cases, corroborating the existence of barcoding gaps between the diploid genomes/subgenomes (Fig. 3). Nonetheless, the detection of some interspecific ITS and GI recombinant sequences in *B. hybridum* (López-Alvarez et al. 2012) pointed towards to the occurrence of frequent genomic rearrangements within the hybrid nucleus. This agreed with cytogenetic CCP evidence demonstrating the existence of structural rearrangements in the *B. hybridum* chromosomes with respect to the *B. distachyon* and *B. stacei* ones (Idziak et al. 2011). However, despite the plausible existence of rapid structural changes in the allotetraploid genome, the integrity of the respective subgenomes is still prevalent. The two subgenomes have kept the same or similar signatures as those of the parental genomes, supporting the recent origin of *B. hybridum* in the Pleistocene (Catalan et al. 2012; Diaz-Perez et al. unpub. data).

A striking finding of our study was the demonstration of the existence of different directional crosses that likely gave rise to the new allotetraploid species (López-Alvarez et al. 2012, unpub. data). This was confirmed through the analysis of the maternally inherited plastid haplotypes in the *B. hybridum* sampling. Though the majority of the surveyed *B. hybridum* individuals show the inheritance of a *B. stacei*-like plastid genome, in a few cases some *B. hybridum* individuals have inherited a *B. distachyon*-type plastome (López-Alvarez et al. 2012; Fig. 5). Consequently, it is assumed that the former derived from a cross between maternal *B. stacei* and paternal *B. distachyon* parents whereas the second resulted from a converse cross, with maternal *B. distachyon* and paternal *B. stacei* parents. The fact that *B. hybridum* plants derived from reciprocal crosses occurred in different Mediterranean localities supports the multiple and polytopic origins of the allotetraploid *B. hybridum*. The recurrent formation of allopolyploid plant species has been widely documented in angiosperms (Soltis and Soltis 1999; Soltis et al. 2010), with speciation occurring from bidirectional crosses in some grasses (e.g., *Aegylops*, Meimberg et al. 2009). In some instances, similar directional crosses have even led to distinct allopolyploid species (Meimberg et al. 2009), probably as a result of different genomic rearrangements and losses in the stabilizing allopolyploid genomes. An inspection of the more variable ITS and GI networks and phylogenetic trees also revealed distinct relationships of the *B. hybridum* sequences to different parental geographic haplotypic groups, corroborating the polyphyletic origin of the *B. hybridum* samples. Complementary or unique parental haplotypic clusters have been found for some western and eastern Mediterranean *B. hybridum* groups (López-Alvarez et al. 2012). Nonetheless, all the studied hybrids correspond to what is considered to be the same allopolyploid species.

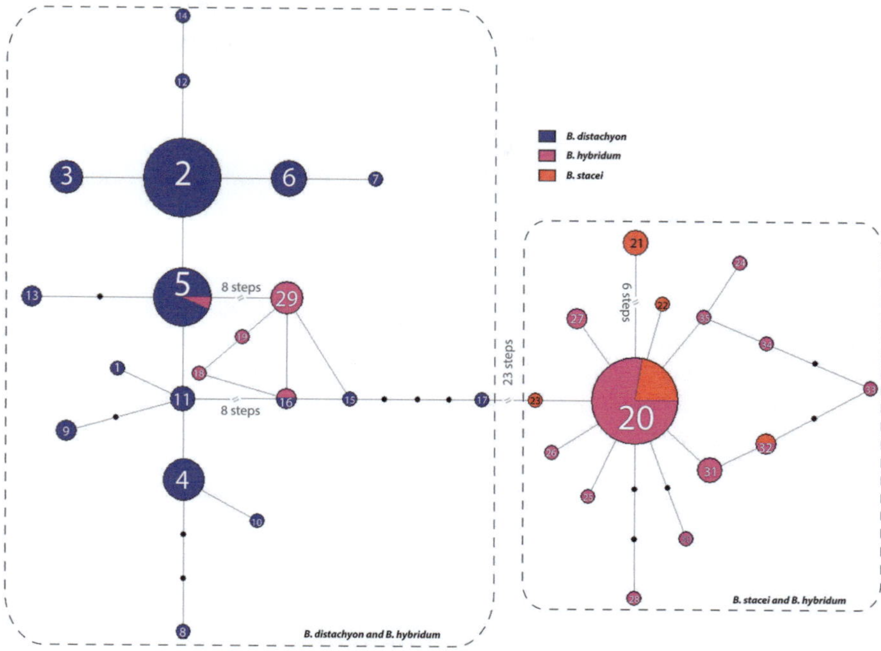

Fig. 5 Plastid haplotypic statistical parsimony network of the annual *Brachypodium distachyon–B. stacei–B. hybridum* complex, based on maternally inherited cpDNA sequences, showing the polyphyletic origin of *B. hybridum*, arisen from bidirectional crosses (♀ *B. stacei* × ♂ *B. distachyon*; ♀ *B. distachyon* × ♂ *B. stacei*) of its progenitors. Species colors are indicated in the charts. Partially adapted from López-Alvarez et al. (2012; Fig. 2a) and unpub. data

Other interesting conclusions about the dispersal capabilities of these species were also drawn from our studies (López-Alvarez et al. 2012, 2015). Despite their abundant distributions in the circumMediterranean region, the intraspecific genetic diversities of the parental *B. distachyon* and *B. stacei* sequences were low. This was manifested in the sharing of their respective most common plastid and nuclear haplotypes by individuals from populations located far apart in disparate parts of the Mediterranean region (Figs. 2 and 5). In contrast, individuals from geographically close populations, or even intraindividual clones, showed different haplotypes. A similar scenario was also recovered for the intra-parental *B. distachyon*-type and *B. stacei*-type sequences found in *B. hybridum*. Because the three annuals are self-fertile plants, the observed pattern could only be explained by long distance dispersal (LDD) of their seeds (Vogel et al. 2009; Mur et al. 2011; López-Alvarez et al. 2012). Selfing species are expected to show low within-population and high among-population genetic diversities (Hamrick and Godt 1996). However, the autogamous *B. distachyon*, *B. stacei* and *B. hybridum* samples show overall low geographical structuring of genetic diversity. This might be a consequence of LDD of seeds coupled with the high capability of these annuals to adapt to different environmental conditions (Manzaneda et al. 2012). Seed dispersal mediated by ants has been observed in *B. distachyon* (Catalán pers. obs.). Ants have been shown to be important seed dispersers in several plant species, though the dispersal distance does not usually exceed 100 m (Gómez and Espadaler 1998). Thus, the suspected LDD of *B. distachyon*, *B. stacei* and *B. hybridum* seeds might have occurred through other dispersal vectors, including anthropogenic mediated dispersal, like the cases documented for the recent introduction of *B. hybridum* in non-native ranges (Bakker et al. 2009; Catalán et al. 2012). Despite the ample distribution of common haplotypes in the native range, the three taxa show, however, some traces of geographic isolation between the western and eastern Mediterranean regions, evidenced by the detection of regional haplotypic clades (e.g. western Mediterranean, Iberian-Balearic, Turkish and Middle East—SW Asian; cf. López-Alvarez et al. 2012). Phylogeographic studies based on next generation genotypic RAD data are currently under way, aiming to dissect the origins and the spatio-temporal colonization routes of populations of the three species in their native circumMediterranean region.

The intriguing question of the potential existence of different environmental-adaptive speciation events among these three annual species was examined through niche overlap tests and niche breadth comparisons (López-Alvarez et al. 2015). Environmental niche modeling (ENM) analysis was conducted for *B. distachyon*, *B. stacei* and *B. hybridum* across their wide circumMediterranean native range as a way to predict the potential distribution of species across time and space and to investigate the roles of geography and environment in the evolutionary diversification of species. A large data set of contrasted occurrence data from 561 accessions and 19 bioclimatic variables plus altitude was used to construct the species' niche models under current and past [e.g., Mid-Holocene (MH), Last Glacial Maximum (LGM), and Last Interglacial (LIG)] climatic scenarios using maximum entropy probability distribution methods (MAXENT). The ENM models obtained under current climate conditions showed a potential distribution of *B. distachyon* in higher altitudinal areas and more northern latitudes, of *B. stacei* in coastal and lowland areas, and of *B. hybridum* in both mountain and lowland and coastal areas, though more commonly in the latter settings (Fig. 6a). A comparative analysis of overlapping areas in these ENMs detected ranges of potential shared occupancy between the two diploids, the allopolyploid and both diploids, and the allopolyploid and one or the other diploid. The environmental data indicate that *B. distachyon* grows in higher, cooler and wetter places than *B. stacei*, which grows in lower, warmer and drier environments, whereas *B. hybridum* grows in zones with intermediate values but also in low altitudinal warmer and drier places, like its *B. stacei* progenitor (López-Alvarez et al. 2015). These results fit well the ecophysiological requirements of the species (e.g., vernalization for most of the *B. distachyon* accessions and lack of it for the *B. hybridum* and *B. stacei* accessions; cf. Vogel et al. 2009, López-Alvarez and Catalán unpubl. data), which are crucial for the germination of seeds and survival of these annual species in their respective Mediterranean niches.

Interestingly, the paleoclimatic MH, LGM and LIG models also showed potential overlapping of environmental niches of the two diploid parents and its hybrid,

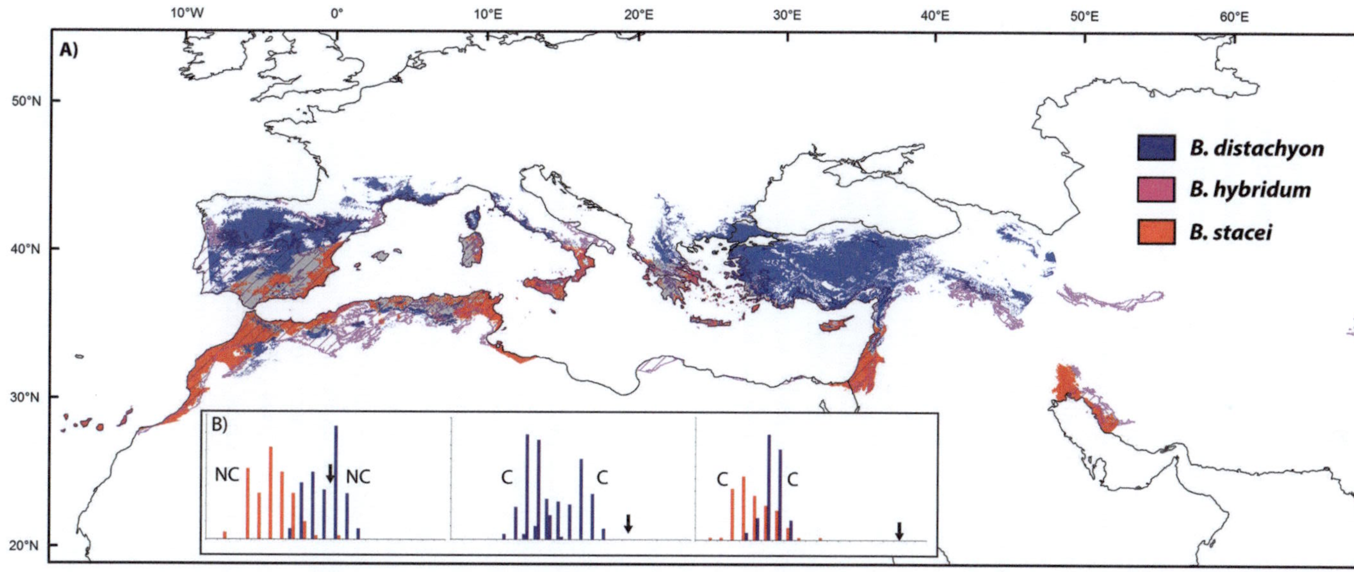

Fig. 6 (**a**) Environmental niche models of *Brachypodium distachyon*, *B. stacei* and *B. hybridum* in their native circumMediterranean range under current climate conditions showing the overlapping areas of their respective niches. *B. distachyon* (*blue*), *B. stacei* (*red*) and *B. hybridum* (*purple*). (**b**) Plots of niche divergence and conservatism tests. Overlap values larger that or similar to the null distribution support niche conservatism (C) or are inconclusive (NC), respectively. Adapted from López-Alvarez et al. (2015; Figs. 2, 5)

with larger overlaps in the LGM, more restricted in the LIG and similar to present in the MH (López-Alvarez et al. 2015). These results support the Mediterranean basin and its adjacent areas as long-term refugia for *B. distachyon* and *B. stacei*, suggesting the existence of multiple potential hybrid zones along the Pleistocene and Holocene, which might have favored the recurrent origin of *B. hybridum*. The amount of significant environmental differences found between *B. distachyon* and *B. stacei* for 18 environmental variables, and their divergence in niche equivalence, suggest distinct adaptations to different ecological tolerances in these diploids. However, the reciprocal niche similarity tests were non-significant (López-Alvarez et al. 2015; Fig. 6b), implying that niche divergence was not the major driver of speciation for these species, but rather reproductive isolation or other biological or life-history traits. Surprisingly, despite the apparent recurrent origin of *B. hybridum* in the past, artificial crossing experiments have failed to generate the synthetic hybrid or have only produced sterile F1 hybrids (Chalhoub, pers. com.), confirming the strong reproductive isolation shown by the two diploids today. The pairwise niche similarity comparison tests showed evidence of niche conservatism for *B. hybridum* and each of its parents (Fig. 6b). Most evolutionary ecological studies conducted with sister or with less-related lineages have accumulated evidences of niche conservatism (Warren et al. 2008). In our case, niche conservatism is predicted for the recent allopolyploid, which shares niche occupancy with both progenitors (Fig. 6a) but is also reproductively isolated from them.

A further outcome of our study is that *B. hybridum* shows the largest niche overlap compared to its two diploid progenitors (Fig. 6a), being phenotypically more variable (Catalán et al. 2012; López-Álvarez and Catalán unpubl. data), but a niche breadth smaller than that of *B. distachyon* and only slightly greater than that of *B. stacei* (López-Alvarez et al. 2015). Niche competition with its diploid progenitors could be invoked to explain the observed restricted range distribution and niche breadth of *B. hybridum* in its native area, overlapping with but without displacing them. Conversely, *B. hybridum* is the only species of the complex that has apparently successfully colonized other non-native world regions (Catalán et al. 2012; López-Alvarez et al. 2015). This suggests a greater ecological tolerance of the allotetraploid compared to the diploids. This could be explained by the high potential of highly heterozygous allopolyploids to large genome rearrangements and to genomic and epigenetic expressions, increasing genetic diversity to buffer against inbreeding depression and to boost diversifying selection (Bakker et al. 2009; Meimberg et al. 2009). The environmental success of the young *B. hybridum* colonizers in America, S Africa and Oceania might also be related to rapid shifts in physiological and adaptive traits, such as changes in flowering time related to photoperiod and weediness (cf. Bakker et al. 2009). Yet the underlying factors causing the apparent different adaptive capabilities of *B. hybridum* in its native range and in the allochthonous areas should be tested through more detailed studies. It would also require the corroboration of the apparent but still unconfirmed absence of the diploid *B. distachyon* and *B. stacei* parents in those areas.

Future Perspectives: Evolutionary Comparative Genomic Studies of *Brachypodium*

The impact of the new model plant *Brachypodium distachyon* on grass genomic research gathered pace since the publication of the full genome sequence of the diploid genotype Bd21 by the International Brachypodium Initiative (IBI 2010). Over the last decade, more than 400 laboratories worldwide have worked on investigating the genomics, transcriptomics and metabolomics of *B. distachyon* (Vain 2011; Mur et al. 2011). *Brachypodium* represents an excellent resource for comparative evolutionary genomic studies (Gordon et al. 2014), and recent work has identified the important role played by hybridization in the history and ecology of its species (Betekhtin et al. 2014; Diaz-Perez et al. unpub. data). The small genome sizes, compact genomes (e.g. low levels of repetitive DNA), diverse ecological tolerances, ready propagation under controlled growth conditions, and considerable existing molecular and genomic resources make this genus an excellent candidate for addressing fundamental questions in ecological and comparative genomics. The new taxonomic and phylogenetic findings, and the advent of inexpensive next generation sequencing technology, has set the stage for high definition investigation of the unusual genomic diversity and evolutionary relationships in *Brachypodium* (Catalán et al. 2014). The demonstration that the model plant was not one but three species (Catalán et al. 2012) opened the way to a thoroughly comparative genomic study of this diploid-polyploid complex. The nuclear and organellar genomes of *B. stacei* and *B. hybridum* are being sequenced and will serve as a model for the origins and consequences of the speciation and polyploidization events that might parallel those of economically important cereals (e.g., wheats; Marcussen et al. 2014). The analysis of the intraspecific diversity in *B. distachyon* is also under way through the nuclear and organellar resequencing of 56 diverse natural accessions (Gordon et al. 2014, unpub. data).

Genomic resources have been also developed for the perennial species *B. sylvaticum* (Steinwand et al. 2013) and the sequencing of other perennial genomes is on the way (John Vogel pers. com.). Comparative genomics of annual vs. perennial species of *Brachypodium* aim to identify the genome donors and the hybridization processes involved in the origin of allopolyploid species (approximately half of the studied taxa) and the switches from perenniality to annuality. A valuable set of metatranscriptomic data would enlarge the scope of the phylogenomic analysis of *Brachypodium*, facilitating the analysis of the transcriptomic contents across the *Brachypodium* phylogeny. Comparative genomic-transcriptomic analysis would allow us to reconstruct the gene content and the non-coding content evolution in *Brachypodium*. It is of particular interest to comparatively examine the evolution of genes and of regulatory elements in non-coding regions and how this might affect the expression of the transcriptomes. The *Brachypodium* data could be contrasted with 970 single or low copy nuclear genes analysed across monocots and a larger number of single or low copy nuclear genes studied in grasses. Since the various *Brachypodium* species and *B. distachyon* accessions are native to a wide geographic region with varied climates, these new sequence resources will facilitate genome-wide association mapping of genes controlling tolerance to drought and other abiotic stresses (Catalán et al. 2014) and to phenotypic and biological traits that might have triggered different speciation processes.

References

Bakker EG, Montgomery B, Nguyen T, Eide K, Chang J, Mockler TC, et al. Strong population structure characterizes weediness gene evolution in the invasive grass species *Brachypodium distachyon*. Mol Ecol. 2009;18:2588–601.

Betekhtin A, Jenkins G, Hasterok R. Reconstructing the evolution of Brachypodium genomes using comparative chromosome painting. PLoS One. 2014;9(12):e115108.

Bouchenak-Khelladi Y, Salamin N, Savolainen V, Forest V, Van der Bank M, Chase MW, et al. Large multi-gene phylogenetic trees of the grasses (Poaceae): progress towards complete tribal and generic level sampling. Mol Phylogenet Evol. 2008;47:488–505.

Bouchenak-Khelladi Y, Verboom GA, Savolainen V, Hodkinson TR. Biogeography of the grasses (Poaceae): a phylogenetic approach to reveal evolutionary history in geographical space and geological time. Bot J Linn Soc. 2010;162:543–57.

Cai D, Rodriguez F, Teng Y, Ane C, Bonierbale M, Mueller LA, et al. Single copy nuclear gene analysis of polyploidy in wildpotatoes (*Solanum* section *Petota*). BMC Evol Biol. 2012;12:70.

Catalán P, Olmstead RG. Phylogenetic reconstruction of the genus *Brachypodium* P. Beauv. (Poaceae) from combined sequences of chloroplast *ndhF* gene and nuclear ITS. Plant Syst Evol. 2000;220:1–19.

Catalán P, Shi Y, Amstrong L, Draper J, Stace CA. Molecular phylogeny of the grass genus *Brachypodium* P. Beauv based on RFLP and RAPD analysis. Bot J Linn Soc. 1995;177:263–80.

Catalán P, Kellogg EA, Olmstead RG. Phylogeny of Poaceae subfamily Pooideae based on chloroplast *ndh*F gene sequencing. Mol Phylogenet Evol. 1997;8:1–18.

Catalán P, Müller J, Hasterok R, Jenkins G, Mur LA, Langdon T, et al. Evolution and taxonomic split of the model grass *Brachypodium distachyon*. Ann Bot. 2012;109:385–405.

Catalán P, Chalhoub B, Chochois V, Garvin DF, Hasterok R, et al. Update on the genomics and basic biology of Brachypodium International Brachypodium Initiative (IBI). Trends in Plant Science. 2014;19:414–8.

Clayton WD, Vorontsoba M, Harman KT, Williamson H. World checklist of Poaceae. Kew: Royal Botanic Gardens. 2015. http://www.kew.org/data/grasses-db.

Davis JI, Soreng RJ. A preliminary phylogenetic analysis of the grass subfamily Pooideae (Poaceae), with attention to structural features of the plastid and nuclear genomes, including an intron loss in *GBSSI*. Aliso. 2007;23:335–48.

Diaz-Pérez A, Sharifi-Tehrani M, Inda LA, Catalán P. Polyphyly, gene-duplication and extensive allopolyploidy framed the evolution of the ephemeral *Vulpia* grasses and other fine-leaved Loliinae (Poaceae). Mol Phylogenet Evol. 2014;79:92–105.

Garvin DF, Gu YQ, Hasterok R, Hazen SP, Jenkins G, Mockler TC, et al. Development of genetic and genomic research resources for *Brachypodium distachyon*, a new model system for grass crop research. Crop Sci. 2008;48:69–84.

Giraldo P, Rodriguez-Quijano M, Vazquez JF, Carillo JM, Benavente E. Validation of microsatellite markers for cytotype discrimination in the model grass *Brachypodium*. Genome. 2012;55:1–5.

Gómez C, Espadaler X. Myrmecochorous dispersal distances: a world survey. J Biogeogr. 1998;25:573–80.

Gordon SP, Priest H, Marais DLD, Schackwitz W, Figueroa M, Martin J, et al. Genome diversity in *Brachypodium distachyon*: deep sequencing of highly diverse inbred lines. Plant J. 2014;79:361–74.

GPWG. The Grass Phylogeny Working Group. Phylogeny and subfamilial classification of the grasses (Poaceae). Ann Mo Bot Gard. 2001;88:373–457.

Hammami R, Jouve N, Cuadrado A, Soler C, Gonzalez JM. Prolamin storage proteins and alloploidy in wild populations of the small grass Brachypodium distachyon (L.) P. Beauv. Plant Syst Evol. 2011;297:99–111.

Hamrick JL, Godt MJW. Effects of life history traits on genetic diversity in plant species. Philos Trans R Soc Lond B Biol Sci. 1996;351:1291–8.

International Brachypodium Initiative (IBI). Genome sequencing and analysis of the model grass *Brachypodium distachyon*. Nature. 2010;463:763–8.

Idziak D, Betekhtin A, Wolny E, Lesniewska K, Wright J, Febrer M, et al. Painting the chromosomes of Brachypodium: current status and future prospects. Chromosoma. 2011;120:469–79.

Idziak D, Hazuka I, Poliwczak B, Wiszynska A, Wolny E, Hasterok R. Insight into the karyotype evolution of *Brachypodium* species using comparative chromosome barcoding. PLoS One. 2014;9:e93503.

Jaaska V. Isozyme variation and differentiation of morphologically cryptic species in the Brachypodium distachyon complex. Biochem Syst Ecol. 2014;56:185–90.

Jacques-Félix H. Les Graminées d'Afrique Tropicale. Paris: Institut des Recherches Agronomiques Tropicales et des Cultures Vivrières; 1962. p. 168–78.

Jenkins G, Hasterok R, Draper J. Building the molecular infrastructure of a new model grass. In: Zwierzykowski Z, Surma M, Kachlicki P, editors. Applications of novel cytogenetic and molecular techniques in genetics and breeding of the Grasses. Poznan: Institute of Plant Genetics PAS; 2003. p. 77–84.

Jones G, Sagitov S, Oxelman B. Statistical inference of allopolyploid species networks in the presence of incomplete lineage sorting. Syst Biol. 2013;62:1–12.

Khan MA. Biosystematic studies in *Brachypodium* (Poaceae). PhD thesis. University of Leicester; 1984.

Khan MA. Seed-protein electrophoretic pattern in *Brachypodium* P. Beauv species. Ann Bot. 1992;70:61–8.

Khan MA, Stace CA. Breeding relationships in the genus *Brachypodium* (Poaceae: Pooideae). Nord J Bot. 1999;19:257–69.

López-Alvarez D, López-Herranz ML, Betekhtin A, Catalan P. A DNA barcoding method to discriminate between the model plant *Brachypodium distachyon* and its close relatives *B. stacei* and *B. hybridum* (Poaceae). PLoS One. 2012;7(12):e51058.

López-Alvarez D, Manzaneda AJ, Rey PJ, Giraldo P, Benavente E, Allainguillaume J, et al. Environmental niche variation and evolutionary diversification of the *Brachypodium distachyon* grass complex species in their native circum-Mediterranean range. Am J Bot. 2015;102:1–16.

Manzaneda AJ, Rey PJ, Bastida JM, Weiss-Lehman C, Raskin E, Mitchell-Olds T. Environmental aridity is associated with cytotype segregation and polyploidy occurrence in *Brachypodium distachyon* (Poaceae). New Phytol. 2012;193:797–805.

Marcussen T, Sandve SR, Heier L, Spannagl M, Pfeifer M, IWGSC, et al. Ancient hybridizations among the ancestral genomes of bread wheat. Science. 2014;345(6194). doi:10.1126/science.1250092.

Marcussen T, Heier L, Brysting AK, Oxelman B, Jakobsen KS. From gene trees to a dated allopolyploid network: insights from the Angiosperm Genus Viola (Violaceae). Syst Biol. 2015;64:84–101.

Meimberg H, Rice KJ, Milan NF, Njoku CC, McKay JK. Multiple origins promote the ecological amplitude of allopolyploid *Aegilops* (Poaceae). Am J Bot. 2009;96:1262–73.

Minaya M, Pimentel M, Mason-Gamer R, Catalan P. Distribution and evolutionary dynamics of *Stowaway* Miniature Inverted repeat Transposable Elements (MITEs) in grasses. Mol Phylogenet Evol. 2013;68:106–18.

Minaya M, Díaz-Pérez AJ, Mason-Gamer R, Pimentel M, Catalán P. Evolution of the *beta-amylase* gene in the temperate grasses: non-purifying selection, recombination, semiparalogy, homeology and phylogenetic signal. Mol Phylogenet Evol. 2015;91:68–85.

Mur LA, Allainguillaume J, Catalan P, Hasterok R, Jenkins G, et al. Exploiting the Brachypodium Tool Box in cereal and grass research. New Phytol. 2011;191:334–47.

Robertson IH. Chromosome numbers in *Brachypodium* Beauv. (Gramineae). Genetica. 1981;56:55–60.

Shi Y, Draper J, Stace CA. Ribosomal DNA variation and its phylogenetic implication in the genus *Brachypodium* (Poaceae). Plant Syst Evol. 1993;188:125–38.

Schippmann U. *Brachypodium boissieri* Nyman. An endemic grass species of southern Spain. Lagascalia. 1990;15:179–88.

Schippmann U. Revision der europäischen Arten der Gattung *Brachypodium* Palisot de Beauvois (Poaceae). Boissiera. 1991;45:1–250.

Scholz H. On the identity of *Brachypodium firmifolium* (Poaceae) from Cyprus. Willdenowia. 2007;37:215–20.

Schneider J, Winterfeld G, Hoffmann MH, Röser M. Duthieeae, a new tribe of grasses (Poaceae) identified among the early diverging lineages of subfamily Pooideae: molecular phylogenetics, morphological delineation, cytogenetics, and biogeography. Syst Biodiver. 2011;9:27–44.

Soltis DE, Soltis PS. Polyploidy: recurrent formation and genome evolution. Trends Ecol Evol. 1999;14:348–52.

Soltis DE, Buggs RJA, Doyle JJ, Soltis PS. What we still don't know about polyploidy. Taxon. 2010;59:1387–403.

Steinwand MA, Young HA, Bragg JN, Tobias CM, Vogel JP. *Brachypodium sylvaticum*, a model for perennial grasses: transformation and inbred line development. PLoS One. 2013;8(9): e75180.

Vain P. *Brachypodium* as a model system for grass research. J Cereal Sci. 2011;54:1–7.

Valdés B, Scholz H.. *Brachypodium genuense* (DC.) Roem. & Schult. Euro+Med PlantNase—the information resource for Euro-Mediterranean plant diversity. 2009. http://ww2.bgbm.org/euroPlusMed/PTaxonDetail.asp?UUID=142AB795-174E-40A3-9787-9B1D435403FA.

Vogel JP, Tuna M, Budak H, Huo NX, Gu YQ, Steinwand MA. Development of SSR markers and analysis of diversity in Turkish populations of *Brachypodium distachyon*. BMC Plant Biol. 2009;9:88.

Warren DL, Glor RE, Turelli M. Environmental niche equivalency versus conservatism: quantitative approaches to niche evolution. Evolution. 2008;62:2868–83.

Watson L, Dallwitz MJ. The grass genera of the world. Wallingford, Oxon: CAB International; 1992.

Wolny E, Hasterok R. Comparative cytogenetic analysis of the genomes of the model grass *Brachypodium distachyon* and its close relatives. Ann Bot. 2009;104:873–81.

Wolny E, Lesniewska K, Hasterok R, Langdon T. Compact genomes and complex evolution in the genus *Brachypodium*. Chromosoma. 2011;120:199–212.

Zucol AF, Brea M, Bellosi E. Phytolith studies in Gran Barranca (central Patagonia, Argentina): middle and Late Eocene. In: Madden RH, Carlini AA, Vucetich MG, Kay RF, editors. The paleontology of Gran Barranca: evolution and environmental change through the Middle Cenozoic of Patagonia. Cambridge: Cambridge University Press; 2010. p. 317–40.

Molecular Cytogenetics in the Genus *Brachypodium*

Robert Hasterok, Alexander Betekhtin, Natalia Borowska-Zuchowska, Agnieszka Braszewska-Zalewska, Dominika Idziak-Helmcke, Ewa Robaszkiewicz, and Elzbieta Wolny

Abstract Cytogenetics is the part of biology that focuses on the study of nuclear genomes at the microscopic level. In its modern incarnation, which is known as molecular cytogenetics, it represents a multidiscipline that amalgamates the various methodological approaches of cytology and molecular genetics as well as advanced microscopy and digital image processing. One of the most spectacular molecular cytogenetic techniques is fluorescence in situ hybridisation, which offers unprecedented insights into the various aspects of nuclear genome organisation at the level of mitotic and meiotic chromosomes and in the interphase nucleus. A complementary cytomolecular approach that utilises fluorescent antibodies that target methylated DNA or chemically modified histones in situ gives in-depth insights into the epigenetic modifications of chromatin that govern the modulation of the gene expression.

Brachypodium genus in general and *B. distachyon*, in particular, represent exceptionally good model systems for cytomolecular studies, which is due to their highly desirable 'model' biological features and rapidly and continuously growing repertoire of various experimental tools. This chapter outlines the present status of molecular cytogenetic analyses of the plant genome structure, dynamics and evolution using *B. distachyon* and some of its relatives, focusing on important scientific questions, such as: What mechanisms shape the karyotypes? Is the distribution of individual chromosomes within an interphase nucleus random or determined? Which epigenetic processes play crucial roles in the *B. distachyon* embryo and plant development by regulating tissue-specific gene expression and cell differentiation?

Keywords *Brachypodium* • *B. distachyon* • Chromosome • Epigenetic modifications • Fluorescence in situ hybridisation • Genome • Immunostaining techniques • Interphase nucleus • Model grass • Molecular cytogenetics

---

R. Hasterok (✉) • A. Betekhtin • N. Borowska-Zuchowska • A. Braszewska-Zalewska
D. Idziak-Helmcke • E. Robaszkiewicz • E. Wolny
Faculty of Biology and Environmental Protection, Department of Plant Anatomy and Cytology, University of Silesia in Katowice, 28 Jagiellonska Street, 40-032 Katowice, Poland
e-mail: robert.hasterok@us.edu.pl

J. Vogel (ed.), *Genetics and Genomics of Brachypodium*,
Plant Genetics and Genomics: Crops Models, DOI 10.1007/7397_2015_1
Published online: 21 October 2015

Cytogenetics is the part of biology that focuses on the study of nuclear genomes at the microscopic level. In its modern incarnation, which is known as molecular cytogenetics, it goes far beyond mere chromosome counting and measurement and represents a multidiscipline that amalgamates the various methodological approaches of cytology and molecular genetics as well as advanced microscopy and digital image processing. This is ensured by powerful methodology. One of the most spectacular molecular cytogenetic techniques is in situ hybridisation, which is the kinetically controlled renaturation of labelled nucleic acid (DNA or RNA) molecules that are called probes along with their complementary DNA or RNA targets in a cytological substrate. Its most common variant, DNA:DNA fluorescence in situ hybridisation (FISH), with evermore sophisticated probes, empowered by cutting-edge microscopes and digital image acquisition and processing systems, offers unprecedented insights into the various aspects of nuclear genome organisation. This is done not only at the level of mitotic and meiotic chromosomes but even in the interphase nucleus, thus permitting three-dimensional reconstructions of the nucleus structure whose results can be presented in a convincing and self-explanatory way on the one hand and in a colourful and visually attractive one on the other. A good example is so-called chromosome painting (CP), which permits the tracking of individual chromosomes and their specific segments during virtually all of the phases of the cell cycle. A complementary cytomolecular approach that utilises fluorescent antibodies that target methylated DNA or chemically modified histones in situ gives in-depth insights into the epigenetic modifications of chromatin that govern the modulation of the expression of sets of tissue-specific genes in various developmental programmes. Other antibodies may also target various proteins that compose specific biological structures such as the synaptonemal complex or those that are responsible for various enzymatic processes, for example, recombination events during meiosis.

In contrast to that of animals, the organisation of plant genomes at the cytomolecular level is still relatively poorly studied and understood. However, the *Brachypodium* genus, in general, and *B. distachyon*, in particular, represent exceptionally good model systems for such studies. This is not only due to their highly desirable ‘model’ biological features, such as a small nuclear genome, low chromosome number and complex phylogenetic relations, but also due to the rapidly and continuously growing repertoire of experimental tools, such as the large collections of accessions, whole genome sequence information, large-insert (BAC) libraries of genomic DNA, etc. This chapter outlines the present status of molecular cytogenetic analyses of the plant genome structure, dynamics and evolution using *B. distachyon* and some of its relatives. The current projects focus on important scientific questions, such as: What mechanisms shape the karyotypes? Is the distribution of individual chromosomes within an interphase nucleus random or determined? Which epigenetic processes play crucial roles in the *B. distachyon* embryo and plant development by regulating tissue-specific gene expression and cell differentiation?

Karyotype Structure and Evolution

The genus *Brachypodium* comprises less than 20 species with different basic chromosome numbers (x) of 5, 7, 8, 9 and 10 as well as different ploidy levels. The karyotypes of *Brachypodium* also display some differences in their chromosome size and morphology (Robertson 1981; Khan and Stace 1999; Catalan et al. 2012). These features make the species of this genus an amenable model system for studying the evolution and divergence of grass genomes at the level of the chromosome. The first karyological analyses comprised *B. distachyon* cytotypes with three different putative ploidy levels and somatic chromosome numbers (2n) of 10, 20 and 30, respectively (Robertson 1981), which have traditionally been regarded as constituting an autopolyploid series. However, more recent cytogenetic studies that were based on fluorescence in situ hybridisation (FISH) with genomic DNA (GISH) and ribosomal DNA and BAC (bacterial artificial chromosome) clones as probes (Hasterok et al. 2004, 2006) indicated that the 2n = 10 and 2n = 20 chromosome cytotypes may, in fact, correspond to two different diploid taxa. This assumption was further confirmed by the comparative phenotypic, cytogenetic and molecular studies of Catalan et al. (2012), who taxonomically separated these cytotypes into three distinct species, namely, diploid *B. distachyon* (2n = 10) and *B. stacei* (2n = 20), which are the putative evolutionary ancestors for the third species, an allotetraploid *B. hybridum* (2n = 30).

A cytogenetic analysis of the 2n = 28 allopolyploid *B. pinnatum* and *B. phoenicoides* using GISH with DNA that had been isolated from different *Brachypodium* diploids indicated that *B. distachyon* is a likely candidate for one of the ancestral species for these polyploids (Wolny and Hasterok 2009). Moreover, the chromosome number of *B. distachyon* fits the hypothesis that allopolyploids with 2n = 28 chromosomes arose through hybridisation between 2n = 10 and 2n = 18 species. Surprisingly, the subsequent molecular phylogenetic data excluded *B. distachyon* as a putative parent of *B. pinnatum* 2n = 28 and *B. phoenicoides* (Wolny et al. 2011). The phylogeny based on a single copy nuclear gene sequence analysis pointed to diploid (2n = 18) *B. pinnatum* and *B. rupestre* as components of these polyploids. These data are in concordance with chromosome barcoding (CB) using single-locus BAC clones that had been derived from *B. distachyon* genomic libraries (Hasterok et al. 2006; Febrer et al. 2010; International Brachypodium Initiative 2010; Idziak et al. 2014).

CB, which utilises FISH with low-repeat BAC clones as probes, has the advantage of the direct visualisation and analysis of the chromosomal rearrangements that are the foundation of karyotype differentiation in related genomes. The feasibility of using single-locus BAC clones from *B. distachyon* genomic libraries in the comparative mapping of *Brachypodium* species was first shown by Hasterok et al. (2006), thus contributing to resolving the true phylogenetic relationships within the *B. distachyon* complex (Catalan et al. 2012). Later, cytomolecular studies, which were based on heterologous BAC-FISH mapping, involved other *Brachypodium* representatives with varying chromosome numbers, such as diploid perennials with 2n = 18, i.e. *B. sylvaticum*, *B. pinnatum* and *B. rupestre* and the allopolyploids *B. pinnatum* and *B. phoenicoides*. The chromosome complements of these species were studied in reference to the *B. distachyon* karyotype structure. The bioinformatic data indicated that the present *B. distachyon* karyotype was likely derived from a 12-chromosome intermediate grass ancestor through seven nested insertions of whole chromosomes into centromeres, thus resulting in x = 5 (International Brachypodium Initiative 2010).

All of the clones that were used for the studies of the karyotype structure and evolution were derived from the physical maps that had previously been developed for *B. distachyon* (Hasterok et al. 2006; Febrer et al. 2010). Pairs of representative clones that mapped to a common *B. distachyon* chromosome were differentially labelled and hybridised in various combinations. The number of chromosomes that were highlighted by a given combination of clones varied among the species that were investigated. In the case of clones from *B. distachyon* chromosomes 1–4 (Bd1–Bd4), the number of homoeologous chromosomes (homoeologues) that were revealed by comparative mapping always exceeded the number of their counterparts in *B. distachyon*. This number was also correlated with the ploidy level; for example, clones from Bd3 mapped to two chromosome pairs in diploids (*B. sylvaticum* and *B. pinnatum* 2n = 18), whereas four pairs of Bd3 homoeologues were found in the allotetraploids (*B. pinnatum* 2n = 28 and *B. phoenicoides*) (Fig. 1a, b, Idziak et al. 2014). A different situation was observed for the clones that had been derived from Bd5, which consistently mapped to only one chromosome pair in diploids and two pairs in allopolyploids.

A comparison of the distribution pattern of BAC signals indicated that some of the nested chromosome insertions that were identified in *B. distachyon* were shared by other diploids with x = 9. This suggested that these chromosomal fusions had already appeared in their common ancestor. Two nested chromosome insertions were found that were specific only for *B. distachyon* (Wolny et al. 2011; Idziak et al. 2014). The results of comparative CB in allotetraploids (Wolny et al. 2011; Idziak et al. 2014) combined with earlier phylogenetic studies (Wolny and Hasterok 2009) allowed to assume that allopolyploid *B. pinnatum* and *B. phoenicoides* resulted from the hybridisation between two species with 2n = 18 chromosomes and then underwent a progressive reduction in chromosome number from 2n = 36 to 2n = 28.

Chromosome painting (CP) is an approach that is complementary to chromosome barcoding and is one of the most advanced and spectacular tools of modern molecular cytogenetics. It refers to the method of the selective visualisation of entire chromosomes or their segments using FISH with chromosome-specific DNA probes (Lichter et al. 1988; Pinkel et al. 1988). CP permits the discrimination and tracking of chromosomes and specific chromosome regions at various stages of mitosis and meiosis as well as in interphase nuclei. Initially, this technique was successfully applied only to the study of animal chromosomes, especially of primates (Muller et al. 1999), and as an important tool in the clinical diagnostics of chromosomal aberrations that are linked to various human diseases and malignancies (Bishop 2010; Xu et al. 2010).

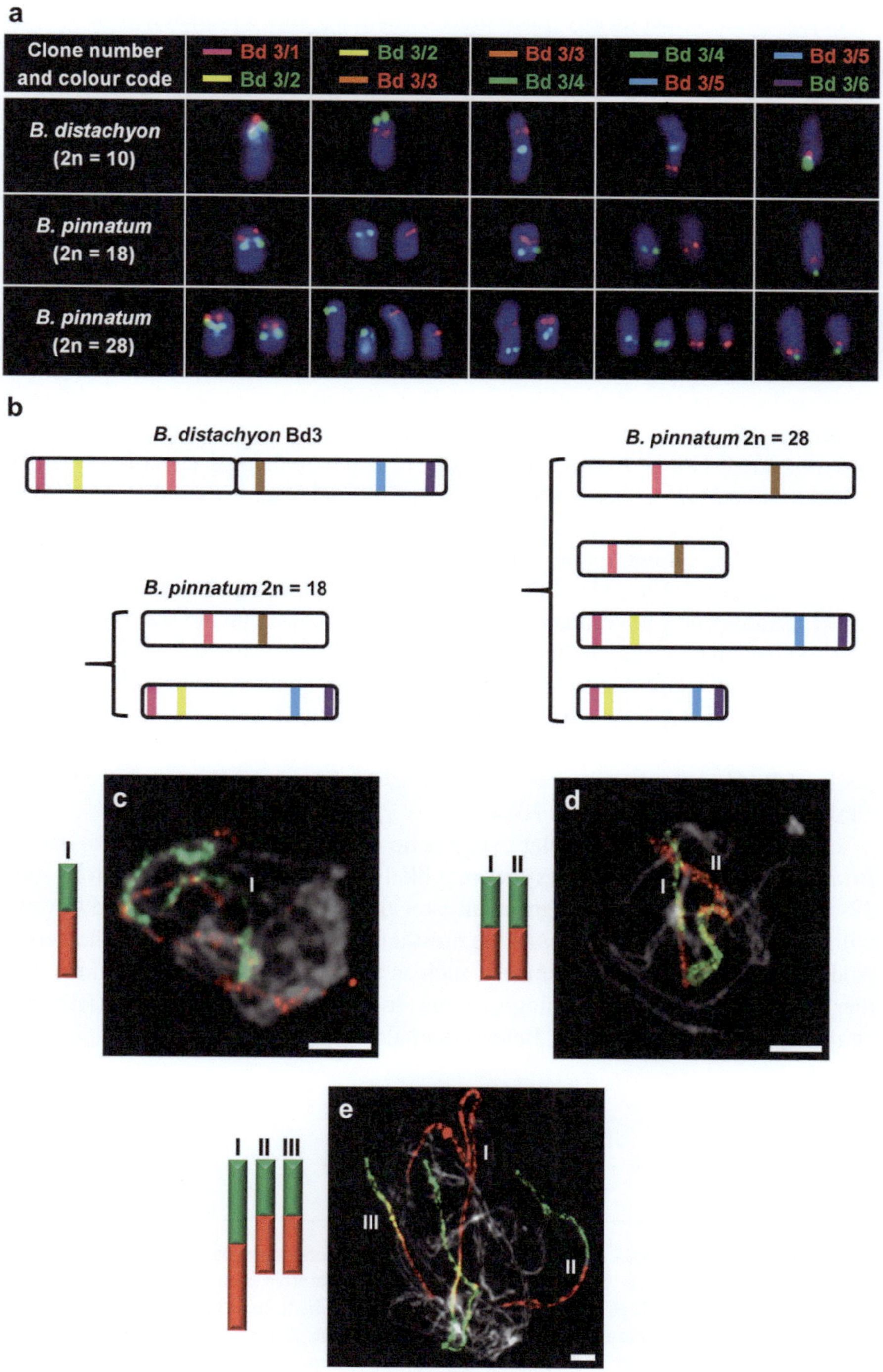

Fig. 1 (**a**) Comparative BAC-FISH mapping of clones from the Bd3 chromosome to various species of *Brachypodium*. Only one chromosome of a homologous pair is shown. The colour of the text label in the *first row* indicates the fluorochrome used (*red*, TRITC; *green*, FITC). Chromosomes counterstained with DAPI (*blue*). The *coloured bars* assigned to specific clones correspond

In plants, CP is technically much more demanding and delimited to a very small number of species such as the model dicot plant *Arabidopsis thaliana* and some of its relatives (Mandakova and Lysak 2008; Lysak et al. 2010) and several species of *Cucumis* (Lou et al. 2014). In the case of *B. distachyon*, this method was developed by exploiting large and ordered pools of BAC clones (Idziak et al. 2011). Comparative chromosome painting (CCP) with pools of low-repeat BAC clones that represent individual *B. distachyon* chromosomes (Fig. 1c–e) was also proved to be an effective tool with which to decipher the evolutionary history of its close relatives in the genus *Brachypodium*, both diploids (*B. arbuscula*, *B. pinnatum* 2n = 18, *B. stacei* and *B. sylvaticum*) and allotetraploids (*B. hybridum*, *B. phoenicoides* and *B. pinnatum* 2n = 28) as well as in two species that have an unknown ploidy level (*B. retusum* 2n = 38 and *B. mexicanum* 2n = 40). The results suggested that the genomes of *Brachypodium* diploids may have originated from a *B. mexicanum*-like or *B. stacei*-like genome through multiple dysploidy-related chromosome fission or fusion events in one of two hypothetical scenarios. The first model postulates *B. distachyon* and the second diploid perennials with 2n = 18 to be intermediate species in the evolution of the genus *Brachypodium*. CCP has also confirmed the allotetraploid origin and nature of *B. pinnatum* 2n = 28, *B. phoenicoides* and *B. hybridum* and indicated that one of their putative ancestors was *B. distachyon*. The results of CCP using BAC clones that are specific for Bd2, Bd4 and Bd5 in *B. pinnatum* 2n = 28 and *B. phoenicoides* clearly demonstrated that the painted bivalents represent the sum of their counterparts that have been observed in *B. distachyon* and in diploid *B. pinnatum* or *B. sylvaticum*, respectively. A similar situation was demonstrated in *B. hybridum* using Bd1- and Bd5-derived pools of BAC clones with respect to the *B. distachyon*- and *B. stacei*-like painted chromosomes (Betekhtin et al. 2014).

In summary, recent molecular cytogenetic data indicate that the karyotypes of *Brachypodium* representatives were likely to have evolved from the 12-chromosome intermediate grass ancestor through a series of dysploidy events with *B. mexicanum* or *B. stacei* as the most ancestral representatives of the genus. Additional structural rearrangements, such as translocations and duplications, further contributed to the increasing genome diversification in the genus (Hasterok et al. 2006; Wolny et al. 2011; Betekhtin et al. 2014; Idziak et al. 2014).

Fig. 1 (continued) to their positions as marked on cytogenetic maps (**b**) of *B. distachyon*, *B. pinnatum* 2n = 18 and *B. pinnatum* 2n = 28. (**c–e**) Comparative chromosome painting in (**c**) *B. distachyon* 2n = 10, (**d**) *B. stacei* 2n = 20 and (**e**) *B. hybridum* 2n = 30 using BAC pools that span the short (*green*) and long (*red*) arm of Bd1. Yellow fluorescence on (**d**) and (**e**) is caused by the cross-hybridisation of non-specific repeats that are common to the two chromosome arms. Painted bivalents in the photomicrographs are numbered arbitrarily using *Roman numerals*, which correspond to those on the ideograms. Chromosomes counterstained with DAPI (*grey*). Bars, 5 μm

Organisation of the Interphase Nucleus

In spite of the fact that the most commonly analysed objects at the cytomolecular level are chromosomes, for most of cell cycle, at the interphase, they appear in the form of decondensed chromatin fibres. To date, in all of the eukaryotic organisms that have been studied, chromatin was not found to be distributed randomly but was organised into specific, distinct 3-D areas of the nucleus, which are known as chromosome territories (CTs) (Cremer and Cremer 2010). The size and structure of each CT primarily depend on such factors as the size of a given chromosome and the transcriptional status of the genes that it carries (Mahy et al. 2002).

The territorial organisation of chromosomes was initially proposed by Carl Rabl (1885), who observed that centromeres and telomeres were clustered at the opposite poles of a nucleus in *Salamandra*. However, the direct visualisation of the CT arrangement within nuclei only became technically feasible almost a century later, along with the development of the FISH-based chromosome painting methodology (Lichter et al. 1988; Pinkel et al. 1988) and 3-D imaging microscopy (Cremer and Cremer 2001). Based on analyses of human cells, two models of CT distribution have been proposed (Parada and Misteli 2002). The first one assumes a radial arrangement of chromosomes, with gene-rich chromosomes or their regions distributed more internally than the gene-poor ones. This model reflects the CT distribution in many vertebrates, in which this type of arrangement seems to be evolutionarily conserved (Cremer and Cremer 2001; Mahy et al. 2002; Neusser et al. 2007; Berchtold et al. 2011). Another model postulates that chromosomes occupy specific positions within the nucleus in relation to each other. This type of chromosome arrangement was described in human quiescent fibroblast cells and Sertoli cells (Chandley et al. 1996; Nagele et al. 1999).

Plants are considerably more difficult to investigate than animals at the cytomolecular level, mainly because of the large amounts of repetitive DNA in their nuclear genomes. This hampered the application of the most precise methods for the analysis of CT distribution at the interphase, which is the painting of individual chromosomes using FISH, for a long time. The first and only complex model of CT arrangement in plants was proposed in 2004 by Pecinka et al. for *A. thaliana*. In this study, BAC clones that contained a small amount of repetitive DNA that was specific to each chromosome/chromosome arm were used, which permitted the observation of a predominantly random CT arrangement and somatic homologous pairing in the interphase nuclei of roots and leaves. The exceptions were the NOR-bearing chromosomes 2 and 4, which were associated more often rather than random, probably because of their involvement in nucleolus formation. Additionally, it was later demonstrated by the same group that CT distribution was not influenced by nuclear shape, volume or the level of endopolyploidy and was similar in meristematic and differentiated cells (Berr and Schubert 2007; Schubert et al. 2012).

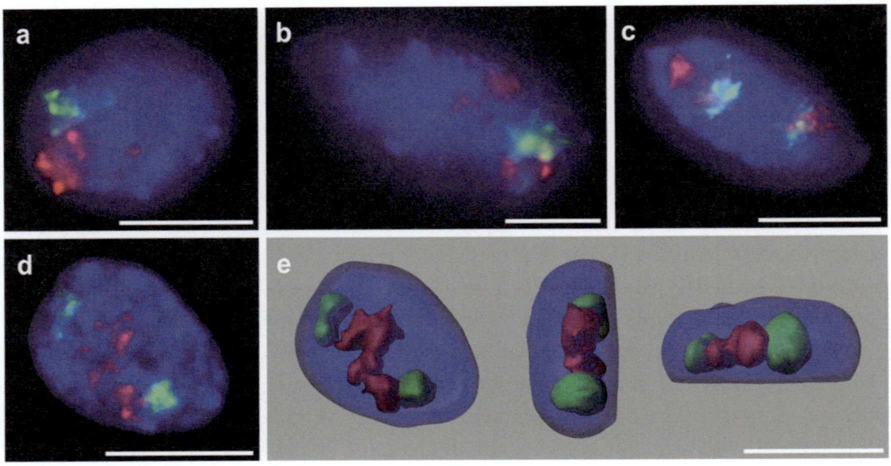

Fig. 2 Different arrangements of the Bd2 homologous chromosome territories at the interphase nuclei in *B. distachyon* root cells (*green*, top arm; *red*, bottom arm). (**a**) Complete association of homologues, (**b**) association of top arms only, (**c**) complete separation, (**d**) association of bottom arms only. (**e**) Nuclei shown in (**d**) modelled with Imaris software (Bitplane) in three different planes. Chromatin counterstained with DAPI (*blue*). Bars, 5 μm

Recently, for the first time in the monocots, CT visualisation has also been feasible in *B. distachyon*, due to the favourable combination of its small nuclear genome with a low repeat content and low chromosome number, which is complemented by the presence of the necessary methodological infrastructure that has been ensured by the completion of the whole genome sequencing project (International Brachypodium Initiative 2010; Idziak et al. 2011). The preliminary study of the CP that was applied for Bd2 in 3-D preserved nuclei that had been isolated from roots (Fig. 2a–e) revealed the presence of four types of CT associations, which were observed at different frequencies. Most commonly (38 %) both homologues were associated along their entire length (Fig. 2a), while the opposite situation, i.e. a complete separation of homologues, was observed in 27 % of the nuclei that were analysed (Fig. 2c). In other two nuclei types, only the top (18 %, Fig. 2b) or bottom arm (17 %, Fig. 2d) of the homologues was associated. This study points out a new and potentially very intriguing area in *Brachypodium* cytogenetics, since some fundamental biological processes, for example, DNA replication, transcription and repair, require interactions within and between chromatin regions. Further analyses of the CT distribution in *Brachypodium* may address some important questions, such as whether the differentiation processes or tissue and cell-specific gene expression is reflected in the spatial arrangement of chromatin and whether the CT arrangement is species specific and dependent on the ploidy level (Catalan et al. 2014).

DNA Methylation Patterns of Chromosomes

DNA methylation at position 5 of cytosine is widespread in plant nuclear genomes. It is well known that DNA methylation is linked to the stability of the chromatin structure and that it plays a crucial role in the plant genome defence system against mobile elements and invading alien DNA (Miura et al. 2001; Casacuberta and Santiago 2003). Most importantly, the methylation of cytosine residues provides a mechanism for the control of gene expression (Gehring and Henikoff 2007; Vaillant and Paszkowski 2007). There are several molecular methods that permit the mapping of 5-methylcytosine (5-MeC) at specific DNA sites in the genome (Gupta et al. 2010). Cytological techniques that enable the analysis of 5-MeC distribution at the chromosomal level have also been developed, and the DNA methylation patterns of metaphase chromosomes have been studied in different plant species (Ruffini Castiglione et al. 2002).

In 2011, the 5-MeC immunopatterns of *B. distachyon* metaphase chromosomes were determined by Borowska et al. (2011). The immunostaining technique with monoclonal antibodies that had been raised against 5-MeC was followed by FISH with chromosome-specific BAC clones in order to identify each chromosome in the complement. The anti-5-MeC signals were additionally quantified and averaged along the longitudinal axes of all of the chromosomes. Only two acrocentric chromosomes, Bd4 and Bd5, were found to have the characteristic 5-MeC distribution. It is known that Bd4 bears a 5S rDNA locus, which is proximally located in its long arm, while Bd5 has a nucleolar organising region (NOR), which is found in the terminal part of the short arm (Fig. 3a). In the case of Bd5, the most prominent anti-5-MeC signals were visible in the pericentromeric regions, and they decreased in their intensity abruptly towards both chromosome ends. Moreover, chromosomes Bd5 that had visible satellite regions were characterised by a lower methylation level at the 35S rDNA loci, which most probably reflected the transcriptional activity of the rRNA genes (Fig. 3b, c). Chromosome Bd4 displayed two sites that were characterised by strong signals that corresponded to 5-MeC. The first one spanned the pericentromeric region, while the second one was located interstitially in the long arm. The proximal region of Bd4, which is the site of 5S rDNA, was characterised by a lower density of 5-MeC foci (Borowska et al. 2011). It is well known that only a portion of rRNA genes in a particular locus is expressed (Birch and Zomerdijk 2008) and that transcriptionally active and repressed rRNA genes have distinct epigenetic patterns. A higher level of DNA methylation has been attributed to a silenced fraction of the rRNA genes (Lawrence et al. 2004). It was found by Borowska et al. (2011) that there is a decrease in methylation levels in both types of rDNA loci in *B. distachyon* chromosomes and these data corroborate the results of studies on other plant species, e.g. *Zingeria biebersteiniana* (Cremonini et al. 2003), *Haplopappus gracilis* (Ruffini Castiglione et al. 2008) and *Colpodium versicolor* (Castiglione et al. 2009).

In contrast to the smallest chromosomes in the complement, pairs of the metacentric chromosomes Bd1-Bd3 did not have any specific DNA methylation

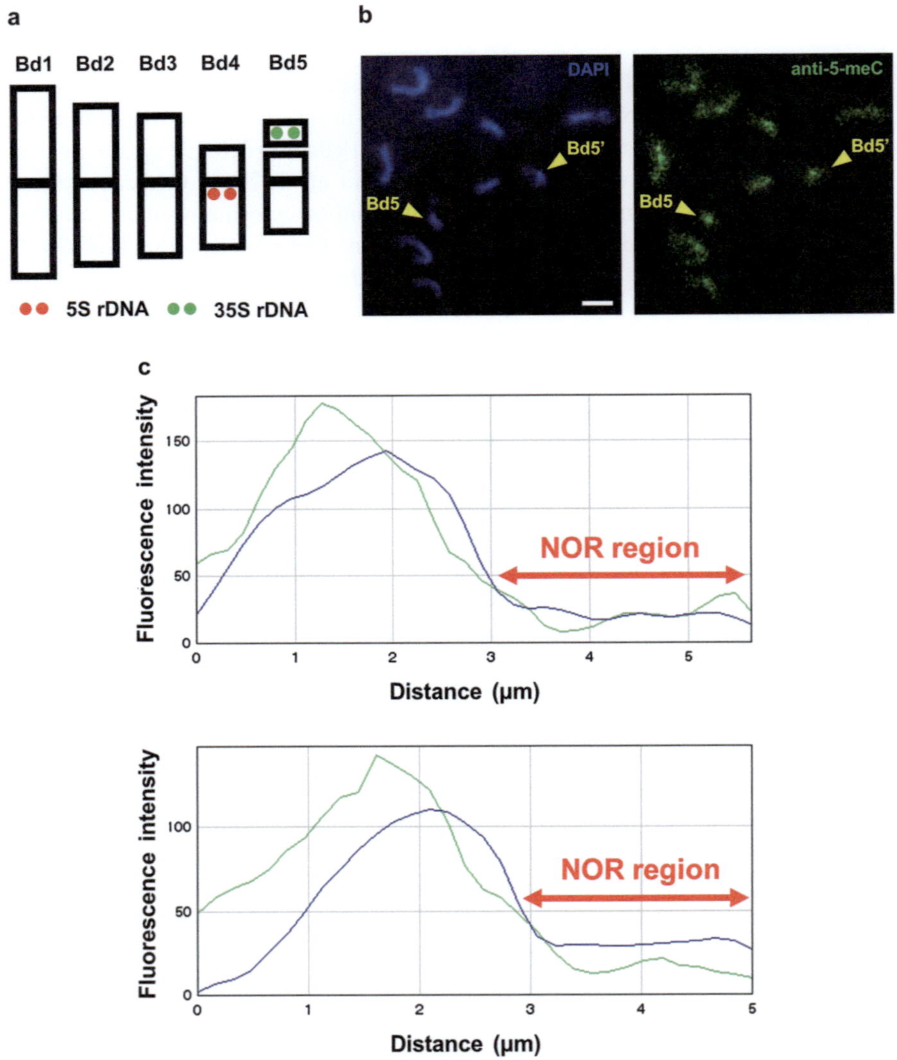

Fig. 3 DNA methylation patterns along the Bd5 chromosomes with a visible satellite region. (**a**) Idiogram of *B. distachyon* chromosomes with the positions of rDNA loci indicated. (**b**) Distribution of 5-MeC foci (*green*) along the metaphase chromosomes. Chromatin counterstained with DAPI (*blue*). Bar, 5 μm. (**c**) DNA methylation profiles of Bd5 and Bd5′ homologues shown on (**b**). Chromosomes oriented with their long arm to the *left*. The length (μm) of the chromosomes is visualised on the X-axis and signal intensity on the Y-axis (arbitrary units). *Blue curves* reflect the DAPI signal and *green* represent the anti-5-MeC signal

patterns. The heterogeneity of both the anti-5-MeC signal distribution and fluorescence intensity was observed not only among the chromosomes of different meristematic cells but also between the homologues from the same cell and between the arms of the same chromosome. The different 5-MeC foci distributions between two homologues can be explained by the distinct chromatin organisation, which can be attributed to the differential transcriptional activity of homologous chromosomes (Borowska et al. 2011). Distinct 5-MeC patterns were also found between *Melandrium album* sex chromosomes, where one X chromosome in the female homozygote was hypermethylated along its entire length, while its homologue showed a higher density of 5-MeC foci only on its short arm (Siroky et al. 1998).

The most distinct anti-5-MeC signals were found in the pericentromeric regions of all of the *B. distachyon* chromosomes. It was documented that LTR elements, which belong mostly to the Ty3-*gypsy* superfamily, are located within these regions (International Brachypodium Initiative 2010). An accumulation of 5-MeC in these chromosome segments is most probably the result of the genome defence system against the activation of mobile elements. In contrast, the terminal chromosome regions were often unmethylated or showed a lower density of 5-MeC foci as compared with the other chromosome segments. An analysis of the *B. distachyon* genomic sequence revealed that genes are evenly distributed along the chromosome arms with the exception of the regions that are near centromeres (International Brachypodium Initiative 2010). A decrease in the methylation level at the distal parts of the chromosomes suggests that transcriptionally active genes may be located predominantly within these regions (Borowska et al. 2011).

In order to analyse the effect of the induced hypomethylation on the 5-MeC distribution along *B. distachyon* chromosomes, the seeds were treated with 5-azacytidine (5-AzaC) at three different concentrations: 0.1, 0.01 and 0.001 mmol/L. This hypomethylating agent led to a significant decrease in the DNA methylation level in the *B. distachyon* genome, especially in the case of the two highest concentrations that were used. Interestingly, some chromosome pairs retained strong signals that corresponded to 5-MeC, while the others seemed to undergo the hypomethylation process more readily. However, no evidence for the specific hypomethylation of particular chromosome regions was observed. Moreover, the application of 5-AzaC apparently caused the loss of chromosome-specific DNA methylation patterns in Bd4 and Bd5, and this finding is in concordance with the work of Siroky et al. (1998) in which treatment with this hypomethylating agent resulted in the loss of specific 5-MeC distribution along the sex chromosomes of *M. album*.

Epigenetic Modifications During Seed Development

The seed and embryo are important stages of plant development that have a critical influence on the life cycle of a new plant. The development of a seed after fertilisation is divided into three main phases, namely, seed maturation, desiccation and storage and seed germination. Following morphogenesis, the developing seed enters the maturation stage. This stage commences with the transition phase during which the switch from maternal to filial control occurs. The seed then undergoes a period of embryo growth, the so-called reserve accumulation, reorganisation of metabolism and synthesis of storage compounds (Angelovici et al. 2010). Maturation ends with the desiccation phase after which the embryo enters into a quiescent state (Gutierrez et al. 2007). This stage is associated with a major loss of water, which leads to a dry seed in preparation for the quiescent period, dormancy and thereafter germination. Three phases can also be distinguished during germination: (1) seed imbibition and reinitiation of the metabolic processes, (2) limited water uptake and (3) increased water uptake, which results in the emergence of the radicle (Holdsworth et al. 2008).

Plants have the ability to react to seasonal changes by synchronising their life cycle transitions with the environmental conditions. The switch from one developmental phase to the next requires significant changes in both the spatial and temporal patterns of gene expression. The transcriptional reprogramming of these genes involves an active modification of the chromatin structure. Many studies have indicated that chromatin organisation is a dynamic process and that it undergoes considerable rearrangement during plant development. There are several factors that can greatly influence the chromatin structure, one of which is a chemical modification of the histone proteins. Core histones are structurally conserved through evolution and contain flexible N-terminal tails that may be subject to numerous posttranslational covalent modifications, including acetylation, methylation, phosphorylation, ubiquitination, ribosylation, glycosylation and sumoylation (Jenuwein and Allis 2001). There is much data on the dynamics of the turnover of epigenetic modifications during plant development. Through the developmental regulation of these mechanisms, plants undergo epigenetic reprogramming in various cell types and developmental stages.

The distribution of two epigenetic markers, i.e. the histone H4 acetylation of lysine K5 (H4K5ac) and the histone H3 dimethylation of lysine K4 (H3K4me2), was analysed in tissue sections of 'matured', 'dry' and 'germinating' embryos of *B. distachyon* (Wolny et al. 2014). Coupled with immunostaining, this provided a unique opportunity to analyse the distinct marks of epigenetic modifications in the nuclei of particular embryo tissues and organs with their preserved topographical context. The results indicated that the abundance of these modifications differed considerably in the various organs and tissues of the three types of embryos. Moreover, the patterns of epigenetic modifications varied not only between particular tissues of the same embryo type but also between the different types of embryos that were analysed. The scutellum, coleorhiza and coleoptiles were the most variable organs in terms of histone H4 acetylation and histone H3 methylation. It was revealed in detail that histone H4 acetylation and H3 methylation in 'matured' embryos were most prominent in the root apical meristem (Fig. 4a, b), whereas it occurred in the shoot apical meristem and leaf primordia in 'dry' ones

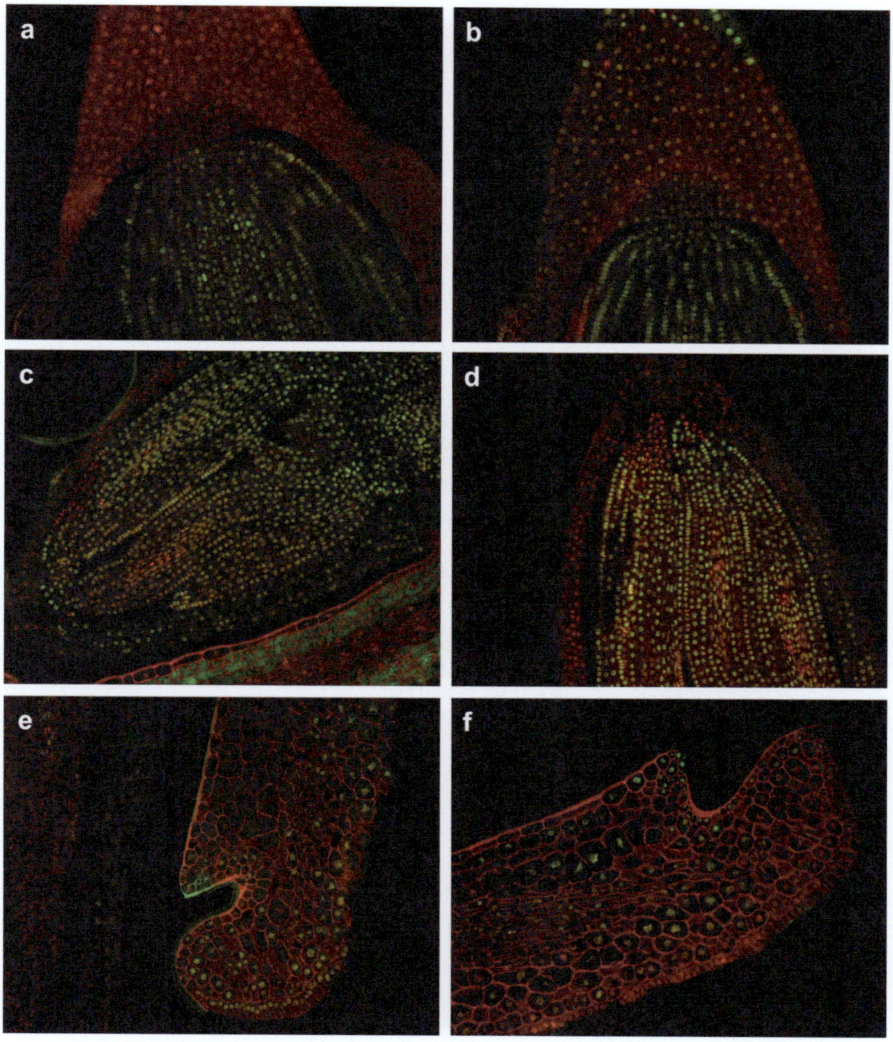

Fig. 4 Immunodetection (*green*) of chemically modified histones in *B. distachyon* embryos at the various stages of seed development. (**a**, **b**) 'Matured', (**c**, **d**) 'dry', (**e**, **f**) 'germinating' embryos. (**a**, **c**, **e**) H4K5ac, (**b**, **d**, **f**) H3K4me2. Sections counterstained with DAPI and visualised in *red* (artificial colour). Bars, 50 μm

(Fig. 4c, d). Conversely, histone H4 acetylation and H3 methylation were prevalent in the scutellum in the 'germinating' embryos (Fig. 4e, f). The distinct patterns of epigenetic modifications that were observed may be involved in the switch of the gene expression profiles in specific organs of the developing embryo and may be linked to the physiological changes that accompany seed desiccation, imbibition and germination (Wolny et al. 2014).

Conclusions

Although the first molecular cytogenetic analyses in *Brachypodium* date back to 2000, there is no doubt that a real breakthrough in this field came a decade later with the sequencing of *B. distachyon* nuclear genome (International Brachypodium Initiative 2010). The integration of sequence data and DNA fingerprinting of deep-coverage large-insert BAC libraries (Febrer et al. 2010) laid the foundation for construction of much more sophisticated FISH probes. Their ultra-high region and chromosome specificity enabled the successful application of advanced cytomolecular approaches in plants, such as CB and CP, not only in *B. distachyon* but also in other representatives of the genus, thereby accelerating the research and contributing to our better understanding of the various aspects of grass nuclear genome organisation at the microscopic level (Idziak et al. 2011, 2014; Betekhtin et al. 2014). Considering that sequencing of some other *Brachypodium* nuclear genomes has already reached the final stages of annotation, and also appreciating the recent advances in the immunolocalisation of epigenetic modifications of chromatin (Borowska et al. 2011; Wolny et al. 2014, 2015), it can be anticipated that the most exciting time for cytomolecular analyses in *Brachypodium* is still to come.

Acknowledgements This work was supported by the Polish National Science Centre (grants no. DEC-2012/04/A/NZ3/00572 and DEC-2011/01/B/NZ3/00177) and by the Polish Ministry of Science and Higher Education (grant no. N303 570738).

References

Angelovici R, Galili G, Fernie AR, Fait A. Seed desiccation: a bridge between maturation and germination. Trends Plant Sci. 2010;15:211–8.

Berchtold D, Fesser S, Bachmann G, Kaiser A, Eilert JC, Frohns F, et al. Nuclei of chicken neurons in tissues and three-dimensional cell cultures are organized into distinct radial zones. Chromosome Res. 2011;19:165–82.

Berr A, Schubert I. Interphase chromosome arrangement in *Arabidopsis thaliana* is similar in differentiated and meristematic tissues and shows a transient mirror symmetry after nuclear division. Genetics. 2007;176:853–63.

Betekhtin A, Jenkins G, Hasterok R. Reconstructing the evolution of *Brachypodium* genomes using comparative chromosome painting. PLoS One. 2014;9:e115108.

Birch JL, Zomerdijk JC. Structure and function of ribosomal RNA gene chromatin. Biochem Soc Trans. 2008;36:619–24.

Bishop R. Applications of fluorescence *in situ* hybridization (FISH) in detecting genetic aberrations of medical significance. Biosci Horizons. 2010;3:85–95.

Borowska N, Idziak D, Hasterok R. DNA methylation patterns of *Brachypodium distachyon* chromosomes and their alteration by 5-azacytidine treatment. Chromosome Res. 2011;19:955–67.

Casacuberta JM, Santiago N. Plant LTR-retrotransposons and MITEs: control of transposition and impact on the evolution of plant genes and genomes. Gene. 2003;311:1–11.

Castiglione MR, Kotseruba V, Cremonini R. Methylated-rich regions and tandem repeat arrays along the chromosome complement of *Colpodium versicolor* (Stev.) Schmalh. Protoplasma. 2009;237:13–8.

Catalan P, Muller J, Hasterok R, Jenkins G, Mur LA, Langdon T, et al. Evolution and taxonomic split of the model grass *Brachypodium distachyon*. Ann Bot. 2012;109:385–405.

Catalan P, Chalhoub B, Chochois V, Garvin DF, Hasterok R, Manzaneda AJ, et al. Update on the genomics and basic biology of *Brachypodium*. Trends Plant Sci. 2014;19:414–8.

Chandley AC, Speed RM, Leitch AR. Different distributions of homologous chromosomes in adult human Sertoli cells and in lymphocytes signify nuclear differentiation. J Cell Sci. 1996;109(Pt 4):773–6.

Cremer T, Cremer C. Chromosome territories, nuclear architecture and gene regulation in mammalian cells. Nat Rev Genet. 2001;2:292–301.

Cremer T, Cremer M. Chromosome territories. Cold Spring Harb Perspect Biol. 2010;2:a003889.

Cremonini R, Castiglione MR, Grif VG, Kotseruba VV, Punina EO, Rodionov AV, et al. Chromosome banding and DNA methylation patterns, chromatin organisation and nuclear DNA content in *Zingeria biebersteiniana*. Biol Plantarum. 2003;46:543–50.

Febrer M, Goicoechea JL, Wright J, McKenzie N, Song X, Lin J, et al. An integrated physical, genetic and cytogenetic map of *Brachypodium distachyon*, a model system for grass research. PLoS One. 2010;5:e13461.

Gehring M, Henikoff S. DNA methylation dynamics in plant genomes. Biochim Biophys Acta. 2007;1769:276–86.

Gupta R, Nagarajan A, Wajapeyee N. Advances in genome-wide DNA methylation analysis. Biotechniques. 2010;49:iii–xi.

Gutierrez L, Van Wuytswinkel O, Castelain M, Bellini C. Combined networks regulating seed maturation. Trends Plant Sci. 2007;12:294–300.

Hasterok R, Draper J, Jenkins G. Laying the cytotaxonomic foundations of a new model grass, *Brachypodium distachyon* (L.) Beauv. Chromosome Res. 2004;12:397–403.

Hasterok R, Marasek A, Donnison IS, Armstead I, Thomas A, King IP, et al. Alignment of the genomes of *Brachypodium distachyon* and temperate cereals and grasses using bacterial artificial chromosome landing with fluorescence *in situ* hybridization. Genetics. 2006;173:349–62.

Holdsworth MJ, Finch-Savage WE, Grappin P, Job D. Post-genomics dissection of seed dormancy and germination. Trends Plant Sci. 2008;13:7–13.

Idziak D, Betekhtin A, Wolny E, Lesniewska K, Wright J, Febrer M, et al. Painting the chromosomes of *Brachypodium*: current status and future prospects. Chromosoma. 2011;120:469–79.

Idziak D, Hazuka I, Poliwczak B, Wiszynska A, Wolny E, Hasterok R. Insight into the karyotype evolution of *Brachypodium* species using comparative chromosome barcoding. PLoS One. 2014;9:e93503.

International Brachypodium Initiative. Genome sequencing and analysis of the model grass *Brachypodium distachyon*. Nature. 2010;463:763–8.

Jenuwein T, Allis CD. Translating the histone code. Science. 2001;293:1074–80.

Khan MA, Stace CA. Breeding relationships in the genus *Brachypodium* (Poaceae: Pooideae). Nord J Bot. 1999;19:257–69.

Lawrence RJ, Earley K, Pontes O, Silva M, Chen ZJ, Neves N, et al. A concerted DNA methylation/histone methylation switch regulates rRNA gene dosage control and nucleolar dominance. Mol Cell. 2004;13:599–609.

Lichter P, Cremer T, Borden J, Manuelidis L, Ward DC. Delineation of individual human chromosomes in metaphase and interphase cells by *in situ* suppression hybridization using recombinant DNA libraries. Hum Genet. 1988;80:224–34.

Lou Q, Zhang Y, He Y, Li J, Jia L, Cheng C, et al. Single-copy gene-based chromosome painting in cucumber and its application for chromosome rearrangement analysis in *Cucumis*. Plant J. 2014;78:169–79.

Lysak MA, Mandakova T, Lacombe E. Reciprocal and multi-species chromosome BAC painting in crucifers (Brassicaceae). Cytogenet Genome Res. 2010;129:184–9.

Mahy NL, Perry PE, Gilchrist S, Baldock RA, Bickmore WA. Spatial organization of active and inactive genes and noncoding DNA within chromosome territories. J Cell Biol. 2002;157:579–89.

Mandakova T, Lysak MA. Chromosomal phylogeny and karyotype evolution in x = 7 crucifer species (Brassicaceae). Plant Cell. 2008;20:2559–70.

Miura A, Yonebayashi S, Watanabe K, Toyama T, Shimada H, Kakutani T. Mobilization of transposons by a mutation abolishing full DNA methylation in Arabidopsis. Nature. 2001;411:212–4.

Muller S, Stanyon R, O'Brien PC, Ferguson-Smith MA, Plesker R, Wienberg J. Defining the ancestral karyotype of all primates by multidirectional chromosome painting between tree shrews, lemurs and humans. Chromosoma. 1999;108:393–400.

Nagele RG, Freeman T, McMorrow L, Thomson Z, Kitson-Wind K, Lee H. Chromosomes exhibit preferential positioning in nuclei of quiescent human cells. J Cell Sci. 1999;112(Pt 4):525–35.

Neusser M, Schubel V, Koch A, Cremer T, Muller S. Evolutionarily conserved, cell type and species-specific higher order chromatin arrangements in interphase nuclei of primates. Chromosoma. 2007;116:307–20.

Parada L, Misteli T. Chromosome positioning in the interphase nucleus. Trends Cell Biol. 2002;12:425–32.

Pecinka A, Schubert V, Meister A, Kreth G, Klatte M, Lysak MA, et al. Chromosome territory arrangement and homologous pairing in nuclei of *Arabidopsis thaliana* are predominantly random except for NOR-bearing chromosomes. Chromosoma. 2004;113:258–69.

Pinkel D, Landegent J, Collins C, Fuscoe J, Segraves R, Lucas J, et al. Fluorescence *in situ* hybridization with human chromosome-specific libraries: detection of trisomy 21 and translocations of chromosome 4. Proc Natl Acad Sci U S A. 1988;85:9138–42.

Rabl C. Über Zelltheilung. Morphol Jahrb. 1885;10:214–330.

Robertson IH. Chromosome numbers in *Brachypodium* Beauv. (Gramineae). Genetica. 1981;56:55–60.

Ruffini Castiglione M, Cremonini R, Frediani M. DNA methylation patterns on plant chromosomes. Caryologia. 2002;55:275–82.

Ruffini Castiglione M, Frediani M, Venora G, Cremonini R. Cytological investigation of *Haplopappus gracilis* (Nutt.) Gray: 5-methylcytosine-rich regions, fluorochrome banding and chromatin sensitivity to DNase I digestion. Protoplasma. 2008;233:107–13.

Schubert V, Berr A, Meister A. Interphase chromatin organisation in Arabidopsis nuclei: constraints versus randomness. Chromosoma. 2012;121:369–87.

Siroky J, Castiglione MR, Vyskot B. DNA methylation patterns of *Melandrium album* chromosomes. Chromosome Res. 1998;6:441–6.

Vaillant I, Paszkowski J. Role of histone and DNA methylation in gene regulation. Curr Opin Plant Biol. 2007;10:528–33.

Wolny E, Hasterok R. Comparative cytogenetic analysis of the genomes of the model grass *Brachypodium distachyon* and its close relatives. Ann Bot. 2009;104:873–81.

Wolny E, Lesniewska K, Hasterok R, Langdon T. Compact genomes and complex evolution in the genus *Brachypodium*. Chromosoma. 2011;120:199–212.

Wolny E, Braszewska-Zalewska A, Hasterok R. Spatial distribution of epigenetic modifications in *Brachypodium distachyon* embryos during seed maturation and germination. PLoS One. 2014;9:e101246.

Wolny E, Braszewska-Zalewska A, Kroczek D, Hasterok R. *In situ* analysis of epigenetic modifications in the chromatin of *Brachypodium distachyon* embryos. Plant Signal Behav. 2015;10(5):e1011948.

Xu W, Li JY, Liu Q, Zhu Y, Pan JL, Qiu HR, et al. Multiplex fluorescence *in situ* hybridization in identifying chromosome involvement of complex karyotypes in *dę novo* myelodysplastic syndromes and acute myeloid leukemia. Int J Lab Hematol. 2010;32:e86–95.

The *Brachypodium distachyon* Reference Genome

Robert VanBuren and Todd C. Mockler

Abstract Grasses provide the bulk of human calories but improvement in grass yields is hindered by the characteristically large and complex genomes of these species; the genomes of wheat, maize, and sugar cane are 17,000, 2300, and 10,000 Mb, respectively. *Brachypodium distachyon* has one of the smallest genomes of all grasses at 272 Mb, and a number of key traits that make it a good model grass. *Brachypodium* was the fourth sequenced grass genome, after rice, *Sorghum*, and maize, and was the first sequenced in the Pooideae subfamily, a diverse group that includes wheat, barley, oat, and rye. The *Brachypodium* genome was sequenced using a whole genome shotgun approach with Sanger sequencing and is nearly complete with 99.6 % of the sequences anchored to five chromosomes. Sequencing of *Brachypodium* enabled comparative genomic analysis of grass genomes and shed light on processes involved in chromosome fusions and maintenance of a small genome. The high-quality *Brachypodium* genome sequence provides a framework for gene expression atlases, resequencing, quantitative trait loci (QTL) mapping, GWAS, and ENCODE datasets. The wealth of *Brachypodium* genomic resources have cemented its utility as a model organism and will facilitate translational work for improving the grasses that feed the world.

Keywords *Brachypodium* • Sanger sequencing • WGS • BAC • Comparative genomics • Polyploidy • Gene families • Resequencing

Abbreviations

BAC	Bacterial artificial chromosome
cM	centiMorgan
contigs	contiguous pieces
ENCODE	Encyclopedia of DNA Elements
EST	Expressed sequence tags
GWAS	Genome-wide association study
indel	Insertion/deletion

R. VanBuren • T.C. Mockler (✉)
Donald Danforth Plant Science Center, St. Louis, MO 63132, USA
e-mail: tmockler@danforthcenter.org

J. Vogel (ed.), *Genetics and Genomics of Brachypodium*,
Plant Genetics and Genomics: Crops Models, DOI 10.1007/7397_2015_15
Published online: 02 October 2015

mb	Megabase
MYA	Million years ago
NGS	Next generation sequencing
RNAseq	RNA sequencing
SMRT	Single molecule real time
SNP	Single nucleotide polymorphism
SV	Structural variant
TD	Tandem duplications
TILLING	Targeting Induced Local Lesion IN Genome
WGS	Whole genome shotgun

Introduction

More than 70 % of calories consumed by humans are derived from grasses and the grass family, and Poaceae is arguably the most important plant family in terms of global food security (Elert 2014). Roughly 42 % of human calories come directly from maize, wheat, rice, and other grasses (Elert 2014) and many other grasses are used for animal feed. In addition to their roles as food, grasses like bamboo and thatching grasses are used as construction materials, and highly productive grasses like *Sorghum*, sugarcane, and *Miscanthus* show potential as cellulosic biofuel crops. Grasses have some of the highest productivity and photosynthetic efficiency of all plants, traits that are largely attributed to the optimized biochemical and morphological adaptations of C4 photosynthesis, which allows for carbon assimilation under high temperatures and drought (Hatch and Slack 1966). Increasing of grass yields, an essential task given the challenges of population growth, climate change, and disease, is hindered by holes in our basic understanding of grass biology (Godfray et al. 2010, 2011). Studying grasses is complicated by their large genomes, mature plant size, and long generation times (Jung et al. 2008). Despite these challenges, great progress has been made toward understanding cell wall composition, C4 photosynthesis, plant architecture, and grain composition based on work primarily in rice and maize. Though substantial, these advances pale in comparison to work from the *Arabidopsis* research community and fall short of knowledge needed for translational work aimed at crop improvement.

Recently, the small temperate C3 grass *Brachypodium distachyon* and the panicoid C4 grass *Setaria virdis* have emerged as promising model systems for the grasses (Brkljacic et al. 2011; Brutnell et al. 2010). *Brachypodium* has a short generation time (3 months from seed to seed), small stature, compact genome (272 megabases, Mb), and simple growth requirements making it possible to grow in virtually any lab space that can support *Arabidopsis* research. An efficient transformation system, over 20,000 T-DNA lines, several ethyl methanesulfonate mutant populations, and over 200 inbreed lines can be used to support gene-based discovery and quantitative trait loci mapping of desired traits. The characteristics that make *Brachypodium* a good model combined with plans to develop these rich resources drove the desire to sequence the *Brachypodium* genome. *Brachypodium* was the fourth sequenced grass genome (behind rice, *Sorghum*, and maize) and the first from the Pooideae subfamily, the largest grass subfamily, which also includes wheat, barley, oat, and rye. A Sanger-based sequencing approach combined with several large insert bacterial artificial chromosome (BAC) and fosmid libraries, genetic maps, and manual gap filling produced a near complete genome that is currently the third best assembled genome behind *Arabidopsis* and rice (Michael and VanBuren 2015). The *Brachypodium* genome paved the way for comparisons among the three economically important grass subfamilies and provided insights into genome evolution and gene composition in the grasses. In this chapter we discuss the sequencing strategies and approaches used to assemble the *Brachypodium* genome, summarize the major findings of analyses of the genome, and end with the ongoing and future genome-scale projects in the *Brachypodium* field.

Assembling the *Brachypodium* Genome

The relatively long length of Sanger reads (300–1000 base pairs) combined with their high quality and low error rate produced a series of gold standard reference plant genomes in the beginning of the genomics era. The most complete plant genome, that of *Arabidopsis thaliana*, is currently assembled into 110 contiguous pieces (contigs; TAIR10 release), and the only remaining gaps correspond to centromeric and telemetric repeats, rRNA gene clusters, tandem gene arrays, and other highly repetitive regions (Lamesch et al. 2012). Over 100 plant genomes have been sequenced since publication of the *Arabidopsis* genome in 2000 (Arabidopsis Genome Initiative 2000), and numerous more are planned or in progress. The rapid rise in the number of sequenced plant genomes was largely due to the low cost and speed of next generation sequencing (NGS) technologies, which reduced the cost of sequencing by several orders of magnitude compared with Sanger sequencing. Genomic sequences generated with NGS Illumina or 454 technologies are of much lower quality than those generated with Sanger sequencing. Genomic sequences generated with NGS technologies typically consist of thousands of scaffolds with numerous imbedded gaps. Gaps can split single genes, and some gaps span biologically meaningful sequences including cis-regulatory regions, conserved non-coding sequences, and tandem gene arrays. Several important grasses have been sequenced, including wheat, barley, rice, bamboo, tef, maize, *Sorghum*, *Setaria italica*, and *Brachypodium*, as have several non-grass monocots including banana, oil palm, date palm, and duckweed (Table 1; Fig. 1). *Brachypodium* was sequenced using Sanger methods and is currently one of the best assembled and most complete plant genomes (Michael and VanBuren 2015).

Two Sanger-based approaches were used to sequence the first plant genomes. A targeted, physical map based approach was used in the *Arabidopsis* (Arabidopsis Genome Initiative 2000), maize (Schnable et al. 2009), and rice genomes

Table 1 Statistics of sequenced monocot genomes

Species	Common name	Genome size (Mb)	Contig N50 (kb)	Number of genes
Spirodela polyrhiza	Duckweed	158	8.2	19,623
Brachypodium distachyon	*Brachypodium*	272	6400	31,694
Oryza sativa	Rice	370	6900	37,544
Seteria italica	*Setaria*	510	136	29,000
Musa acuminata	Banana	523	43	36,542
Phoenix dactylifera	Date palm	658	6.4	28,890
Sorghum bicolor	*Sorghum*	730	1013	27,640
Egrostis tef	Tef	772	0.8	38,333
Elaeis guineensis	Oil palm	1800	3.1	34,802
Phyllostachys heterocycla	Bamboo	2075	11.8	31,987
Zea mays	Maize	2300	40	32,540
Hordeum vulgare	Barley	5100	1.4	75,258
Triticum aestivum	Wheat	17,000	8.3	124,201

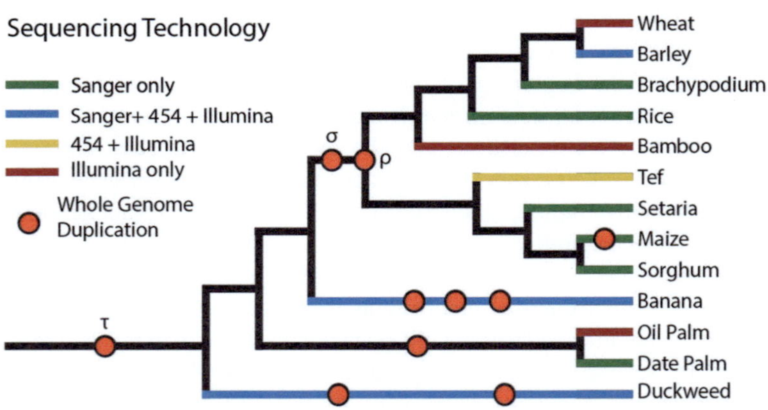

Fig. 1 Summary of sequenced monocot genomes. Branch colors represent sequencing technologies used: Sanger only (*green*), Sanger + 454/Illumina (*blue*), 454 + Illumina (*gold*), and Illumina only (*red*). WGD events are shown *orange*. The ancient WGD event Tau (τ) occurred before the diversification of monocots and the sigma (σ) and rho (ρ) events are Poales specific. Branch length does not correlate with divergence time

(Project IRGS 2005) and a whole genome shotgun (WGS) approach was used in grape (Jaillon et al. 2007), *Brachypodium* (Vogel et al. 2010), *Sorghum* (Paterson et al. 2009), poplar (Tuskan et al. 2006), and papaya (Ming et al. 2008). In a physical map based approach, large pieces of genomic DNA are fragmented into 50–200 kilobase (kb) pieces, inserted into BAC vectors, and transformed into and maintained in *E. coli*. This library of BAC clones is then screened using a variety of approaches to create a map corresponding to the physical location of each BAC along each chromosome. A minimum tiling path of BACs representing the optimized order of BACs that span the genome with minimal overlap is generated and corresponding BACs are sequenced. A BAC-based approach ensures good coverage of the genome, as highly repetitive regions can be assembled separately in each BAC before being combined. The highest quality genomes were sequenced using a BAC-based strategy, but this approach is too cost and labor prohibitive for most projects. The WGS approach is more widely used because it costs less and requires a fraction of the time and resources of BAC-based sequencing. In a WGS approach, total genomic DNA is manually sheared to a desired length (ranging from 500–10,000 base pairs) and cloned into *E. coli*; random clones are then sequenced using Sanger technology. As repetitive sequences are problematic and assembly is computationally challenging, WGS assemblies are often improved using end-sequencing of large insert libraries like BACs or fosmids, which can bridge repeat regions and order contigs into scaffolds. Scaffolds can be joined into chromosome-scale pieces and anchored to specific chromosomes using markers from genetic maps or optical mapping approaches.

The *Brachypodium* genome was sequenced using a Sanger-based WGS approach with the incorporation of end-sequences from BAC and fosmid libraries and a genetic map for generating a chromosome-scale assembly (Vogel et al. 2010). Bd21, the most widely grown *Brachypodium* accession, was chosen for sequencing because it is highly inbreed (eight generations) and has a low level of residual within-genome heterozygosity. A wide range of insert sizes and high physical coverage of large insert BAC libraries allowed bridging of most of the highly repetitive regions and assembly into five chromosomes. Repeat sequences account for 15–90 % of plant genomes and commonly exceed 10–20 kb in length (Schatz et al. 2012). Plant genomes also have large megabase-size rRNA gene clusters and centromere-specific tandem arrays that are impossible to completely assemble using only short WGS Sanger reads. Large insert libraries can span repeats and join flanking assembled regions.

For *Brachypodium*, several large insert libraries were constructed and end-sequenced including ~4× coverage of a 3-kb library, 4.5× of an 8-kb library, 0.6× of two fosmid libraries (average insert size of 33 kb) and 0.5× of six BAC libraries (average insert size of 113 kb). The 4.7 million WGS Sanger reads collectively represent 9.4× genome coverage and were assembled into 2067 contigs distributed in 217 scaffolds using the overlap based assembler Arachne (Batzoglou et al. 2002). Most of the assembly (99.6 %) is in 13 scaffolds exceeding 50 kb in length with a scaffold N50 of 59.3 Mb. Thus, half of the *Brachypodium* genome sequence is in scaffolds 59.3 Mb or larger. The genome only contains 1.1 % gaps, much lower than most other plant genomes. For comparison, *Brachypodium* is better assembled than the *Sorghum bicolor* genome, which was assembled using a similar approach, and has 12,873 contigs and 3304 scaffolds across 738 Mb (Paterson et al. 2009). The *Sorghum* genome is three times larger than that of *Brachypodium* and has a higher repeat content with large pericentric regions spanning as much as half of some chromosomes, which likely explains the lower contiguity. The *Brachypodium* genome is more complete than most other

Sanger-based genomes, including that of papaya, which is slightly larger (400 Mb) and is assembled into 47,483 contigs and 17,764 scaffolds with a scaffold N50 of 1 Mb (Ming et al. 2008), and that of grape, which is assembled into 19,577 contigs and 3514 scaffolds (N50 = 2.07 Mb) totaling 487 Mb (Jaillon et al. 2007). The *Brachypodium* assembly is also more complete than any published NGS-based genome as the shorter read lengths of 454 and Illumina sequencing methods produce more gap-filled assemblies.

The 13 largest scaffolds in the *Brachypodium* genome were assembled into five chromosomes using a high-density map based on single nucleotide polymorphisms (SNPs). A total of 558 SNP-based markers were scored in 476 individuals from an F2 population using the Illumina GoldenGate Genotyping platform (Huo et al. 2011). The 558 markers were placed into five linkage groups with an unusually large combined genetic map length of 1598 centiMorgans, indicating a high recombination rate in *Brachypodium*. Anchoring to the genetic map joined seven scaffolds and helped identify two false scaffold joins that occurred during genome assembly.

The assembly quality of the *Brachypodium* genome was further improved after its initial release using manual curation. Scaffolds were divided into 2-Mb overlapping pieces, and gaps were manually inspected and finished using a variety of approaches. Gaps were filled using primer walking on subclones or fosmid templates, Illumina WGS small insert shatter libraries, and shotgun sequencing of fosmid and BAC clones using Illumina or Sanger sequencing. A total of 119 small scaffolds were integrated into the five chromosomes, and a total of 1217 gaps were closed representing 1.15 Mb of added sequence. The overall contig N50 increased by a factor of 18 from 347.8 kb to 6.4 Mb. The few remaining gaps in the *Brachypodium* genome are currently being filled using third-generation single molecule real time (SMRT) sequencing from Pacific Biosciences (PacBio). SMRT sequencing represents the next shift in DNA sequencing technology; native DNA is sequenced directly in long stretches, currently exceeding 10 kb on average (Chin et al. 2013). The long read length and unbiased nature of SMRT sequencing should allow the closure of most remaining complex repeat structures and GC-rich regions in *Brachypodium*.

Gene Annotation and Gene Family Evolution in the Grasses

Brachypodium gene models were annotated using a set of assembled transcripts from a broad range of tissue types and treatments. A total of 128,000 Sanger-based expressed sequence tags (ESTs), 2.3 million 454-based ESTs, and 289 million Illumina RNAseq reads were generated (Vogel et al. 2006). The assembled 454- and Sanger-based ESTs, Illumina transcripts, and homologous genes from other grasses were aligned to the reference *Brachypodium* genome to generate a training set of 410 high-confidence gene models for *ab initio* gene prediction. The statistical combiner JIGSAW (Allen and Salzberg 2005) was trained using these high-confidence genes to annotate the rest of the genome. EST support, Illumina-based splice site predictions, and gene homologs were used to define alternative splicing patterns and more accurately define intron/exon boundaries. After filtering, the *Brachypodium* genome was predicted to have a total of 25,532 protein coding genes. The version 2.1 annotation release added more genes and refined splicing sites with a total of 31,694 more protein coding loci. This is the same range as *Arabidopsis* (TAIR9, 26,990), rice (RAP2, 28,236) (Tanaka et al. 2008), *Sorghum* (v2.1 33,441) (Paterson et al. 2009), and *Setaria italica* (v2.1 35,471) (Bennetzen et al. 2012) but fewer than wheat (124,201) (Mayer et al. 2014). Gene models were validated using the 289 million Illumina reads, and 92.7 % of the predicted coding genes were supported by Illumina data. A fraction of the gene models (2755) from 97 diverse gene families were manually annotated for improvement and validation. Only 13 % of the gene models required modification, demonstrating the high quality of the original annotation.

Gene duplications form the basis of evolutionary innovation and changes in gene family composition (gains/losses) reflecting adaptive evolution for abiotic stress tolerance, disease resistance, and lineage-specific characteristics (Hanada et al. 2008). The *Brachypodium* gene models were classified into gene families by comparison with gene models of other grasses (rice, *Sorghum*, and ESTs from barley and wheat) and *Arabidopsis* to assess grass-specific, Pooideae-specific, and *Brachypodium*-specific gene family expansions and contractions. Eighty-four percent of the gene families are found in all three economically important grass subfamilies reflecting their conserved gene content and recent diversification. Across the three grass subfamilies, 681 gene families are shared between *Brachypodium* and rice but not *Sorghum*, and 1689 families are shared between *Brachypodium* and *Sorghu*m but not rice. Two hundred sixty-five gene families are Pooideae specific as they are found in *Brachypodium*, wheat, and barley but not *Sorghum* and rice. Grass-specific gene families are enriched in gene ontology (GO) terms for transmembrane receptor proteins, kinases, and ion binding. Pooideae- and *Brachypodium*-specific gene families are enriched for oxidoreductase activity, antioxidant activity, and heme binding, reflecting adaptations to abiotic stresses.

Most of the gene family expansions in the grasses are the products of tandem gene duplications (TDs). TDs occur as a result of ectopic homologous recombination and replication slippage, which can produce large tandem arrays of genes ranging from 2 to 50 or more copies (Leister 2004). TD is the most common source of new genes in plants, and TD facilitates rapid adaptation through partitioning roles in preexisting pathways (subfunctionalization) or adapting new roles (neofunctionalization) (Edger and Pires 2009). Roughly 13.5 % (3452) of *Brachypodium*-specific gene models are tandem duplicates, and 21 % of the grass-specific genes are tandems. These grass-specific tandems are enriched for transcription factors, kinases, and enzymes with antioxidant activity, supporting their role in grass evolution.

Synteny, Colinearity, and Chromosome Evolution of *Brachypodium*

Grasses have a propensity for polyploidy, a trait that likely contributes to their emergent agronomic properties such as nutrition and high yield, and that helped spur their rapid diversification. Whole genome duplication (WGD) and tandem gene duplication provide the raw material for evolutionary innovation (reviewed in (Edger and Pires 2009)). A WGD event likely predates the radiance of flowering plants, and in all plants (Albert et al. 2013) with the exception of *Amborella*, there is evidence for at least one additional, lineage-specific WGD. Evidence for WGD in the grasses was first suggested nearly 20 years ago based on apparently duplicated regions in rice genetic maps (Kishimoto et al. 1994). This duplication event, referred to as rho, was later confirmed in the Ehrhartoideae upon completion of the rice genome (Goff et al. 2002) and extended to include the Panicoideae based on evidence from the *Sorghum* genome (Paterson et al. 2009). Evidence of rho was also found in the *Brachypodium* genome (clade Pooideae), confirming that rho is indeed found in all major grass families. Comparisons of synonymous substitution rates in shared duplicated genes in rice, *Sorghum*, and *Brachypodium* suggest that rho occurred roughly 70 million years ago (MYA) (Vogel et al. 2010), coinciding with the K-T mass extinction event and the diversification of the grasses (Fawcett et al. 2009) (Fig. 2a). The grasses also share a second WGD event, sigma, which occurred roughly 110 MYA (Tang et al. 2010; Paterson et al. 2004), and a third ancient WGD, tau, which is found in all sequenced monocots and occurred not long after the origin of monocots 150–130 MYA (Jiao et al. 2014) (Fig. 1).

Aside from the shared rho and sigma events, *Brachypodium*, *Sorghum* (Paterson et al. 2009), and rice (Yu et al. 2002) have no additional, lineage-specific WGD events, and gene order (or synteny) is highly conserved among the grasses (Fig. 2b). In total, 99 % of the *Brachypodium* genome has detectable colinearity with other grasses, and these large syntenic chromosomal regions have shared gene order and content in grass species. The few non-collinear regions are mostly confined to the pericentric regions, which are highly expanded in *Sorghum* and generally gene poor (Fig. 2c). Most of the conserved regions span 12 large syntenic blocks that can be traced back to the ancestral pan-monocot genome and reveal a series of nested insertions responsible for chromosome reduction in *Brachypodium*. About 92 % of the genome is retained in six major duplication blocks from the rho WGD event. Roughly 42 % and ~45 % of the orthologous gene pairs in *Brachypodium* are collinear with rice and *Sorghum* (Wang et al. 2012), and 58 % of genes are collinear with the D genome of wheat (Mayer et al. 2014). Non-syntenic genes are the products of numerous small-scale rearrangements, small chromosomal and tandem gene duplications, and lineage-specific genes.

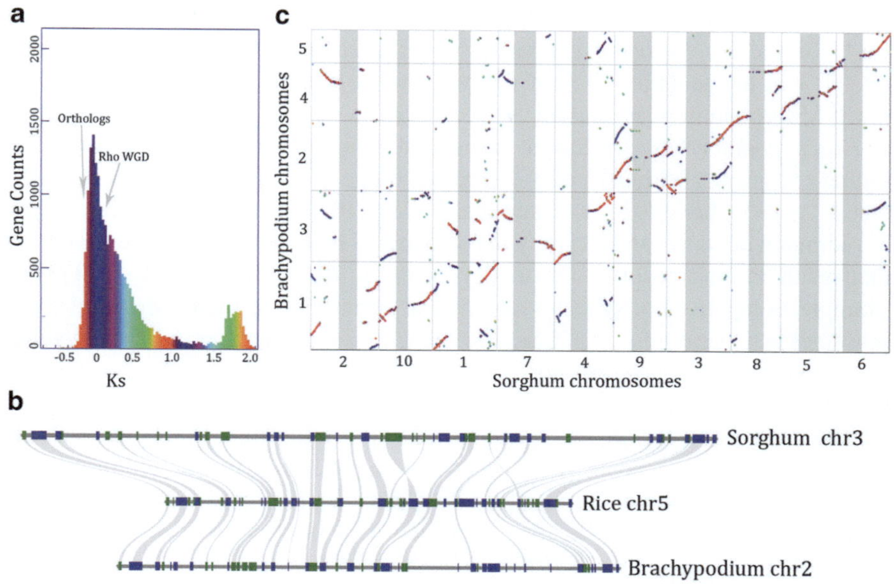

Fig. 2 Comparative genomics and the compact genome structure of *Brachypodium*. (**a**) Distribution of the log10 substitutions per synonymous site (Ks) between homologous gene pairs in *Brachypodium* and *Sorghum*. Peaks for orthologs and rho WGDs are shown. (**b**) Genomic dot plots of syntenic blocks between *Sorghum* and *Brachypodium*. One-to-one syntenic blocks are shown in *red* and regions retained from the rho WGD event are shown in *blue/purple*. *Grey lines* denote the expanded pericentric regions in *Sorghum* that lack orthology to *Brachypodium*. Results were regenerated at: https://genomevolution.org/r/f766. (**c**) Microsynteny between *Sorghum*, rice and *Brachypodium*. Forward genes are denoted in *green* and genes in the reverse orientation are denoted in *blue*. Colinearity and gene content is highly conserved and intergenic expansion in *Sorghum* is apparent

The Compact Genome Structure of *Brachypodium*

Brachypodium has the smallest sequenced grass genome to date at 272 Mb and is only slightly larger than the smallest known grass genome of *Oropetium thomaeum* (244 Mb) (Brandham and Bennett 1995). For comparison, the rice genome is ~400 Mb, *Setaria* 500 Mb, *Sorghum* 850 Mb, maize 2300 Mb, sugarcane ~10,000 Mb, and wheat 17,000 Mb. The compact genome structure of *Brachypodium* makes it desirable as a model organism but also allows insight into genome size reduction and maintenance. Despite its small size, gene content, order, and intragenic spacing (introns and exons) are highly conserved between *Brachypodium* and other grasses (Fig. 2b). The compact structure of the *Brachypodium* genome is largely due to its low repeat content and comparatively small pericentric regions. Retrotransposons compromise only 21.4 % of the *Brachypodium* genome, compared to 25 % in rice, 54 % in *Sorghum*, 85 % in maize, and ~90 % in wheat (Yu et al. 2002; Paterson et al. 2009; Schnable et al. 2009; Mayer et al. 2014). The distribution of transposable element classes in *Brachypodium* is similar to that of other grasses; LTR retrotransposons make up 21.4 % of the *Brachypodium* genome (76 % of repeats) and DNA transposons make up 4.7 % of the genome (17 % of repeats), but the overall proportion of repeats is much lower than in other grasses.

Brachypodium has the lowest monoploid chromosome number among the sequenced grasses with a 1*n* of 5 compared to rice (1*n* = 12), *Sorghum* (1*n* = 10), wheat (1*n* = 7), barley (1*n* = 7), and maize (1*n* = 10). Chromosomal fusion events are common in plants, and a variety of mechanisms have been proposed or observed that result in chromosomal fusion (Lysak et al. 2006; Xiong et al. 2011). For example, comparison of syntenic regions that correspond to shared ancestral chromosomes can identify fusions. The reduced chromosome number in *Brachypodium* is a product of repeated nested chromosomal fusions into centromeric regions (Vogel et al. 2010). Centromeric repeats, peaks in retrotransposons, and low gene densities are observed in the regions flanking the insertion sites. The current *Brachypodium* chromosome 1, for instance, is a product of two nested insertions; the ancestral equivalent of modern rice chromosome 6 was inserted into the centromeric region of chromosome 7, which then inserted into the centromeric region of rice chromosome 3. *Brachypodium* chromosomes 1, 3, and 4 each have two nested insertions, chromosome 2 has one, and chromosome 5 has no insertions. This pattern of nested insertions in *Brachypodium* contrasts with the reciprocal translocations, inversions, and deletions responsible for the *Arabidopsis* karyotype of *n* = 5 from its ancestral karyotype of eight chromosomal pairs (Lysak et al. 2006). Perhaps most interestingly, reduction in chromosome number in wheat and *Brachypodium* occurred independently, as none of the fusion events are shared between the species (Vogel et al. 2010).

Bioinformatics Resources for *Brachypodium* Research

The vast and continually growing genomic resources for *Brachypodium* are publically available in a suite of online databases and tools (Table 2). These resources are were initially centered around brachypodium.org, an online portal that was similar to the TAIR database for *Arabidopsis* research (Huala et al. 2001). Brachypodium.org played a critical role in the establishment and adoption of *Brachypodium* as a model but has recently been taken down due to resource limitations. The tools and datasets from brachypodium.org can be easily accessed through other portals. Phytozome has a number of useful resources and tools including a BLAST portal, T-DNA line information, genome assembly, annotation and a highly integrated genome browser

SNPs, insertions/deletions (indels), and structural variants (SV) are available for six diverse, deep-sequenced *Brachypodium* accessions: Bd1-1, Bb21-3, Bd3-1, Bd30-1, BdTR12c, and Koz-3 (Gordon et al. 2014). These six re-sequenced accessions are highly divergent and display a wide range of phenotypes with respect to flowering time, response to abiotic stresses, disease resistance, plant morphology,

Table 2 Available bioinformatics resources for *Brachypodium*

Resource	Institution	URL	Description
Phytozome	DOE Joint Genome Institute	http://phytozome.jgi.doe.gov/	Genome browser, natural diversity data, T-DNA insertion sites
BrachyBio	Boyce Thompson Institute for Plant Research	http://bti.cornell.edu/brachybio	TILLING population, resources for teachers
Brachypodium genome information	Munich Information Center for Protein Sequences	http://mips.helmholtz-muenchen.de/plant/brachypodium/	Genome browser, synteny browser, BLAST database, repeat database
Brachypodium resources	DOE Joint Genome Institute	http://jgi.doe.gov/our-science/science-programs/plant-genomics/brachypodium/	T-DNA lines, germplasm, and methods for crossing, mutagenesis, and transformation
CoGe	University of California, Berkeley	http://synteny.cnr.berkeley.edu/CoGe/	Comparative genomic tools
ELEMENT	Danforth Plant Science Center	http://element.cgrb.oregonstate.edu/	Promoter searching tool
Gramene	Cold Spring Harbor Laboratories	http://www.gramene.org	Comparative genomics tools
PlantGDB	Iowa State University	http://www.plantgdb.org/	Comparative genomic tools
PlexDB	Iowa State University	http://www.plexdb.org/	RNAseq and microarray gene expression data
TILLING database	Unité de Recherche en Génomique Végétale	http://urgv.evry.inra.fr/UTILLdb	TILLING collection
USDA NPGS	USDA	http://www.ars-grin.gov/npgs/	*Brachypodium* germplasm collection

and growth habits. The collection of variants includes 2.5 million non-redundant SNPs with 152,920 SNPs in protein-coding genes, 586,206 small indels, and 2237-8413 SVs per accession ranging from 65 base pairs to 20 kb in size. Sequence variants have a high concordance with altered gene expression of hundreds of genes including genotype-dependent treatment responses to abiotic stresses (Gordon et al. 2014). Fifty six additional *Brachypodium* accessions are currently being resequenced and will soon be available. The density and location of variants as well as associated changes in protein sequences can be viewed in the Phytozome genome browser (JBrowse). Several tracks of variants in other accessions and available T-DNA accessions can be displayed in the genome browser simultaneously, allowing for in-depth analyses of individual loci or genomic regions of interest. The *Brachypodium* genome assembly and annotations, resequencing data, can also be downloaded in bulk from Phytozome using a secure ftp server.

Over 23,000 *Brachypodium* T-DNA insertion lines are publically available from the DOE Joint Genome Institute (http://Brachypodium.pw.usda.gov). The T-DNA lines have tagged more than 9000 genes, and a number of co-segregating phenotypes have been observed (Bragg et al. 2012). T-DNA insertion lines of interest can be ordered through the JGI Brachypodium resources website. A sodium azide-induced mutant population of 5530 lines and a Targeting Induced Local Lesion IN Genome (TILLING) platform have also been established in *Brachypodium* (Dalmais et al. 2013). Many of these 5530 lines have an obvious phenotype, and genes screened using the TILLING platform can be searched. Mutant lines are also being screened through the BrachyBio project, a collaborative outreach project with Boyce Thomas Institute for Plant Research and high schools (http://bti.cornell.edu/education/educational-resources/brachybio/) reference or unpub.

Brachypodium genomic resources are also available through Plant Genome and Systems Biology (PGSB) (http://pgsb.helmholtz-muenchen.de/plant/recat/index.jsp. Accessed 8 March 2015), which is a useful portal for *Brachypodium* genomics and comparisons with other grasses. PGSB has a built-in BLAST database, integrated genome browser, comparative genomics browser, and the PGSB Repeat Element Database (PGSB-REdat) for annotated grass repeats. Other comparative genomics tools for *Brachypodium* and other sequenced genomes include CoGe, Gramene, PlantGDB, and GrainGenes (Table 2). The promoter searching tool ELEMENT has a searchable database of promoter sequences from a number of species including *Brachypodium* (Mockler et al. 2007). The currently available and in-development resources for *Brachypodium* provide the raw materials for a thriving research community.

Current Prospects and the Future of *Brachypodium* Genomics

The number of research groups using *Brachypodium* as a model grass has risen steadily since plans to sequence the genome were announced in 2006 (Brkljacic et al. 2011). Requests for seed stocks of both germplasm and T-DNA mutant lines increase yearly, and current numbers are similar to stock requests in the early days of *Arabidopsis* research. The wide adoption of *Brachypodium* as a model is reflected in the number of *Brachypodium*-centered publications. As of March 2015, the *Brachypodium* genome paper has been cited over 670 times, and there had been 417 publications with "Brachypodium" in the title, including 389 in the past 5 years. Funding for *Brachypodium* research is substantial, and several large-scale *Brachypodium* research projects are currently funded.

The complete reference genome was simply a starting point; more in-depth analyses are needed to link the genome to the biology. Six diverse *Brachypodium* accessions have been resequenced and 56 accessions are currently being sequenced. Future plans include resequencing the entire USDA-maintained *Brachypodium* germplasm and phenotyping all of the lines for a wide range of traits using high-throughput phenotyping. Variants from resequencing can be combined with phenotyping data in a Genome Wide Association Studies (GWAS) to identify the underlying loci controlling interesting traits. A number of recombinant inbreed line populations have been established for *Brachypodium* and allow for mapping of loci involved in flowering time, plant architecture, and response to abiotic stresses (Cui et al. 2012). Reduced representation sequencing strategies such as genotyping by sequencing are being applied to these populations to identify high-confidence SNP-based markers for QTL mapping.

Whole genome resequencing projects capture small variants, such as SNPs and indels, but are ill suited for accurately identifying large-scale structural variations such as complex rearrangements, regions missing from the reference, and copy number variation. De novo assembly of three diverse rice accessions identified several Mbps of novel sequences in each accession that were missing from the reference rice genome (Schatz et al. 2014). Some of these rice sequences contained actively transcribed genes and are likely biologically meaningful but were missed entirely using a typical resequencing approach. Several high-quality *Brachypodium* genomes are needed to capture all of the structural variation found in the germplasm. PacBio sequencing is particularly well suited for de novo assembly of more accessions as the long reads can resolve complex repeat structures and distinguish haplotype specific differences.

Microarray platforms are well developed for *Brachypodium* and have been used extensively to investigate transcriptional responses to abiotic stresses, circadian processes, and light (Priest et al. 2014). RNAseq has largely replaced microarrays, and RNAseq-based gene expression profiles of virtually all tissue types, stresses, and temporal samples have either been generated or are planned (http://jgi.doe.gov/a-brachypodium-encode-like-project/). Tools and methods for combining results from multiple datasets are also in development. Paired-end RNAseq has allowed identification of extensive alterative splicing events that are likely involved in processes such as pathogen responses (Mandadi and Scholthof 2015). Deep sequencing of microRNAs, which function in gene regulation and post-transcriptional silencing, has also been done in *Brachypodium*. MicroRNAs involved in cold stress (Zhang et al. 2009), drought (Budak and Akpinar 2011), reproductive processes (Wei et al. 2009) have been surveyed. Tissue-specific microRNAs have also been identified by analysis of 17 tissues in plants subjected to various stresses (Jeong et al. 2013).

The next phase of crop genomics is to completely elucidate the role of functional non-coding elements in plant genomes as has been done for other model systems like humans, mice, and *Drosophila* in a plant ENCODE (Encyclopedia of DNA Elements) project (Lane et al. 2014). The *Brachypodium* ENCODE project was the first formal plant ENCODE project funded by the DOE/USDA (http://pgsb.helmholtz-muenchen.de/plant/recat/index.jsp. Accessed 8 March 2015). In the *Brachypodium* ENCODE project, multiple datasets of DNA methylation, small RNA, and mRNA transcriptome profiles will be surveyed under progressive drought stress to assess changes in epigenetics, chromatin dynamics, and gene regulatory networks. These datasets will provide foundational insights into basic plant processes and help identify candidates for translational work to increase grass feedstock productivity.

The development of *Brachypodium* as a model plant followed a unique trajectory: genomic resources were established in parallel with other key resources such as T-DNA and TILLING mutant populations, an efficient transformation system, and collection of natural accessions. The high-quality *Brachypodium* genome provides a platform for uncovering the basic aspects of grass biology as well as translational work for improving the grasses that feed the world.

References

Albert VA, Barbazuk WB, Der JP, Leebens-Mack J, Ma H, Palmer JD, et al. The Amborella genome and the evolution of flowering plants. Science. 2013;342(6165):1241089.

Allen JE, Salzberg SL. JIGSAW: integration of multiple sources of evidence for gene prediction. Bioinformatics. 2005;21(18):3596–603.

Arabidopsis Genome Initiative. Analysis of the genome sequence of the flowering plant Arabidopsis thaliana. Nature. 2000;408(6814):796.

Batzoglou S, Jaffe DB, Stanley K, Butler J, Gnerre S, Mauceli E, et al. ARACHNE: a whole-genome shotgun assembler. Genome Res. 2002;12(1):177–89.

Bennetzen JL, Schmutz J, Wang H, Percifield R, Hawkins J, Pontaroli AC, et al. Reference genome sequence of the model plant Setaria. Nat Biotechnol. 2012;30(6):555–61.

Bragg JN, Wu J, Gordon SP, Guttman ME, Thilmony R, Lazo GR, et al. Generation and characterization of the Western Regional Research Center Brachypodium T-DNA insertional mutant collection. PLoS One. 2012;7(9):e41916.

Brandham P, Bennett S. Nuclear DNA amounts in diploid and tetraploid Oropetium (Poaceae). Kew Bulletin. 1995;50:601–4.

Brkljacic J, Grotewold E, Scholl R, Mockler T, Garvin DF, Vain P, et al. Brachypodium as a model for the grasses: today and the future. Plant Physiol. 2011;157(1):3–13.

Brutnell TP, Wang L, Swartwood K, Goldschmidt A, Jackson D, Zhu X-G, et al. Setaria viridis: a model for C4 photosynthesis. Plant Cell. 2010;22(8):2537–44.

Budak H, Akpinar A. Dehydration stress-responsive miRNA in Brachypodium distachyon: evident by genome-wide screening of microRNAs expression. Omics. 2011;15(11):791–9.

Chin C-S, Alexander DH, Marks P, Klammer AA, Drake J, Heiner C, et al. Nonhybrid, finished microbial genome assemblies from long-read SMRT sequencing data. Nat Methods. 2013;10(6):563–9.

Cui Y, Lee MY, Huo N, Bragg J, Yan L, Yuan C, et al. Fine mapping of the Bsr1 barley stripe mosaic virus resistance gene in the model grass Brachypodium distachyon. PLoS One. 2012;7(6):e38333.

Dalmais M, Antelme S, Ho-Yue-Kuang S, Wang Y, Darracq O, d'Yvoire MB, et al. A TILLING platform for functional genomics in Brachypodium distachyon. PLoS One. 2013;8(6):e65503.

Edger PP, Pires JC. Gene and genome duplications: the impact of dosage-sensitivity on the fate of nuclear genes. Chromosome Res. 2009;17(5):699–717.

Elert E. Rice by the numbers: a good grain. Nature. 2014;514:50.

Fawcett JA, Maere S, Van de Peer Y. Plants with double genomes might have had a better chance to survive the Cretaceous–Tertiary extinction event. Proc Natl Acad Sci. 2009;106(14):5737–42.

Godfray HCJ, Beddington JR, Crute IR, Haddad L, Lawrence D, Muir JF, et al. Food security: the challenge of feeding 9 billion people. Science. 2010;327(5967):812–8.

Godfray H, Pretty J, Thomas S, Warham E, Beddington J. Linking policy on climate and food. Science. 2011;331(6020):1013–4.

Goff SA, Ricke D, Lan T-H, Presting G, Wang R, Dunn M, et al. A draft sequence of the rice genome (Oryza sativa L. ssp. japonica). Science. 2002;296(5565):92–100.

Gordon SP, Priest H, Des Marais DL, Schackwitz W, Figueroa M, Martin J, et al. Genome diversity in Brachypodium distachyon: deep sequencing of highly diverse inbred lines. Plant J. 2014;79(3):361–74.

Hanada K, Zou C, Lehti-Shiu MD, Shinozaki K, Shiu S-H. Importance of lineage-specific expansion of plant tandem duplicates in the adaptive response to environmental stimuli. Plant Physiol. 2008;148(2):993–1003.

Hatch MD, Slack CR. Photosynthesis by sugar-cane leaves. Biochem J. 1966;101:103–11.

http://Brachypodium.pw.usda.gov.

http://pgsb.helmholtz-muenchen.de/plant/recat/index.jsp. Accessed 8 March 2015.

Huala E, Dickerman AW, Garcia-Hernandez M, Weems D, Reiser L, LaFond F, et al. The Arabidopsis Information Resource (TAIR): a comprehensive database and web-based information retrieval, analysis, and visualization system for a model plant. Nucleic Acids Res. 2001;29(1):102–5.

Huo N, Garvin DF, You FM, McMahon S, Luo M-C, Gu YQ, et al. Comparison of a high-density genetic linkage map to genome features in the model grass Brachypodium distachyon. Theor Appl Genet. 2011;123(3):455–64.

Jaillon O, Aury J-M, Noel B, Policriti A, Clepet C, Casagrande A, et al. The grapevine genome sequence suggests ancestral hexaploidization in major angiosperm phyla. Nature. 2007;449(7161):463–7.

Jeong D-H, Schmidt SA, Rymarquis LA, Park S, Ganssmann M, German MA, et al. Parallel analysis of RNA ends enhances global investigation of microRNAs and target RNAs of Brachypodium distachyon. Genome Biol. 2013;14(12):R145.

Jiao Y, Li J, Tang H, Paterson AH. Integrated syntenic and phylogenomic analyses reveal an ancient genome duplication in monocots. Plant Cell. 2014;26(7):2792–802.

Jung K-H, An G, Ronald PC. Towards a better bowl of rice: assigning function to tens of thousands of rice genes. Nat Rev Genet. 2008;9(2):91–101.

Kishimoto N, Higo H, Abe K, Arai S, Saito A, Higo K. Identification of the duplicated segments in rice chromosomes 1 and 5 by linkage analysis of cDNA markers of known functions. Theor Appl Genet. 1994;88(6–7):722–6.

Lamesch P, Berardini TZ, Li D, Swarbreck D, Wilks C, Sasidharan R, et al. The Arabidopsis Information Resource (TAIR): improved gene annotation and new tools. Nucleic Acids Res. 2012;40(D1):D1202–10.

Lane AK, Niederhuth CE, Ji L, Schmitz RJ. pENCODE: a plant encyclopedia of DNA elements. Annu Rev Genet. 2014;48:49–70.

Leister D. Tandem and segmental gene duplication and recombination in the evolution of plant disease resistance genes. Trends Genet. 2004;20(3):116–22.

Lysak MA, Berr A, Pecinka A, Schmidt R, McBreen K, Schubert I. Mechanisms of chromosome number reduction in Arabidopsis thaliana and related Brassicaceae species. Proc Natl Acad Sci U S A. 2006;103(13):5224–9.

Mandadi KK, Scholthof K-BG. Genome-wide analysis of alternative splicing landscapes modulated during plant-virus interactions in Brachypodium distachyon. Plant Cell. 2015;27(1):71–85.

Mayer KF, Rogers J, Doležel J, Pozniak C, Eversole K, Feuillet C, et al. A chromosome-based draft sequence of the hexaploid bread wheat (Triticum aestivum) genome. Science. 2014;345(6194):1251788.

Michael TP, VanBuren R. Progress, challenges and the future of crop genomes. Curr Opin Plant Biol. 2015;24:71–81.

Ming R, Hou S, Feng Y, Yu Q, Dionne-Laporte A, Saw JH, et al. The draft genome of the transgenic tropical fruit tree papaya (Carica papaya Linnaeus). Nature. 2008;452(7190):991–6.

Mockler T, Michael T, Priest H, Shen R, Sullivan C, Givan S, et al. The DIURNAL project: DIURNAL and circadian expression profiling, model-based pattern matching, and promoter analysis. In: Cold Spring Harbor Symposia on quantitative biology. Cold Spring Harbor Laboratory Press; 2007;72:353–63.

Paterson A, Bowers J, Chapman B. Ancient polyploidization predating divergence of the cereals, and its consequences for comparative genomics. Proc Natl Acad Sci U S A. 2004;101 (26):9903–8.

Paterson AH, Bowers JE, Bruggmann R, Dubchak I, Grimwood J, Gundlach H, et al. The Sorghum bicolor genome and the diversification of grasses. Nature. 2009;457(7229):551–6.

Priest HD, Fox SE, Rowley ER, Murray JR, Michael TP, Mockler TC. Analysis of global gene expression in Brachypodium distachyon reveals extensive network plasticity in response to abiotic stress. PLoS One. 2014;9(1):e87499.

Project IRGS. The map-based sequence of the rice genome. Nature. 2005;436(7052):793–800.

Schatz MC, Witkowski J, McCombie WR. Current challenges in de novo plant genome sequencing and assembly. Genome Biol. 2012;13(4):243.

Schatz MC, Maron LG, Stein JC, Wences AH, Gurtowski J, Biggers E, et al. Whole genome de novo assemblies of three divergent strains of rice, Oryza sativa, document novel gene space of aus and indica. Genome Biol. 2014;15:506.

Schnable PS, Ware D, Fulton RS, Stein JC, Wei F, Pasternak S, et al. The B73 maize genome: complexity, diversity, and dynamics. Science. 2009;326(5956):1112–5.

Tanaka T, Antonio BA, Kikuchi S, Matsumoto T, Nagamura Y, Numa H, et al. The rice annotation project database (RAP-DB): 2008 update. Nucleic Acids Res. 2008;36(Supp 1):D1028–33.

Tang H, Bowers JE, Wang X, Paterson AH. Angiosperm genome comparisons reveal early polyploidy in the monocot lineage. Proc Natl Acad Sci. 2010;107(1):472–7.

Tuskan GA, Difazio S, Jansson S, Bohlmann J, Grigoriev I, Hellsten U, et al. The genome of black cottonwood, Populus trichocarpa (Torr. & Gray). Science. 2006;313(5793):1596–604.

Vogel JP, Gu YQ, Twigg P, Lazo GR, Laudencia-Chingcuanco D, Hayden DM, et al. EST sequencing and phylogenetic analysis of the model grass Brachypodium distachyon. Theor Appl Genet. 2006;113(2):186–95.

Vogel JP, Garvin DF, Mockler TC, Schmutz J, Rokhsar D, Bevan MW, et al. Genome sequencing and analysis of the model grass Brachypodium distachyon. Nature. 2010;463(7282):763–8.

Wang Y, Tang H, DeBarry JD, Tan X, Li J, Wang X, et al. MCScanX: a toolkit for detection and evolutionary analysis of gene synteny and collinearity. Nucleic Acids Res. 2012;40(7):e49.

Wei B, Cai T, Zhang R, Li A, Huo N, Li S, et al. Novel microRNAs uncovered by deep sequencing of small RNA transcriptomes in bread wheat (Triticum aestivum L.) and Brachypodium distachyon (L.) Beauv. Funct Integr Genomics. 2009;9(4):499–511.

Xiong Z, Gaeta RT, Pires JC. Homoeologous shuffling and chromosome compensation maintain genome balance in resynthesized allopolyploid Brassica napus. Proc Natl Acad Sci. 2011;108 (19):7908–13.

Yu J, Hu S, Wang J, Wong GK-S, Li S, Liu B, et al. A draft sequence of the rice genome (Oryza sativa L. ssp. indica). Science. 2002;296(5565):79–92.

Zhang J, Xu Y, Huan Q, Chong K. Deep sequencing of Brachypodium small RNAs at the global genome level identifies microRNAs involved in cold stress response. BMC Genomics. 2009;10 (1):449.

Brachypodium Paleogenomics: From Genome Evolution to Translational Research in Grass Crops

Jérôme Salse

Abstract The modern *Brachypodium distachyon* genome consisting in 25,532 genes, 271 Megabases covering five chromosomes derives from a founder grass ancestor structured in 7/12 protochromosomes containing 10,000–15,000 protogenes with a minimal physical coding space size of <50 Megabases. From the grass ancestor, the modern *Brachypodium* genome evolved through a polyploidization event followed by a diploidization mechanism at the genome level, consisting in centromere-oriented ancestral chromosome fusions leading to chromosome number reduction, and at the gene level, consisting in the deletion of the duplicated gene copies leading to a subgenome dominance. Finally, the *Brachypodium* genome can be used as a guide for translational research in grasses for applied research in dissecting and improving traits of agricultural relevance.

Keywords Synteny • Duplication • Evolution • Genome • Ancestor • Rearrangement • Translational research

Brachypodium Paleogenomics and Evolution

Since 2010 with the access to the *Brachypodium distachyon* genome sequence, with 25,532 genes, 271 Megabases and 5 chromosomes (Fig. 1a, IBI 2010), it has been possible to gain insights into grass genome evolution to an unprecedented resolution (Abrouk et al. 2010, Bolot et al. 2009). Comparison of *Brachypodium* genome with that of rice (IRGSP 2005), maize (Schnable et al. 2009) and sorghum (Paterson et al. 2009) using proposed community standards for accurate comparative genomics studies (Salse et al. 2009a) unravelled 8533 *Brachypodium* orthologs consisting in 12 syntenic blocks and involving the following conserved synteny chromosome-to-chromosome relationships with ***os*** for rice, ***sb*** for sorghum, ***zm*** for maize and ***bd*** for *Brachypodium*: os1/sb3/zm3-8/bd2 (pink), os2/sb4/zm4-5/bd3

J. Salse (✉)
INRA/UBP UMR 1095 GDEC 'Génétique, Diversité et Ecophysiologie des Céréales', Institut National de la Recherche Agronomique (INRA), Domaine de Crouelle, 234 avenue du Brézet, 63100 Clermont Ferrand, France
e-mail: jsalse@clermont.inra.fr; http://www4.clermont.inra.fr/umr1095/ppav

J. Vogel (ed.), *Genetics and Genomics of Brachypodium*,
Plant Genetics and Genomics: Crops Models, DOI 10.1007/7397_2015_2
Published online: 02 October 2015

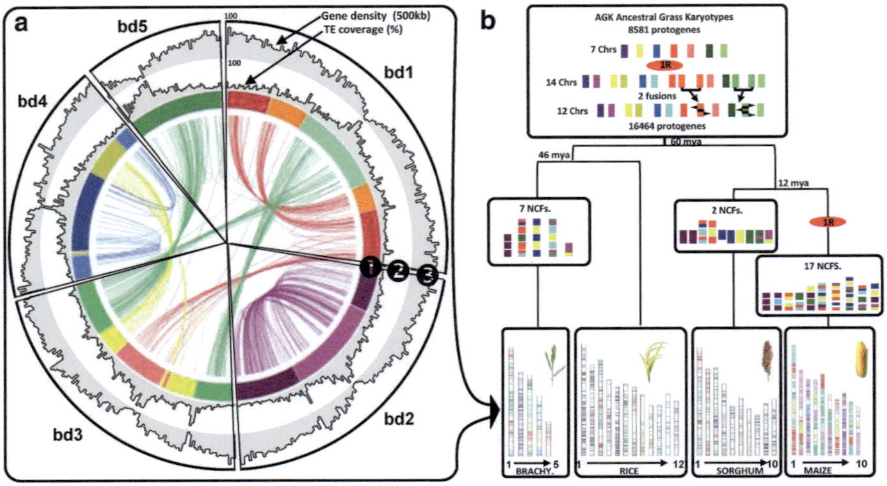

Fig. 1 *Brachypodium* genome paleo-evolution (Adapted from Murat et al. 2010). (**a**) *Brachypodium* genome circle. The *Brachypodium* genome is shown as *circles* with the five modern chromosome (layer 1) associated with a color code illuminating the ancestral chromosome origins, the gene (layer 2, expressed in % within 500 Kb windows) and repeat (layer 3, expressed in percentage) density as well as the ancestral retained duplicated gene linked with colored connecting lines at the center of the circle. (**b**) Grass paleo-evolutionary scenario. The present day grass genomes (*bottom*) are represented with color codes to illustrate the evolution of segments from their founder ancestors (*top*) with 7 (pre-paleoWGD) and 12 (post-paleoWGD) protochromosomes. The WGD events that have shaped the structure of the different grass genomes during their evolution from the common ancestor are indicated as red dots. The geological periods (Jurassic, Cretaceaous, Paleogene and Neogene) are indicated on the *left* along with the time scale (in mya). Evolutionary genome shuffling events such as chromosomal fusions and fissions are illustrated in *boxes*

(light green), os3/sb1/zm1-5-9/bd1 (red), os4/sb6/zm2-10/bd5 (dark green), os5/sb9/zm6-8/bd2 (violet), os6/sb10/zm5-6-9/bd1 (green), os7/sb2/zm2-7/bd1 (orange), os8/sb7/zm1-4-6-10/bd3 (yellow), os9/sb2/zm2-7/bd4 (light yellow), os10/sb1/zm1-5-9/bd3 (brown), os11/sb5/zm2-4/bd4 (blue), os12/sb8/zm1-3-10/bd4 (light blue), Fig. 1b.

Similarly, interchromosomal duplications defined 642 *Brachypodium* paralogs consisting in seven duplicated blocks and involving the following chromosome pair combinations: os11-12/sb5-8/zm2-4-1-3-10/bd4-4, os5-1/sb9-3/zm6-8-3/bd2-2, os10-3/sb1-1/zm1-5-9/bd1-3, os7-3/sb2-1/zm2-7-1-9-5/bd1-1, os4-2/sb6-4/zm2-10-4-5/bd3-5, os9-8/sb2-7/zm2-7-1-4-10-6/bd3-4, and os2-6/sb4-10/zm4-5-6-9/bd1-3 (Fig. 1a). Time divergence estimate of such orthologs and paralogs highlighted rice as the closest grass relative (*Brachypodium*/rice speciation dating back to 46 million years ago, *i.e.* mya) and the confirmation of a shared ancestral whole genome duplication event (*i.e.* paleoWGD dating back to 90 mya, Salse et al. 2008, 2009b, Tang 2010) among grasses (Fig. 1b).

Previous comparative genomic data aiming at identifying conserved (orthologous) and duplicated (paralogous) genes allowed the reconstruction of founder ancestors (*i.e.* post- and pre-paleoWGD karyotypes) in terms of protochromosome and protogene numbers as well as physical size, founding the field of paleogenomic research. This field is devoted to proposing the most likely evolutionary scenario based on the following assumptions where: (1) ancestor modelling is based on duplications (or any shuffling events) found at orthologous positions between modern species, and thus considered as ancestral, (2) evolutionary history is based on the smallest number of shuffling operations (including inversions, deletions, fusions, fissions, translocations) which explain evolution from the ancestral genome to modern karyotypes (Salse et al. 2009a, Salse and Feuillet 2011).

Comparison of *Brachypodium*, rice, maize and sorghum unveiled an AGK (Ancestral Grass Karyotype, pre-paleoWGD) structured in seven protochromosomes containing 8581 protogenes with a minimal gene space physical size of 33 Mb (Murat et al. 2010, 2014a). This ancestor went through a paleotetraploidization event (consisting of seven duplicated blocks shared by the modern grasses, Salse et al. 2008, 2009b) followed by two reciprocal translocations to reach a 12 chromosome ancestor (*i.e.* post-paleoWGD) intermediate containing 16,464 protogenes (Fig. 1b) (Murat et al. 2010, 2014a). Furthermore, it is possible to represent the chromosome number for any modern grass species by a karyotypic equation reflecting the evolutionary events, such as duplications-fusions-fissions, that took place in the course of evolution. To this regards, the *Brachypodium* karyotypic equation is $[[7 \times 2] - 2 -] + 7 - 14$, reflecting that the $n = 7$ AGK went through a WGD ($\times 2$) with 2 reciprocal translocations (-2) to reach a $n = 12$ AGK intermediate followed by 14 fusions and 7 fissions (to reach the modern *Brachypodium* genome structure of five chromosomes). The five *Brachypodium* chromosomes (bd1 to bd5) consist in a mosaic of the $n = 12$ AGK intermediate (A1 to A12) with bd1 = A3(red) + A6(green) + A7(orange), bd2 = A1(pink) + A5 (violet), bd3 = A2(light green) + A8(yellow) + A10(brown), bd4 = A9(light yellow) + A11(blue) + A12(light blue), bd5 = A4(dark green), Fig. 1b.

Brachypodium Genome Architecture Inherited from Paleoevolution

According to the previous evolutionary scenario, the *Brachypodium* genome has been duplicated at least once in the course of its paleohistory (for reviews see Salse 2012). Such polyploidization event constituted a genomic shock and leaded to a diploidization process at both the whole-genome and gene levels. At the whole-genome level, such diploidization of the *Brachypodium* genome consisted in intense chromosome reshuffling resulting in 7 fissions followed by 14 chromosome fusions (also referenced as seven nested chromosome fusions (NCF) in Luo et al. 2009 and Murat et al. 2010 or asymmetric/symmetric reciprocal translocations in Schubert and Lysak 2011), then resulting in a reduced number of chromosomes in today's *Brachypodium* (five chromosomes) compared to the reconstructed common paleopolyploid ancestor (n = 12 chromosomes for the post-paleoWGD progenitor), as illustrated by the dotplot between AGK and modern grasses in Fig. 2a.

Chromosome fusion boundaries (also referred as synteny breakpoints) correspond in *Brachypodium* to regions considered as 'fragile' genomic structures as they became hotspots of structural rearrangements such as inversions, repeat amplifications and tandem gene duplications (Murat et al. 2010). When intervals comprising ancestral chromosome fusion points are compared in their structures they may correspond to (1) meiotic recombination hotspots, (2) high sequence turn over loci through repeat invasion, and (3) hotspots of evolutionary novelty that could putatively act as a reservoir for producing novel phenotypes (Murat et al. 2010). Finally, modern genomes harbour traces of their evolutionary history in their current chromosomal architecture, particularly their specific patterns of ancestral chromosome rearrangement events (Fig. 2a). For *Brachypodium* chromosome 2, as an illustration case example, that went through an NCF event (dotted rectangle and black arrow in Fig. 2a), the gene density pattern indicates that gene hotspots precisely corresponds to subtelomeric regions of either the modern *Brachypodium* chromosome (external red arrows) or the ancestral fused chromosome (internal red arrows), Fig. 2b. As a result, gene density (or TE density that is negatively correlated), was observed as a prominent evolutionary marker of the NCF pattern of any of the grass genomes including *Brachypodium*, Fig. 2b.

Brachypodium genome diploidization also occurs at the gene level through duplicated gene redundancy reduction. However, duplicated gene deletion is not performed at random as it has been shown that duplicated gene redundancy is eliminated through a so-called subgenome dominance (Salse 2012 in which only one of the duplicated blocks has preferentially retained the majority of ancestral copies of the duplicates. However, an exception to this general subgenome dominance pattern has been characterised within *Brachypodium* chromosome 4 (corresponding to fused ancestral chromosomes A11–A12 and also observed between rice (11–12), sorghum (5–8), foxtail millet (7–8) and maize (1–2–3–4–10) chromosomes) where a highly conserved region has been retained during 50–70 million years of evolution by an active and recurrent gene conversion phenomenon (Jacquemin et al. 2011, Wang et al. 2011, Murat et al. 2014a). Is has been clearly established that protein coding gene functions do not behave in the same way regarding this diploidization phenomenon depending on their roles in macromolecular complexes or networks with transcription factors (TFs) or regulators (TRs), part of product-to-product connectivity interactions, retained as duplicated copies following WGDs. Despite TF and TR, the vast majority of the duplicated genes return to a singleton status following the subgenome dominance process (Murat et al. 2010, 2014a). This diploidization phenomenon then leads at the whole chromosome level to dominant (low duplicated gene loss and retention of the ancestral function) and sensitive (high duplicated gene loss or neo- / sub-functionalization) subgenomes in paleopolyploids such as *Brachypodium* as well as neopolyploids relatives such as wheat (Pont et al. 2011; 2013).

Comparing orthologous blocks os1-bd2-sb3-zm3-zm8 (corresponding to the ancestral grass chromosome A1) and os5-bd2-sb9-zm6-zm8 (corresponding to the

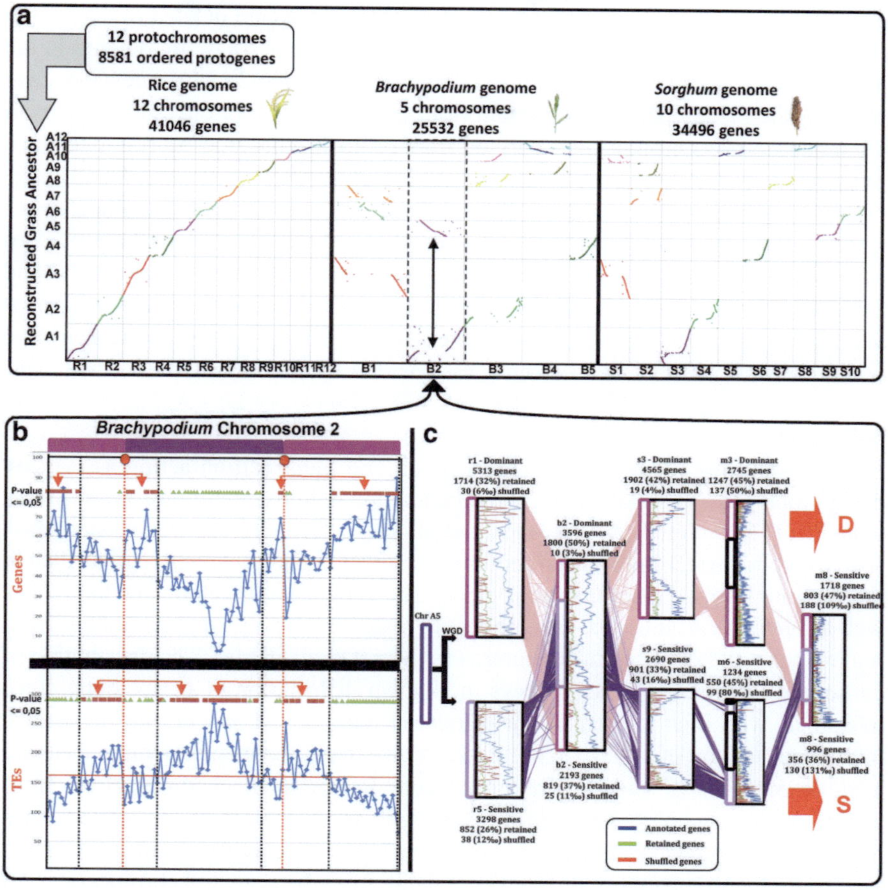

Fig. 2 *Brachypodium* genome diploidization following paleoWGD (Adapted from Murat et al. 2014a). (**a**) Dotplot comparison between AGK and modern grasses. The reconstruction of the AGK ancestor (n = 7) from 16,464 (or 8959 excluding maize) orthologous relationships leaded to a post-duplication ancestor (n = 12) with 6246 (or 8581 excluding maize) ordered protogenes. Dotplot representation of the n = 12 ancestor structured with 6246 ordered protogenes aligned with the rice (12 chromosomes with 41,046 genes), *Brachypodium* (5 chromosomes with 25,532 genes) and sorghum (10 chromosomes with 34,496 genes) modern genomes. (**b**) Diploidization at the whole-genome level leading to fused ancestral chromosomes. Illustration of *Brachypodium* chromosome 2 (bd2) resulting from the nested fusion of A1 (*pink*) and A5 (*violet*) protochromosomes leading to bias in gene and TE density (*blue curves*). Statistically-based enrichment of genes (*top*) and TEs (*bottom*) are highlighted with red rectangles. Synteny breakpoints resulting from the ancestral chromosome fusions are shown with red dots. (**c**) Diploidization at the gene level leading to dominant and sensitive subgenomes. Illustration of paralogous chromosomes in rice (Chr 1–5), *Brachypodium* (Chr 2), sorghum (Chr 3–9) and maize (Chr 3–6–8) originating from a single ancestral pre-duplication chromosome A5 (*left*). The orthologous genes are illustrated with colored lines between chromosomes. The distribution of annotated genes (*blue curve*), ancestral retained paralogous genes (*green curve*) and shuffled duplicated genes (*red curve*) are shown at the right side of the chromosomes. The total numbers of annotated, retained (*i.e.* ortholog) and shuffled genes are shown for each chromosome (*top*) defining dominant (referenced as D) and sensitive (referenced as S) chromosomal segments

ancestral grass chromosome A5), 58 % of ancestral genes are deleted in the former (among 17,937 genes with 7466 retained genes), whereas 67 % are deleted in the latter (among 10,411 genes with 3478 retained). Figure 2c illustrates the gene shuffling frequency between dominant (top) and sensitive (bottom) subgenomes in the modern grass chromosomes derived from protochromosome A5 (*i.e.* rice chromosomes 1–5, *Brachypodium* chromosome 2, sorghum chromosomes 3–9 and maize chromosomes 3–6–8). Annotated genes (blue curves), ancestral retained genes (green curves) and shuffled genes (red curves) have been mapped on these modern chromosomes. At the whole genome level, grass genomes are then a mosaic of D and S blocks inherited from the diploidization process following the ancestral shared WGD where A5–6–7–8–10–12 are dominant and A1–2–3–4–9–11 are sensitive making *Brachypodium* chromosomes as: bd1 = A3(D) + A6(S) + A7(S), bd2 = A1 (D) + A5(S), bd3 = A2(D) + A8(S) + A10(S), bd4 = A9(D) + A11(D) + A12(S), bd5 = A4(D).

Gene shuffling (gene duplication, deletion and transposition acting as synteny decay) took place in distal and high-recombination regions of chromosomes (Massa et al. 2011, Murat et al. 2014a, Wicker et al. 2010). Synteny between related grass species eroded faster in the distal, high-recombination chromosomal regions than in the proximal, low-recombination regions. The subtelomeric regions of today's as well as ancient chromosome arms showed greater erosion of synteny in their distal regions than in their proximal counterparts (Murat et al. 2014a). The distribution of conserved or syntenic genes follows the known distribution of genes along chromosomes, with high density in sub-telomeric regions and low density in centromeric ones (Murat et al. 2010). Figure 2c shows that lower gene retention (with an average of 43 % of ancestral retained genes in dominant genomes *vs* 35 % in sensitive ones) and higher gene movement (with an average 34 ‰ of shuffled genes in dominant genomes *vs* 50 ‰ in sensitive ones) are observed in dominant (D) and sensitive (S) chromosomes (or blocks) deriving from the ancestral chromosome A5 in all four species (including two regions in maize) compared. This suggests that genome sensitivity is driven by lower gene conservation, involving gene deletion, movement (such as duplicated gene transposition) or diversity (such as the accumulation of synonymous and non-synonymous substitutions). In this scenario, subgenome dominance is driven either by massive loss, transposition or/and enhanced diversification of genes from the S compartment.

Brachypodium as a Guide for Translational Research in Grass Crops

A translational genomics approach can be accurately applied from *Brachypodium* to grass crops in (1) marker development, (2) improving conserved gene annotation or (3) candidate gene selection for traits of interest (either conserved or lineage-specific ones) (Salse 2012, Murat et al. 2014b). The paleogenomics data reported here, in terms of ancestral genome structures (*i.e.* protochromosome characterisation as well as protogene order inference), associated with a robust comparison of modern grass genomes, can now be considered as an applied tool to navigate accurately between genomes and transfer genomic information (*i.e.* gene structures and functions) from *Brachypodium* to grass crops.

To do so, we have recently provided a user friendly web tool named 'PlantSyntenyViewer' [http://urgi.versailles.inra.fr/synteny-cereal], allowing access to the orthologous, paralogous and ancestral relationships described in the current chapter (Fig. 3a). Using this tool it is possible to navigate from one genome to another using a gene name, a modern chromosome nomenclature or ancestral protochromosome references. This tool offers for the first time in the same screen shot the complete set of identified orthologs and paralogs from the sequenced monocot genomes for any considered region or gene of interest. 'PlantSyntenyViewer' thus provides information about the ancestral grass gene set that can be used as a platform for the development of COS (Conserved Orthologous Set) markers (Quraishi et al. 2009, Pont et al. 2013) to support cross genome map-based cloning strategies in grasses.

Paleogenomics data can greatly simplify and accelerate the identification of useful markers or more interestingly candidate genes showing that comparative genomics-based paleogenomics data available in the web tool 'PlantSyntenyViewer' represents a valuable resource for marker development and ultimately trait dissection in grasses. With this tool/strategy, since 2010, *Brachypodium* as be used as a pivotal genome to dissect in wheat traits related to yield (*i.e.* NUE for Nitrogen Use Efficiency, Quraishi et al 2011a; FZP for FriZzy Panicle, Dobrovolskaya et al. 2015) or bread making quality (GFC for Grain fiber content, Quraishi et al 2011b; Carotenoid content Dibari et al. 2012).

As a case example, the Fig. 3b illustrates the translational-based cloning of FZP genes in bread wheat (Dobrovolskaya et al. 2015). Bread wheat inflorescences, or spikes, are characteristically unbranched and normally bear one spikelet per rachis node. Based on wheat mutants with supernumerary spikelets (SS) and paleogenomics/comparative genomics data between *Brachypodium* and wheat, it has been shown that the orthologous *FRIZZY PANICLE* (FZP) gene, encoding a member of the APETALA2/Ethylene Response Factor (AP2/ERF) transcription factor family, drives the SS trait in both bread wheat and *Brachypodium*. Structural and functional characterization of the three wheat *FRIZZY PANICLE* homoeologous genes (WFZP-A-B-D of the allohexaploid bread wheat) revealed that coding mutations of *WFZP-D* cause the SS phenotype with the most severe effect when *WFZP-D* lesions are combined with a frameshift mutation in *WFZP-A* (Dobrovolskaya et al. 2015).

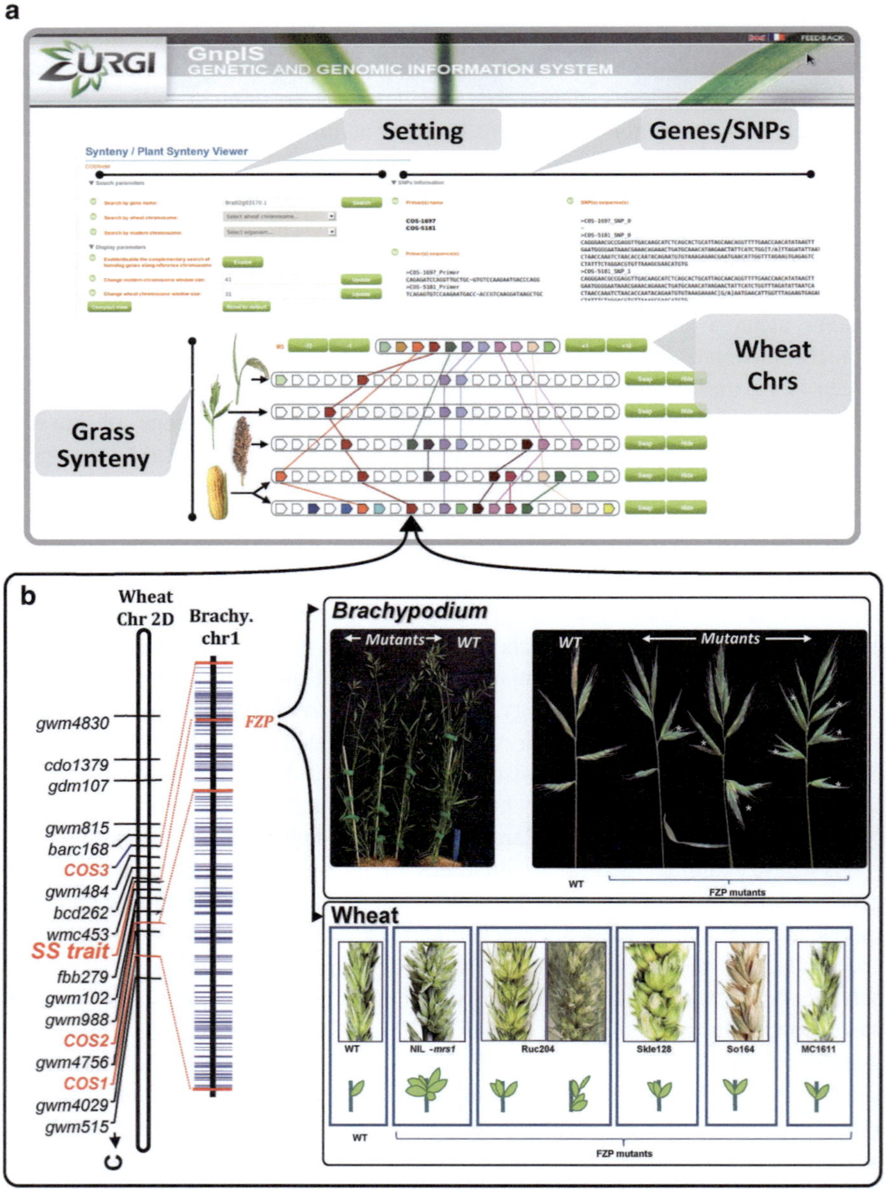

Fig. 3 PlantSyntenyViewer tool and its use for WFZP gene dissection in bread wheat. (**a**) PlantSyntenyViewer tool (Adapted from Murat et al. 2014b). The entry page of the PlantSyntenyViewer tool in which the setting parameters (search by gene name, ancestral or modern chromosomes) are mentioned at the top and the derived paleogenomics data visualization with AGK (A5 in this screen), rice, maize, sorghum, *Brachypodium* gene conservation (*colored connecting lines*) at the *bottom*. The PlantSyntenyViewer tool is available at http://urgi.versailles.inra.fr/synteny-cereal. (**b**) WFZP gene characterization in bread wheat (Adapted from Dobrovolskaya et al. 5015). Composite genetic map of the wheat chromosome 2D, including the supernumerary spikelets (*SS*) locus and flanking COS markers associated with the *Brachypodium* orthologous region (on chromosome 1) highlights FZP as a candidate (*left*) witch function is conserved in wheat and *Brachypodium* mutants (compared to wild type WT) in deriving similar SS phenotypes (*right*)

Conclusion

The modern *Brachypodium* genome has been shaped from a seven ancestral grass genome followed by seven nested chromosome fusions. Its close relationship to both the ancestral grass karyotype and the *Triticeae* clade, makes *Brachypodium* as a pivotal genome for accurate translational research genomics in the economically important wheat, barley and rye crops regarding the transfer of gene annotation/function and ultimately candidate genes of traits.

Acknowledgments This work has been supported by grants from the Agence Nationale de la Recherche (program ANR Blanc-PAGE, *ref*: ANR-2011-BSV6-00801). The author gratefully acknowledges Caroline Pont and Florent Murat (INRA Clermont-Ferrand, France) for their contributions in preparing the article illustrations.

References

Abrouk M, Murat F, Pont C, Messing J, Jackson S, Faraut T, et al. Palaeogenomics of plants: synteny-based modelling of extinct ancestors. Trends Plant Sci. 2010;15(9):479–87.

Bolot S, Abrouk M, Masood-Quraishi U, Stein N, Messing J, Feuillet C, et al. The 'inner circle' of the cereal genomes. Curr Opin Plant Biol. 2009;12(2):119–25.

Dibari B, Murat F, Chosson A, Gautier V, Poncet C, Lecomte P, et al. Deciphering the genomic structure, function and evolution of carotenogenesis related phytoene synthases in grasses. BMC Genomics. 2012;13:221.

Dobrovolskaya O, Pont C, Sibout R, Martinek P, Badaeva E, Murat F, et al. FRIZZY PANICLE drives supernumerary spikelets in bread wheat. Plant Physiol. 2015;167(1):189–99.

International Brachypodium Initiative. Genome sequencing and analysis of the model grass *Brachypodium distachyon*. Nature. 2010;463(7282):763–8.

International Rice Genome Sequencing Project. The map-based sequence of the rice genome. Nature. 2005;436:793–800.

Jacquemin J, Chaparro C, Laudié M, Berger A, Gavory F, Goicoechea JL, et al. Long-range and targeted ectopic recombination between the two homeologous chromosomes 11 and 12 in *Oryza* species. Mol Biol Evol. 2011;28(11):3139–50.

Luo MC, Deal KR, Akhunov ED, Akhunova AR, Anderson OD, Anderson JA, et al. Genome comparisons reveal a dominant mechanism of chromosome number reduction in grasses and accelerated genome evolution in Triticeae. Proc Natl Acad Sci U S A. 2009;106(37):15780–5.

Massa AN, Wanjugi H, Deal KR, O'Brien K, You FM, Maiti R, et al. Gene space dynamics during the evolution of Aegilops tauschii, Brachypodium distachyon, Oryza sativa, and Sorghum bicolor genomes. Mol Biol Evol. 2011;28(9):2537–47.

Murat F, Xu JH, Tannier E, Abrouk M, Guilhot N, Pont C, et al. Ancestral grass karyotype reconstruction unravels new mechanisms of genome shuffling as a source of plant evolution. Genome Res. 2010;20(11):1545–57.

Murat F, Zhang R, Guizard S, Flores R, Armero A, Pont C, et al. Shared subgenome dominance following polyploidization explains grass genome evolutionary plasticity from a seven protochromosome ancestor with 16K protogenes. Genome Biol Evol. 2014a;6(1):12–33.

Murat F, Pont C, Salse J. Paleogenomics in Triticeae for translational research. Curr Plant Biol. 2014b;1:34–9.

Paterson AH, Bowers JE, Bruggmann R, Dubchak I, Grimwood J, Gundlach H, et al. The *Sorghum bicolor* genome and the diversification of grasses. Nature. 2009;457(7229):551–6.

Pont C, Murat F, Confolent C, Balzergue S, Salse J. RNA-seq in grain unveils fate of neo- and paleopolyploidization events in bread wheat (*Triticum aestivum* L.). Genome Biol. 2011;12 (12):R119.

Pont C, Murat F, Guizard S, Flores R, Foucrier S, Bidet Y, et al. Wheat syntenome unveils new evidences of contrasted evolutionary plasticity between paleo- and neoduplicated subgenomes. Plant J. 2013;76(6):1030–44.

Quraishi UM, Abrouk M, Bolot S, Pont C, Throude M, Guilhot N, et al. Genomics in cereals: from genome-wide conserved orthologous set (COS) sequences to candidate genes for trait dissection. Funct Integr Genomics. 2009;9(4):473–84.

Quraishi UM, Abrouk M, Murat F, Pont C, Foucrier S, Desmaizieres G, et al. Cross-genome map based dissection of a nitrogen use efficiency ortho-metaQTL in bread wheat unravels concerted cereal genome evolution. Plant J. 2011a;65(5):745–56.

Quraishi UM, Murat F, Abrouk M, Pont C, Confolent C, Oury FX, et al. Combined meta-genomics analyses unravel candidate genes for the grain dietary fiber content in bread wheat (*Triticum aestivum* L.). Funct Integr Genomics. 2011b;11(1):71–83.

Salse J, Bolot S, Throude M, Jouffe V, Piegu B, Quraishi UM, et al. Identification and characterization of conserved duplications between rice and wheat provide new insight into grass genome evolution. Plant Cell. 2008;20:11–24.

Salse J, Abrouk M, Murat F, Quraishi UM, Feuillet C. Improved criteria and comparative genomics tool provide new insights into grass paleogenomics. Brief Bioinform. 2009a;10 (6):619–30.

Salse J, Abrouk M, Bolot S, Guilhot N, Courcelle E, Faraut T, et al. Reconstruction of monocotelydoneous proto-chromosomes reveals faster evolution in plants than in animals. Proc Natl Acad Sci U S A. 2009b;106:14908–13.

Salse J, Feuillet C. Paleogenomics in cereals: modelling of ancestors for modern species improvement. C R Biol. 2011;334(3):205–11.

Salse J. *In silico* archeogenomics unveils modern plant genome organization, regulation and evolution. Curr Opin Plant Biol. 2012;15(2):122–30.

Schnable PS, Ware D, Fulton RS, Stein JC, Wei F, Pasternak S, et al. The B73 maize genome: complexity, diversity, and dynamics. Science. 2009;326(5956):1112–5.

Schubert I, Lysak MA. Interpretation of karyotype evolution should consider chromosome structural constraints. Trends Genet. 2011;27(6):207–16.

Tang H, Bowers JE, Wang X, Paterson AH. Angiosperm genome comparisons reveal early polyploidy in the monocot lineage. Proc Natl Acad Sci U S A. 2010;107(1):472–7.

Wang X, Tang H, Paterson AH. Seventy million years of concerted evolution of a homoeologous chromosome pair, in parallel, in major Poaceae lineages. Plant Cell. 2011;23(1):27–37.

Wicker T, Buchmann JP, Keller B. Patching gaps in plant genomes results in gene movement and erosion of colinearity. Genome Res. 2010;20(9):1229–37.

Genome Size and the Role of Transposable Elements

Alan H. Schulman

Abstract The lack of correlation between genome size and organismal complexity was early on dubbed the "*C*-value Paradox;" it holds even when gene number is considered instead of overall organismal complexity. The sequencing of large eukaryotic genomes has now conclusively solved this conundrum with the demonstration that most nuclear DNA comprises various classes of repeats, primarily transposable elements (TEs). The inherent and variable capacity of the TEs for mobility and replication explains how genome size can vary so greatly on their account. The Class I TEs or retrotransposons have a replication cycle involving the copying of a transcribed, genomic RNA into dsDNA by reverse transcriptase. As a result of their replicative life cycle, the retrotransposons comprise most of large genomes among plants; differences in their prevalence explain most of the variation in genome size on the monoploid level. However, retrotransposons are not only gained through the propagative life cycle described above, but they also can be lost through a combination of progressive small deletions and truncations. The genome of *Brachypodium distachyon,* at ~372 Mb, is at the lower end of the distribution for flowering plants. The compactness of the *B. distachyon* genome is correlated with a relatively low number of retrotransposons, although it contains many recently inserted transposable elements. The *B. distachyon* genome appears to stay trim through recombinational shedding of retrotransposons, despite their continuing propagation. Nevertheless, the chromosomes show remarkable differences among them regarding the gain and loss of retrotransposons over time and the relative accumulation of the two superfamilies, *Copia* and *Gypsy*.

Keywords Retrotransposon replication • Genome size • Transposable elements • Chromosome dynamics • Genome evolution

---

A.H. Schulman, B.A., M.S., M.Phil., Ph.D. (✉)
Institute of Biotechnology, University of Helsinki, Viikki Biocenter,
P.O. Box 65, 00014 Helsinki, Finland

Green Technology, Luke Natural Resources Institute, Viikki Biocenter,
P.O. Box 65, 00014 Helsinki, Finland
e-mail: alan.schulman@helsinki.fi

J. Vogel (ed.), *Genetics and Genomics of Brachypodium*,
Plant Genetics and Genomics: Crops Models, DOI 10.1007/7397_2015_3
Published online: 02 October 2015

Abbreviations

AP	Aspartic proteinase
BET	Bromodomain and extraterminal domain
C	DNA content of a haploid or monoploid genome
ERV	Endogenous retrovirus
IN	Integrase
LARD	Large retrotransposon derivative
LINE	Long interspersed nuclear element
LTR	Long terminal repeat
MITE	Miniature inverted repeat transposable element
MY	Million years
MYA	Million years ago
NLS	Nuclear localization signal
PAV	Presence-absence variation
RH	RNase H
RT	Reverse transcriptase
SINE	Short interspersed nuclear element
TE	Transposable element
TIR	Terminal inverted repeat
VLP	Virus-like particle

The *C*-value Paradox

Studies on organisms from bacteria to higher plants and animals established by the 1970s that genome size varied enormously across life. Even among the eukaryotes, the range of genome size is enormous: the genomes of *Amoeba dubia*, comprising 670,000 Mbp (Gregory 2001), and that of the microsporidium *Encephalitozoon cuniculi*, containing only 2.9 Mbp (Biderre et al. 1995; Katinka et al. 2001), vary by 200,000-fold. Fungal genomes, which are generally from 10 to 60 Mbp in size, tend to occupy the lower end of this range, topping off at about 200 Mbp (Gregory et al. 2007; Schulman and Wicker 2013). The lack of correlation between genome size (*C*-value) and organismal complexity in terms of tissue and organ numbers was early on dubbed the "*C*-value Paradox" (Gaut and Ross-Ibarra 2008; Rosbash et al. 1974). Weighing in with the largest genomes among the angiosperms are two diploids, the monocot *Trillium hagae* ($1C = 129{,}536$ Mbp) and the eudicot *Viscum album* (Santalaceae) with 100,636 Mbp (Zonneveld 2010), and the octoploid *Paris japonica* ($1C = 148{,}881$ Mbp; (Bennett and Leitch 2011)). These are at least 1600 X larger than the smallest known plant genome, dwarfing that of the carnivorous *Genlisea tuberosa* (61 Mbp; (Fleischmann et al. 2014)). The genomes of *Brachypodium distachyon* and *Arabidopsis thaliana*, at ~372 Mb and ~135 Mb respectively, are nevertheless clearly at the lower end of the distribution for flowering plants.

Within taxonomic groups, where anatomical complexity is expected to be fairly similar, the genomes of many clades of animals and fungi maintain genome size in a fairly narrow range, from around 3000 Mbp for mammals (Gregory et al. 2007; Schulman and Wicker 2013) to about 1000–2000 Mbp for reptiles and birds (Gregory et al. 2007; Krishan et al. 2005). Amphibians and lungfish nevertheless display a greater than a 100-fold genome size variation within their groups (Gregory 2001). Plant genomes, in contrast, can show extreme size variations both over fairly narrow taxonomic ranges, which are not correlated with phylogenetic distance. *B. distachyon* and bread wheat (*Triticum aestivum*) have genomes of 272 Mbp and 5700 Mbp, respectively, yet they diverged only about 35 million years ago (MYA) (Bossolini et al. 2007). Even taking their threefold ploidy difference into account, the genomes of these two species still differ by seven times. Sorghum and maize diverged only 12 MYA (Swigonova et al. 2004), but their genomes are now respectively 727 Mbp and 2500 Mbp (Paterson et al. 2009; Schnable et al. 2009).

As the gene number of organisms in various clades gradually has been clarified, it has also become apparent that the *C*-value Paradox remains valid when gene number is considered instead of overall organismal complexity. While genome size varies widely, gene number shows only about a tenfold variation, from roughly 5000 to 10,000, on the monoploid level in eukaryotic genomes (Michael and Jackson 2013). Clearly, polyploidization multiplies both gene number and genome size by a factor of two (tetraploid) or three (hexaploid), or by more in many highly polyploid plant families. Nevertheless, genome size continues to vary greatly, much more so than gene number, even when considering the DNA content of the basic monoploid set of chromosomes. The upper end of the range in gene number is occupied primarily by animals and plants, which have from about 25,000 to 40,000 genes (Schulman and Wicker 2013). In this regard, the seven-fold difference in monoploid genome size between *B. distachyon* and *T. aestivum* or barley (*Hordeum vulgare*) is also far in excess of the maximum 1.3-fold difference in high-confidence estimates for the number of protein coding genes, respectively 32,000 in *B. distachyon* (International Brachypodium Initiative 2010), 35,000 in *T. aestivum* (International Wheat Genome Sequencing 2014), and 26,159 in barley (International Barley Genome Sequencing Consortium et al. 2012).

Recently, the concept of the "core" and "pan-" genomes has been extended to the plants (Morgante et al. 2007). The core genome is defined as the minimum set of common genes found in a given clade, whereas the pan-genome is the set of all genes found in that clade. The relative proportion of the gene complement that appears to be dispensable, *i.e.*, not part of the core genome, appears to vary from clade to clade and may be related to the predominant breeding system (*e.g.* selfing, outcrossing, or vegetatively propagating). In *Glycine*, 20 % of the ~55,000 genes appear to be dispensable (Li et al. 2014). In maize (*Zea mays*), between 0.5 and 4 % of the annotated genes showed presence-absence variation (PAV) between two inbred lines (Springer et al. 2009), with only 16.4 % of transcribed genes belonging to the core expressed part of the genome (Hirsch et al. 2014). Data is not yet in hand to allow the pan-genome concept to be tested against the *Brachypodium* gene set but, as will become obvious below, the possession of only a small core gene set and a large dispensable set is not needed to explain the compactness of the *B. distachyon* genome.

Transposable Elements Explain the *C*-value Paradox

When one defines the genome as comprising not only the genes ("gene-ome"), but all DNA in the nucleus, the mystery of the *C*-value Paradox clears. The sequencing of large eukaryotic genomes from the beginning of the millennium onward has conclusively shown that most of the nuclear DNA encodes not a diversity of metabolic and regulatory proteins, but rather various classes of repeats, primarily transposable elements (TEs; Bennetzen and Wang 2014; Vitte and Panaud 2005). The TEs can constitute 80 % or more of the total genomic DNA of large genomes, such as those of the cereals (Schnable et al. 2009; Wicker et al. 2009) or gymnosperms (De La Torre et al. 2014). Even compact plant genomes, such as those of *A. thaliana* or *B. distachyon*, are populated by hundreds of different TE families having copy numbers ranging from one or several to many hundreds of copies while others have only one or a few members (International Brachypodium Initiative 2010).

The TEs are best thought of as autonomous or semi-autonomous genetic units or even as intracellular viruses in the sense of being replicating entities within the genome. It is their inherent and variable capacity for mobility and replication that explains how genome size can vary so greatly on their account. They fall into two major groups: the "cut and paste" DNA transposons of Class II (Fig. 1c); the "copy and paste" retrotransposons of Class I (Fig. 1a, b) (Wicker et al. 2007). As described below, in the life cycle of retrotransposons new copies are propagated throughout the genome while the original element remains in place. Transposition of DNA transposons is not, in contrast, inherently replicative, although these elements may be extremely abundant. Within Class I, the LTR (long terminal repeat) retrotransposons are most abundant in plants. They fall into two main superfamilies, *Gypsy* and *Copia*, which are found in almost all eukaryotic lineages. The superfamilies differ both in the order and the sequence affinities of the encoded protein domains described below (Wicker et al. 2007).

Given that intact retrotransposons are generally about 9 kb long, it is therefore unsurprising that, when abundant, they constitute the most extensive class of repetitive DNA by base pairs in plant genomes. Of the retrotransposons, the LTR retrotransposon order (Wicker et al. 2007) contributes the most to genome size variation in plants. *A. thaliana* and sorghum (*Sorghum bicolor*), which have genomes of 135 Mbp and 727 Mbp respectively, contain similar numbers of DNA transposons, while the abundance of LTR retrotransposons largely explains the considerably larger genome of sorghum (Estep et al. 2013). This general pattern has been found repeatedly as the genomes of various plants have been subjected to high-throughput sequencing. For example, the sunflower (*Helianthus annuus*)

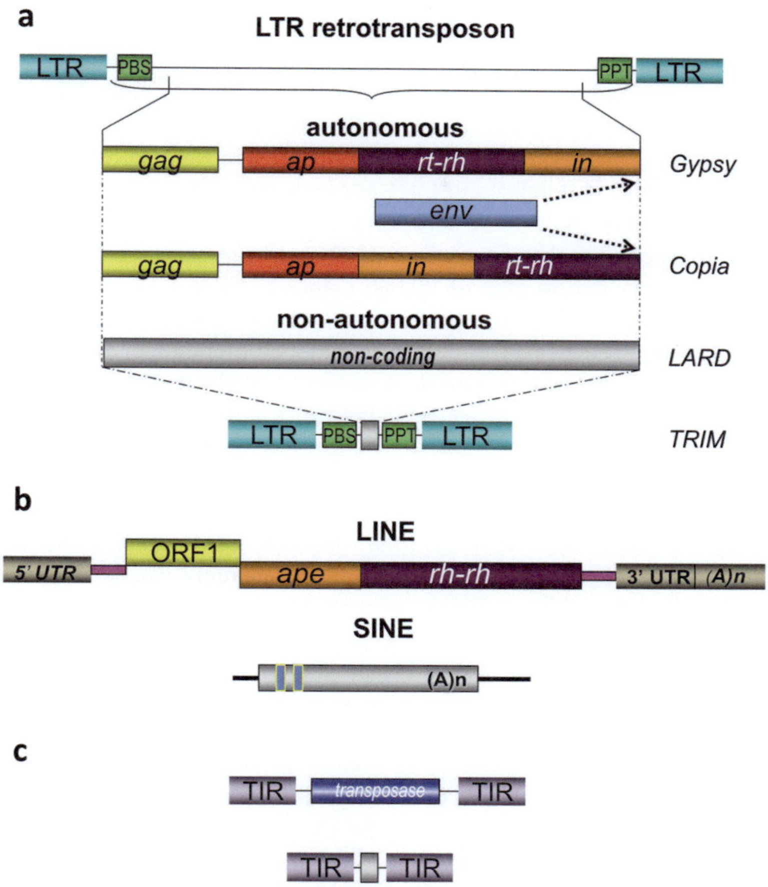

Fig. 1 Main groups of transposable elements. (**a**) Autonomous and non-autonomous LTR retrotransposons. An LTR retrotransposon comprises: the long terminal repeats (LTRs); the (−)-strand primer binding site (PBS) for reverse transcription; the polypurine tract (PPT), which is the (+)-strand priming site for reverse transcription, and a core domain (shown as a *black line above*). Within the core domain, autonomous retrotransposons contain the coding domains for a protein that forms the capsids of the virus-like particles (*gag*), an aspartic proteinase (*ap*), the reverse-transcriptase and RNase H complex (*rt-rh*), and integrase (*in*). The domain orders differ in the two main Superfamilies of LTR retrotransposons, *Copia* and *Gypsy*. The position of the envelope (*env*) domain in those *Gypsy* and *Copia* clades that contain it is shown. *Below*, the non-autonomous LTR retrotransposons. *LARD* elements have a long internal domain with conserved structure but lacking coding capacity. *TRIM* elements have virtually no internal domain except for the PBS and PPT signals. (**a**) Autonomous and non-autonomous non-LTR of the LINE order and the non-autonomous SINE order. A *grey bar* indicates a non-coding domain. (**b**) The non-LTR retrotransposons. LINE elements contain a 5′ untranslated region (UTR) and two open reading frames, ORF1, which specifies an RNA binding protein that can form a ribonucleoprotein particle, and second open reading frame that encodes an apurinic/apyrimidinic-like endonuclease (ape) and the reverse transcriptase and RNase H complex (rt-rh). These are followed by a 3′ untranslated region (3′ UTR) and a polyadenosine tail ((A)n). The SINE elements are non-autonomous, with shared features being the pol III promoter (two *vertical blue stripes*) and the polyadenosine tail ((A)n). (**c**) DNA transposons (Class II transposable elements). The simplest type of Class II element, which belong to the *Tc1/Mariner* superfamily, is diagrammed. These comprise an open reading frame specifying a transposase domain and are bounded by terminal inverted repeats (TIRs). *Below*, a non-autonomous MITE element, in which the coding domain has been replaced by a small non-coding segment (*gray box*). Parts **a** and **b** are modified and reprinted from (Schulman 2013) with permission from Elsevier

genome is comprised of 81 % transposable elements and 77 % LTR retrotransposons, the vast majority of which inserted following the origin of the species (Staton et al. 2012). Comparisons of the genomes of several *Oryza* (rice) species showed that differences in LTR retrotransposon abundance considerably explained variations in genome size in the genus (Chen et al. 2013; Zhang et al. 2007). Spectacularly, the genome of *O. australiensis* doubled in size over three million years due to gain of 90,000 copies of three LTR retrotransposons families, specifically *RIRE1* of superfamily *Copia* and *Kangourou* and *Wallabi* of superfamily *Gypsy* (Piegu et al. 2006). Differential proliferation of retrotransposons in *Gossypium* (cotton) species likewise is well correlated with the expansion of the genomes of some species in this genus in comparison to others.

Comparative analysis of the *B. distachyon* genome is consistent with the general theme of retrotransposon abundance determining genome size. The genome contains over 29,000 DNA transposons, which together cover 4.8 % of the assembly (Table 1). The coverage ratio of 1:4.8 between DNA transposons and the retrotransposons, which comprise 23.3 %, is considerably smaller than that for barley. In barley the ratio is 13.5:1, estimated by annotation of whole-genome shotgun sequence covering 15 % of the genome (International Barley Genome Sequencing Consortium et al. 2012). Hence, the relative expansion of the barley genome compared to *B. distachyon* appears to be due to the growth in retrotransposon abundance in barley.

While the LTR retrotransposons as a whole are generally the most important TEs for genome expansion, particular superfamilies and families among them differ widely in their role depending on the genus and species of plant. In barley, 12 families of retrotransposons account for almost 50 % of the 5.5 Gb genome (Wicker et al. 2009), and overall the superfamily *Gypsy* is 1.5-fold more abundant than that of *Copia* (International Barley Genome Sequencing Consortium et al. 2012). Among the related tribe Triticeae genomes, the *BARE*, *WIS*, and *Angela* families of superfamily *Copia* comprise more than 10 % of the genome (Kalendar et al. 2000; Soleimani et al. 2006; Vicient et al. 1999a; Wicker et al. 2009). Variation in the abundance of the *BARE* retrotransposon family is, moreover, sufficient to explain most of the difference in genome size between two *Hordeum* species (Vicient et al. 1999b). Likewise, in the panacoid grasses, particular families became dominant in particular plant lineages (Estep et al. 2013). In the *B. distachyon* genome, the *Gypsy* superfamily is predominant, comprising 55.4 % of the retrotransposons (Table 1) and forming 19 major clades (International Brachypodium Initiative 2010). Together, the *Gypsy* elements form 70.6 % of the intact LTR retrotransposons and cover 16.1 % of the genome sequence, which is 3.3 times more than the *Copia* elements do. The *Copia* superfamily represents 40.8 % of the retrotransposons in the genome and forms 44 clades.

Beyond the grasses, a similar picture is emerging. Within the enormous (~11 to 20 Gb) genomes of the conifers, depending on the genus, either the *Copia* or the *Gypsy* superfamily played a larger role than the other expanding the genome (Nystedt et al. 2013). In the pepper (*Capsicum annuum*), *Del* elements of the *Gypsy* superfamily are primarily responsible for expansion (Park et al. 2011,

Table 1 *Brachypodium distachyon* transposable element content

Transposable element	Families	Copies	% copy number	Mb	Avg length bp	% of TE bp	% of genome
Total		80,049	100.00	76.091	951	100.00	28.10
Class I: Retroelement (RXX)		*50,419*	*62.99*	*63.168*	*1253*	*83.02*	*23.33*
LTR retrotransposon		47,274	59.06	57.908	1225	76.10	21.39
Full length		690	0.861972	6.468	9373	8.4999	2.3885036
Solo		1814	2.266112	0.685	378	0.900762	0.2531174
Ty1/copia (RLC)	44	12,426	15.52	13.149	1058	17.28	4.86
Full length		282	0.35	1.900	6737	2.50	0.70
Solo		689	0.86	0.332	482	0.44	0.12
Ty3/gypsy (RLG)	19	32,978	41.20	43.464	1318	57.12	16.05
Full length		382	0.48	4.358	11,408	5.73	1.61
Solo		1122	1.40	0.352	313	0.46	0.13
Unclassified LTR (RLX)	9	1870	2.34	1.295	693	1.70	0.48
Full length		26	0.03	0.210	8074	0.28	0.08
Solo		3	0.004	0.002	567	0.002	0.001
Non-LTR retrotransposon (RXX)		3145	3.93	5.259	1672	6.91	1.94
LINE (RIX)		3145	3.93	5.259	1672	6.91	1.94
Class II: DNA Transposon (DXX)		*29,630*	*67.01*	*12.924*	*436*	*16.98*	*4.77*
Superfamily (DTX)		5947	7.43	9.564	1608	12.57	3.53
CACTA (DTC)	14	1523	1.90	5.899	3873	7.75	2.18
hAT (DTA)	76	1220	1.51	1.197	737	1.56	0.44
Mutator (DTM)	65	2854	3.57	1.710	599	2.25	0.63
Tc1/Mariner (DTT)	8	50	0.06	0.177	3542	0.23	0.07
PIF/Harbinger (DTH)	24	862	1.08	1.135	1316	1.49	0.42
MITE (DXX)		23,563	29.44	2.869	122	3.77	1.06
Stowaway (DTT)	21	20,994	26.23	2.394	114	3.15	0.88
Tourist (DTH)	19	2569	3.21	0.475	185	0.62	0.18
Helitron (DHH)	48	120	0.15	0.491	4089	0.64	0.18

The data has been previously reported (International Brachypodium Initiative 2010), with the transposable element groups classified according to (Wicker et al. 2007)

2012) of the genome to its current 2.7 Gb size. One particular *Gypsy* family, *Gorge3*, underwent massive increases in copy number in two particular *Gossypium* lineages (Hawkins et al. 2006). In the legume *Vicia pannonica*, a single, massive (25 kb) *Gypsy* element similar to the family *Ogre* alone comprises 38 % of the genome (Neumann et al. 2006).

In conclusion, the role of retrotransposons in explaining most of the differences in genome size of monoploid chromosome sets has become unassailable. It has not been established for any species why particular retrotransposon families come to predominate. The answer must lie in the dynamics of how each family replicates, the control mechanisms thereof, selective forces acting on newly inserted copies, and the population dynamics and breeding systems of the species in question. The loss of retrotransposons through various recombinational mechanisms, discussed below, clearly has played an important role in shaping plant genomes. Retrotransposons can be activated by abiotic stresses including drought (Kalendar et al. 2000) and UV light (Ramallo et al. 2008) and by biotic stresses set off by pathogens (Anca et al. 2014; Grandbastien et al. 2005). Whereas some of these are abundant, other stress-activated retrotransposons are present at low copy numbers. A short consideration of retrotransposon replication may help to sharpen the issues involved.

LTR Retrotransposon Replication and Growth in Genome Size

Class I transposable elements all share a replication cycle involving the copying of a transcribed, genomic RNA into dsDNA by reverse transcriptase. The LTR retrotransposons form one major order. Two other major orders of retrotransposons (Fig. 1), the LINEs (Long Interspersed Nuclear Elements; Goodier and Kazazian 2008) and SINEs (Short Interspersed Nuclear Elements; Wicker et al. 2007), lack LTRs, differ otherwise in their structures and replication mechanisms from the LTR retrotransposons, and will be discussed briefly below.

Plant LTR retrotransposons are similar by their structure (Fig. 1a) and replication mechanism to the superfamilies *Gypsy* and *Copia* of fungi and animals as well as to retroviruses and endogenous retroviruses (ERVs) of mammals. Transcription initiates from the 5′ LTR, which contains a pol II promoter. LTRs also contain signals for RNA termination and polyadenylation, which are recognized by the transcriptional machinery in the 3′ LTR. The LTRs carry promoter response elements that can modulate transcription and connect the replication cycle to regulatory networks within the plant (Butelli et al. 2012, Grandbastien 2014, McCue et al. 2012). Transcription in different LTR retrotransposon families has been shown to be activated by a variety of biotic and abiotic stresses as well as by tissue culture and plant hormone treatment (Ansari et al. 2007; Cavrak et al. 2014; Grandbastien et al. 2005; Kalendar et al. 2000; Ramallo et al. 2008; Salazar et al. 2007).

Transcription has been examined closely for only a couple of plant retrotransposons (Beguiristain et al. 2001; Chang and Schulman 2008; Hernández-Pinzón et al. 2012). In barley, multiple pools of polyadenylated and non-polyadenylated RNAs are produced, respectively for translation and reverse transcription (Chang et al. 2013).

The transcripts of LTR retrotransposons need to serve both translation and reverse transcription. However, transcription yields RNA with incomplete LTRs, because the promoter and terminator are within the 5′ and 3′ LTRs respectively. Restoration of both LTRs is needed to produce a daughter cDNA competent for both integration and subsequent replication. This problem is resolved by the complex reverse transcription mechanism (Schulman 2013). Following transcription, the retrotransposon RNA is transported to the cytoplasm (Fig. 2). Translation of proteins encoded by the retrotransposon itself is essential for completion of the life cycle. Between the two LTRs, domains encode a capsid protein, Gag, which forms virus-like particles (VLPs), and a polyprotein (Gao et al. 2003; Moisy et al. 2008; Tanskanen et al. 2007). The Gag may be either a part of the polyprotein open reading frame or in a separate one. The polyprotein contains an aspartic proteinase (AP), integrase (IN), RT, and RNase H (RH). For both retrotransposons and retroviruses, the translated and processed Gag binds and encapsidates the retrotransposon RNAs into a VLP together with RT-RNase H and IN (Lee et al. 2012; Schulman 2013). The formation of virus-like particles has, however, been shown only for the native *BARE1* in barley (Jääskeläinen et al. 1999) and the tobacco *Tto1* under an inducible promoter in Arabidopsis (Böhmdorfer et al. 2008).

Completion of the retrotransposon life cycle requires integration into the chromosome, which means that the retrotransposon cDNA must find its way from the cytoplasm back into the nucleus (Fig. 2). Nuclear entry is directed by a nuclear localization signal (NLS), which in various retroviruses can be found in different retroviral proteins (Müllers et al. 2011; Suzuki and Craigie 2007). The NLS signals have not been well studied for plant retrotransposons. While we have evidence that *BARE* Gag contains an NLS (Gómez-Orte et al. unpublished); *gene23*, transcribed under its own promoter from the opposite strand as *gag* and *pol*, encodes a functional NLS in retrotransposon *Grande* of *Zea* species, but its role remains unclear (Gómez-Orte et al. 2013). Once the LTR retrotransposon cDNA is localized to the nucleus, integration is carried out by IN. The enzyme makes staggered cuts at the target site and carries out the reaction by a mechanism highly conserved among retrotransposons and retroviruses (Krishnan and Engelman 2012) that appears to be conserved also with bacteriophage transposases (Hickman et al. 2010; Montaño et al. 2012; Montaño and Rice 2011). Integration site specificity is an interesting issue for plant retrotransposons because, as will be discussed below, retrotransposon density along plant chromosomes is generally highly variable. Integrases in various organisms have local target site preferences, some of them highly specific.

Most work on integration specificity has been done with yeast (*Saccharomyces cerevisiae*) retrotransposons and with retroviruses, excepting one clade of plant retrotransposons. The integrases of yeast *Ty1* and *Ty3* direct integration to a narrow range of sites upstream of genes transcribed by Pol III, such as tRNA and 5S

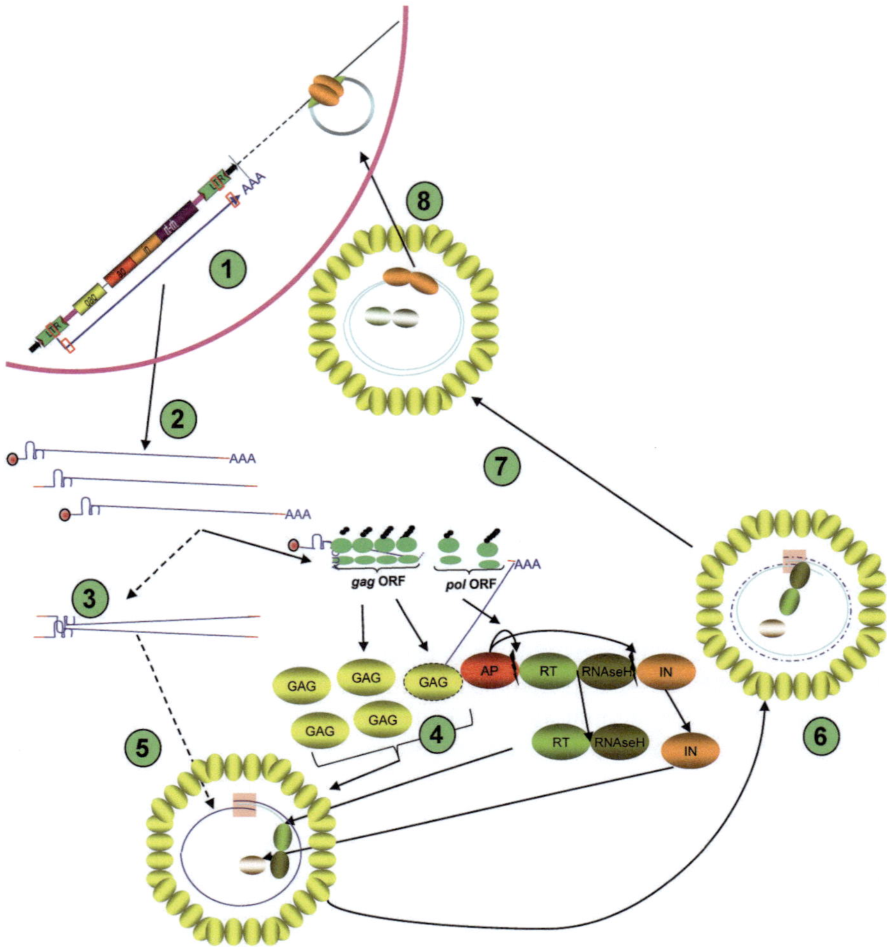

Fig. 2 Retrotransposon life cycle. An element from the *Copia* superfamily is shown within the genome inside the nucleus (*magenta curve*). The successive steps of replication are: (*1*) transcription from the promoter in the LTR (*red boxes* denote the R-domain generated by transcription); (*2*) nuclear export; (*6*) alternative packaging of transcripts into a virus-like particle (VLP) or translation; for the *BARE* retrotransposon, capped (*red balls*) and polyadenylated transcripts are translated, whereas uncapped and unpolyadenylated transcripts are packaged (Chang et al. 2013); (*4*) translation of either distinct *gag* and *pol* open reading frames or of a shared one to produce the capsid protein Gag and a polyprotein containing aspartic proteinase (AP), RT, RNase H (RH), and integrase (IN); (*5*) assembly of a VLP from Gag containing RNA transcripts, IN, RT, RH; (*6*) reverse transcription by RT; (*7*) localization of the VLP to the nucleus; (*8*) passage of the cDNA–IN complex into the nucleus and integration of the cDNA into the genome. The details are essentially as presented earlier (Schulman 2012, 2013). The figure is modified and reprinted from (Schulman 2013) with permission from Elsevier

ribosomal genes, by interacting with transcription factor subunits (Bachman et al. 2005; Yieh et al. 2000), whereas *Ty5* integrase directs integration to heterochromatin through interaction with the heterochromatin protein Sir4 (Brady et al. 2008). The retrovirus HIV targets transcriptionally active regions for integration by interaction with the cellular lens epithelium-derived growth factor LEDGF/p75 (Ciuffi and Bushman 2006), whereas, in parallel, MLV interacts with bromodomain and extraterminal domain (BET) proteins to achieve the same end (Sharma et al. 2013). One widespread group of *Gypsy* retrotransposons, the so-called chromoviruses or CRM clade, is interesting for the presence of chromodomains at the C-terminus of the IN (Gorinsek et al. 2005). The chromodomain is similar to domains of heterochromatin protein 1 (HP1), which may confer particular integration patterns to some chromoviruses (Weber et al. 2013). Phylogenetic and sequence analysis of members of the CRM clade in plants (Neumann et al. 2011), defined one particular group that contains a CR motif similar that found in the chromodomain-bearing MAGGY retrotransposons of fungi (Gao et al. 2008) and is concentrated in centromeric regions.

Non-LTR Retrotransposons

The non-LTR retrotransposons are ubiquitous in the eukaryotes. Although they predominate in vertebrates (Chalopin et al. 2015), LINEs are generally much less abundant in plants (Heitkam et al. 2014), with an exception being the sugar beet, *Beta vulgaris* (Wenke et al. 2009). Probably the most ancient group of Class I elements due to their simple structure, the core of which contains only reverse transcriptase and endonuclease, the LINEs replicate by a rather different mechanism than do the LTR retrotransposons (Schulman 2013). Lacking an integrase gene, RT primes DNA synthesis from the poly-A tail of the transcript directly at the point of insertion and then ligates the end of the cDNA into the insertion point (Yamaguchi et al. 2014). In *B. distachyon* (Table 1), the non-LTR retrotransposons cover only about 2 % of the genome, compared with 21.4 % of the genome by LTR retrotransposons (International Brachypodium Initiative 2010).

Non-autonomous Transposable Elements as the Genome "Dark Matter"

The autonomous members of the foregoing groups of TEs carry out cut-and-paste mobilization (Class II) or copy-and-paste replication (Class I) through the activities of enzymes encoded within the elements themselves. However, perhaps the majority of DNA segments in the genome that identifiably belong to the main groups of

TEs are in fact variously deleted and mutated versions that do not encode all or any of the proteins required for replication or transposition (Fig. 1). As insertions, deletions, and mutations progressively erode the identity of what were once active TEs, these elements increasingly become the "dark matter" of the genome, and in fact most of the unidentifiable sequences in plant genomes probably originate from TEs (Maumus and Quesneville 2014). These are reminiscent of what Ohno referred to as "junk" DNA (Ohno 1972), although many are more accurately fossils of once active TEs that have accumulated inactivating mutations. The *B. distachyon* genome contains only 690 full-length and potentially autonomous retrotransposons (Table 1), which comprise 10.2 % of the total base pairs represented by retrotransposons and 2.39 % of the genome. Fully 19 % of the *B. distachyon* genome is therefore comprised of non-autonomous retrotransposons.

Nevertheless, many of the apparently dead TEs can be re-animated through by parasitizing the proteins of autonomous TEs. Binary pairs of autonomous and non-autonomous TEs were described before their molecular nature was known, by McClintock for the respective controlling elements, *Ac* and *Ds* (Jones 2005; McClintock 1948), which proved to be Class II transposons (Fedoroff et al. 1983). Miniature Inverted Repeat Transposable Elements (MITEs), first identified as insertions in the maize genome, are highly abundant in many plant and other eukaryotic genomes (Fattash et al. 2013; Feschotte and Mouches 2000; Wessler et al. 1995). Evidence that MITEs are derived from, and can be mobilized by, autonomous Class II transposons, was demonstrated by discovery of the *mPing-Pong* system (Jiang et al. 2003); the autonomous partner of the *Stowaway* MITEs was later found as well (Feschotte et al. 2005). The MITEs are highly abundant in the *B. distachyon* genome, being present in 23,500 copies (Table 1), of which 89 % are in the *Stowaway* family. As small elements, the MITEs amount to only 1 % of the genome despite their high copy number.

For MITEs, the minimum requirement for mobility is the possession of terminal inverted repeats (TIRs) that can be recognized by a transposase. For Class I retrotransposons, non-autonomous elements may be blocked at any of the many steps in their complex replicative life cycle (Sabot and Schulman 2006), including transcription, translation, VLP formation and RNA packaging, reverse transcription, and integration. A defective retrotransposon nevertheless could be replicated if a substitute for the non-functional protein (Gag, RT, IN) is available in *trans* from an autonomous element, providing the correct recognition signals are present on the RNA or cDNA. For example, retrotransposon *BARE2* of barley parasitizes *BARE1* for its Gag (Tanskanen et al. 2007). Non-autonomous groups of LTR retrotransposons including the Large Retrotransposon Derivative (*LARD*) elements (Kalendar et al. 2004) and the Terminal-repeat Retrotransposons In Miniature, the TRIMs (Kalendar et al. 2008; Witte et al. 2001), both of which lack protein-coding capacity, are abundant and structurally conserved in plant genomes (Antonius-Klemola et al. 2006; Wu et al. 2012; Yin et al. 2014) and found also in insects (Zhou and Cahan 2012), so therefore appear to have been successful at replication. The SINEs, which unlike the non-autonomous LTR retrotransposons constitute an order of their own (Wicker et al. 2007), comprise the diverse sequences that can be propagated by the enzymatic machinery of the LINEs (Goodier and Kazazian 2008; Vassetzky and Kramerov 2013). Reaching a million copies in mammalian genomes (Kramerov and Vassetzky 2005), SINES are also abundant and appear to be transpositionally active in plants (Ben-David et al. 2013; Deragon and Zhang 2006).

Genome Dynamics and Retrotransposon Gain and Loss

Our grasp of how TEs affect genomes over time is based both on an understanding of their transposition and replication mechanisms, described above, over the time spans of single plant generations and on a view of the current distribution of TEs in genomes, which represents the end result of TE activity over millions of generations. The germ cells of higher plants are formed only after many somatic cell divisions; TEs that have been mobilized or propagated in somatic cells generally can be inherited only if the new insertions occur in a clonal line of cells leading to the floral meristem and ultimately a germ cell. Studies of retrotransposon replication in plants have shown that the process displays tissue specificity (Fukai et al. 2010; Jääskeläinen et al. 2013; Slotkin et al. 2009), which may be one of the keys to clarifying why some families of TEs have become extremely abundant.

Due to the mechanism of reverse transcription (Schulman 2013), the LTRs of an LTR retrotransposon are identical at the time of insertion. These LTRs will accumulate mutations and diverge over time; given a molecular clock for the neutral rate of mutation, the age of an element since insertion can be estimated (SanMiguel et al. 1998). Using this strategy, it was shown that different families of superfamily *Copia* in rice and wheat were active at different times, undergoing "waves" of amplification lasting several hundreds of thousands of years (Wicker and Keller 2007). Dating of Class II insertions is more difficult due both to the lack of an internal "clock" comparable to the LTRs and to the cut-and-paste life cycle, which disrupts the historical connection to the surrounding genome.

A major theme to emerge from genome analysis is that retrotransposons are not only gained through the propagative life cycle described above, but they are also lost through a combination of progressive small deletions and truncations (Devos et al. 2002) and LTR–LTR recombination (Mager and Goodchild 1998; Shirasu et al. 2000; Vitte and Panaud 2005). LTR–LTR recombination is a class of unequal, homologous, intra-strand, recombination that removes most of an element, leaving behind a recombinant solo LTR. The recombination can also occur between LTRs belonging to different individual elements in the genome, removing a large piece of DNA in the process (Vicient et al. 2005). Where genome assemblies are sufficiently long to include large numbers of entire retrotransposons, one can infer an age structure for the families of elements by determining the ages of individual elements according to their LTR sequence divergence. When the numbers of family members for given age classes are plotted, a decay rate can be calculated that gives the half-life of the family. The family abundance appears to decay with time because a truncated element or a solo LTR cannot be dated by its LTR pair and intact elements tend to become increasingly rare with age. In a complementary approach, assigning solo LTRs to particular families gives an idea of how many elements have been lost through recombination, generally an underestimate because solo LTRs themselves can undergo recombinational loss.

Analyses of retrotransposon age structures and relative solo LTR prevalence show great differences among various retrotransposon families and plant genomes. For example, there is only one solo LTR for every nine complete retrotransposons in Norway spruce (*Picea abies*), indicating very slow removal by recombination. The retrotransposons of the related loblolly pine (*Pinus taeda*) genome are very abundant and highly divergent (Kovach et al. 2010; Wegrzyn et al. 2013). Taken together, the data indicate that gymnosperm genomes have enlarged through slow accumulation of retrotransposons with little loss over tens or hundreds of millions years. At the other extreme, cultivated barley has seven solo LTRs for every full-length *BARE* element and wild barley (*Hordeum vulgare* ssp. *spontaneum*) and some other *Hordeum* species even higher ratios (Soleimani et al. 2006; Vicient et al. 1999b). The ratios for *Hordeum* reflect a relatively high turnover rate at least for *BARE* through solo LTR formation, despite the overall abundance of retrotransposons in the barley genome. The other genomes heretofore investigated show ratios of solo LTRs to full-length elements ranging from 0.14:1 in maize to 1.39:1 in rice (El Baidouri and Panaud 2013) and 1.26:1 in soybean (Du et al. 2010). A closer examination in rice revealed a ratio of 1.26:1 recombination-suppressed pericentromeric regions and 1.62:1 outside them. Although it is clear from analyses of solo LTR prevalence that their formation is an important mechanism decreasing genome size through loss of DNA in retrotransposons, it is not the only mechanism. For example, the rice genome contains not only 4937 intact LTR retrotransposons and 7981 solo LTRs, but also 2006 truncated retroelements generated through illegitimate recombination (Tian et al. 2009). In *A. thaliana*, illegitimate recombination also appears to be relatively important (Vitte and Bennetzen 2006).

Brachypodium: A Genome on a Diet

Like every dieter knows, staying slim is a question of the balance between the calories taken in as food and those shed through exercise. In the case of genomes, as demonstrated for the *Gossypium* genus (Hawkins et al. 2009), it is the balance between gain of retrotransposons through propagation and loss through deletions and intrachromosomal LTR–LTR recombination that largely determines the size of the monoploid genome. The relatively few retrotransposons present in the *A. thaliana* genome appear to be largely silent due to chromatin methylation (Tsukahara et al. 2009), with only sporadic activation (Reinders et al. 2013). Not only are *A. thaliana* retrotransposons replicating rarely, but they also are being removed from the genome fairly rapidly. The half-life of superfamily *Copia* elements is 0.648 MY (million years) over the genome as a whole in *A. thaliana* and 0.472 MY outside the peri-centromeric regions (Pereira 2004), compared to 0.790 MY for rice; barley and wheat elements are so persistent that decay curves have been difficult to construct (Wicker and Keller 2007).

The *B. distachyon* genome, being roughly 2.5 times larger than that of *A. thaliana*, is nevertheless still very trim, the retrotransposons comprising only 21.4 % of the sequence, compared to 26 % in rice (International Brachypodium Initiative 2010). The overall half-life for superfamily *Copia* elements in *B. distachyon*, close to that in rice, is 0.859 MY (International Brachypodium Initiative 2010); the intact *Gypsy* elements are older, having a half-life of 1.265 MY (Fig. 3). The slightly longer half-lives in *B. distachyon* than in rice may be an artefact of the quality of the genome assembly, which for *B. distachyon* assembly is

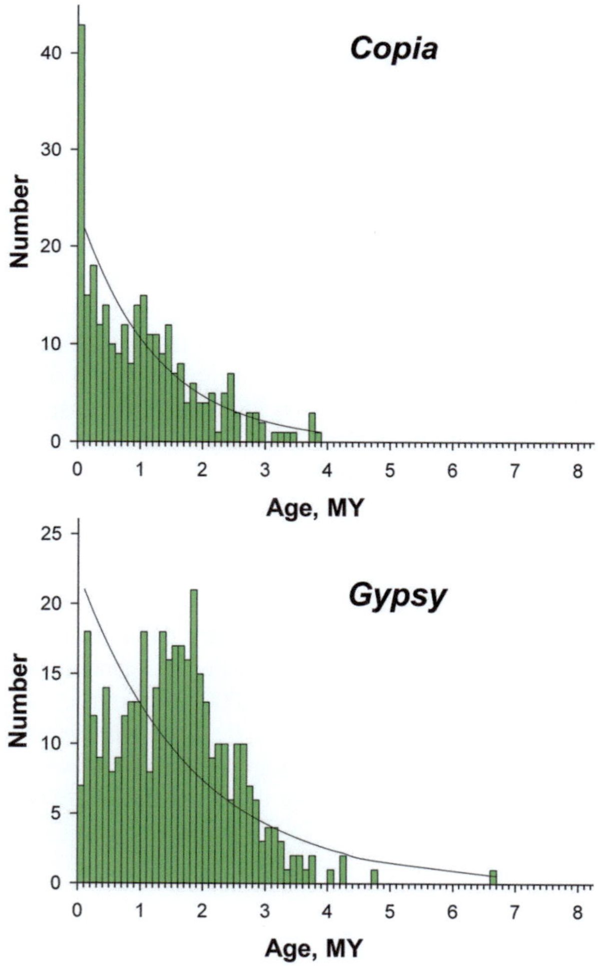

Fig. 3 Age distribution and frequency of intact *Copia* (*above*) and *Gypsy* (*below*) LTR retrotransposons (*green bars*) in the *B. distachyon* genome. The retrotransposons are grouped in age classes of 0.1 MY. Fitted exponential decay curves for the half-life of intact elements are shown

considerably better than that for rice. Generally, retrotransposon clusters create problems in sequence assembly; retrotransposon sequences and clusters cannot be unambiguously placed within the genome and are not incorporated into the final chromosome pseudomolecules. The problem is particularly severe in the pericentromeric regions, which are rich in old retrotransposons. The Japonica rice (*O. sativa* ssp. *japonica*) assembly concurrent with the half-life analysis (Wicker and Keller 2007) comprises 35,047 contigs, whereas the v1.0 *B. distachyon* assembly used for this analysis comprises only 1754 contigs and therefore contains considerably more retrotransposons.

The half-life curve for intact retrotransposons in *Brachypodium* directly translates into sequences shed through the LTR–LTR recombination that generates solo LTRs and through illegitimate recombination. The solo LTRs remaining in the genome after LTR–LTR recombination can be used to give a minimum estimate of how much DNA has been lost. Given an average retrotransposon size of 10 kb, at least 17.4 Mb has been lost from the *B. distachyon* genome. Although this amounts only to roughly 5 % of the genome, it is nevertheless 2.7 times the current genomic coverage by intact elements (6.47 Mb). The estimate assumes recombination between two LTRs of the same element and therefore does not include DNA lost when recombination spans a segment of nested or concatenated retrotransposons. Neither does the analysis include retrotransposon sequences lost by small deletions, or unannotated "orphan" solo LTRs that cannot be associated with an intact family of elements.

The presence of solo LTRs in the *B. distachyon* genome, hence the history of recombinational loss, is not uniform by retrotransposon superfamily or family (International Brachypodium Initiative 2010). Superfamily *Gypsy* solo LTRs are 1.6 times more abundant than are *Copia* solo LTRs, commensurate with the relative abundance of intact members of the superfamilies. Over 69.8 % of the retrotransposon families have no related solo LTRs; one *Gypsy* family has 645 and one *Copia* family has 263. The retrotransposons of *B. distachyon* that are most similar to the *BARE*, *Angela*, and *Wis* families in barley and wheat show the highest number of solo LTRs relative to the age of the intact elements, indicating that this family has a high propensity to form solo LTRs. One member of the family, the Bd2_RLC_14 element, is 20,769 years old and has 35 related solo LTRs. This is consistent with the high turnover seen for *BARE* elements in the *Hordeum* genome (Soleimani et al. 2006; Vicient et al. 1999b). Although the retrotransposon population of the genome, in evolutionary terms, has been removed rapidly, they are nevertheless continuing to propagate. The genome contains at least 13 families of *Copia* elements younger than 20,000 years and 53 that are less than 100,000 years old. The overall picture of the *B. distachyon* genome is of one that stays trim through recombinational shedding of retrotransposons, despite the continuing propagation of these elements and their consequent "fattening" effects.

Retrotransposon Gain and Loss Is Not Uniform Within or Between Chromosomes

On the local level, the distribution of retrotransposons in large genomes is strikingly non-uniform. In the genomes of many diverse plant groups, many retrotransposons are present as nests in which elements have successively integrated one into another (Kronmiller and Wise 2007; Wei et al. 2013; Wicker et al. 2001; Vitte et al. 2013). The process of LTR–LTR recombination within nests can, moreover, lead to nests comprised of solo LTRs (Shirasu et al. 2000). The nests provide a safe "landing pad" for new insertions, protecting genes from disruption. Although nesting patterns within the *B. distachyon* genome have not been analyzed, compact genomes generally have fewer nests; comparisons of specific loci between *B. distachyon* and the syntenic regions in large cereal genomes indicate the accumulation of nested clusters of retrotransposons in the latter (Wang et al. 2010). This nesting and clustering may reflect selection against deleterious insertions into genes and their vicinity (Choulet et al. 2010; SanMiguel et al. 1998; 1996; Shirasu et al. 2000). The 5000 concatenated *BARE* elements derived from recombination between a pair of retrotransposons (Vicient et al. 2005) give some indication of the potential risk for gene loss, were a gene to be present between each pair.

The pericentromeric regions of chromosomes are extremely rich in clustered and nested retrotransposons. This is partly due to the presence of retrotransposons, such as the centromeric "chromoviruses", the CRM clade of *Gypsy* elements (Gorinsek et al. 2005), that tend to insert preferentially into these regions. The centromere of *B. distachyon* is composed of a 156 bp repeat, BdCENT. The centromeres are well assembled in the published genome (International Brachypodium Initiative 2010), and comprise 100–1300 BdCENT repeats, which are interspersed with blocks of retrotransposons. The centromeric retrotransposons here as elsewhere are of the chromovirus type (Qi et al. 2013). Surrounding the centromeres are gene-poor domains consisting almost entirely of *Gypsy* retrotransposons. Together, the 300 kb regions around all *B. distachyon* centromeres contain only 54 genes, none of which are collinear with rice or sorghum. A comparison of "heat maps", which are plots of the relative density of various genome features, shows a spread of the retrotransposon-rich, gene-poor pericentromeric regions towards the telomeres and increases in the relative abundance of retrotransposons in these regions in parallel with growth in genome size, even when comparing *B. distachyon* with the relatively compact (727 Mb) sorghum genome (Fig. 4).

In *B. distachyon*, vast differences in retrotransposon distribution are also found between chromosomes (International Brachypodium Initiative 2010). Chromosome 1 has the lowest density of retrotransposons, which cover 20.3 % of the sequence. Chromosome 4 is deficient in Gypsy elements, which are 2.34 times less abundant than elsewhere. The short arm of chromosome 5 (Bd5S) displays a very high density of retrotransposons, which comprise 28.3 % of the arm, and few genes compared to the other chromosomes. This chromosome also contains the lowest solo LTR density and the youngest (1.37 MY, versus 1.54–1.64 MY elsewhere) and

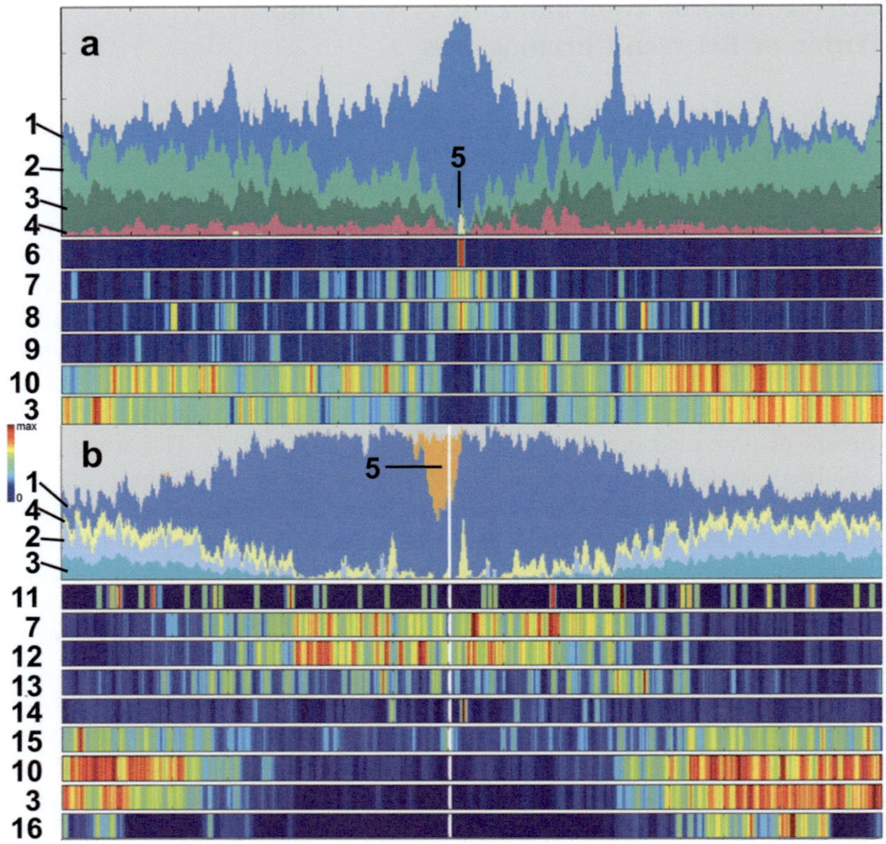

Fig. 4 Comparative distributions of genomic features for *B. distachyon* and *S. bicolor*. (**a**) *B. distachyon* chromosome 2. Relative abundances (*upper*) and heat-map distribution (*lower*) are shown for: track 1, retrotransposons; 2, introns; 3, CDS (exons); 4, DNA transposons; 5 and 6, satellite tandem arrays; 7, full-length LTR retrotransposons; 8, solo LTRs; 9, non-MITE DNA transposons; 10, MITEs. The heat maps (lower) indicate relative abundances by % of bp that differ both by range and level (*blue*, minimum; *red*, maximum) per track: 6, 0–55 % bp, scaled to max. 10 % bp; 7, 0–36 %, max. 20 %; 8, 0–4 %; 9, 0–20 %; 10, 0–22 %; 11, 0–22.3 %. (**b**) *S. bicolor* chromosome 3, which is syntenic to *B distachyon* chromosome 2 *above*. Labels for the relative abundances (*above*) and heat maps (*below*) are as for (**a**) with the following additions; 11, young LTR retrotransposons (<10,000 years old); 12, superfamily *Gypsy* retrotransposons; 13, superfamily *Copia* retrotransposons; 14, superfamily CACTA DNA transposons; 15, CpG islands; 16, paralogues. The heat maps are colored according to the ranges: 11, 0–5.0 %; 7, 0–43.6 %; 12, 0.2–53 %; 13, 0–18.5 %; 14, 0–38.4 %; 15, 0.3–6.3 %; 10, 0–4.6 %; 3, 0–18.1 %; 16, 0–7.1 %. Part (**a**) has been modified from (International Brachypodium Initiative 2010): Part (**b**) has been modified from (Paterson et al. 2009) and used with permission from Macmillan Publishers Ltd

most abundant (2.9 times more than average) *Gypsy* elements among the chromosomes. Nevertheless, chromosome 5 has only four retrotransposons of the 52 in the genome that are younger than 0.1 MY, whereas chromosome 4 has 18.

The distribution of solo LTRs also varies greatly among the chromosomes. The chromosomes have 362 solo LTRs each on average, but the range is from 73, for chromosome 5, to 1016, for chromosome 3. Chromosome 5 has one solo LTR per 389 kb, while chromosome 3 has a 1.6-fold higher density. Chromosome 3 also hosts two most abundant sets of solo LTRs in the genome, both from *Copia* families (International Brachypodium Initiative 2010). Given that solo LTRs are not mobile, the ratio of solo LTRs to intact LTR retrotransposons in a particular region reflects the relative rates of integration and loss there. The *B. distachyon* genome has an overall ratio of 2.6 solo LTRs per intact element, whereas chromosome 5 has a ratio of only 0.89 and chromosome 3 has the highest, at 6.96. Taken together, these facts lead to the conclusion that more retrotransposons have been inserted and fewer lost by recombination from chromosome 5, though not necessarily in the recent past, than elsewhere in the *B. distachyon* genome. The regions syntenic to Bd5S on rice (Os4S and sorghum Sb6S) show the same pattern, suggesting that chromosome-specific retrotransposon dynamics have been maintained for the 50 MYA since divergence of the lines leading to sorghum and *Brachypodium* (Salse et al. 2008). Chromosome 3, on the other hand, appears to be differentially losing elements through LTR–LTR recombination.

Conclusions

The replicative life cycle of retrotransposons has the potential to add on average 9 or 10 kb for every RNA molecule reverse-transcribed into DNA and reintegrated into the genome. Given the ubiquity of retrotransposons throughout the plant kingdom, at any given ploidy level plant genomes grow primarily by gaining retrotransposons. Compact genomes such as that of *B. distachyon* are relatively depauperate of retrotransposons. Genomes can become compact or stay that way either by blocking the replication of retrotransposons by transcriptional and post-transcriptional silencing, through selection against insertions, or through shedding of integrated copies. The copies can be lost by LTR–LTR recombination, which removes either most of one element or long segments of DNA spanning between two elements of the same family, or by illegitimate recombination, which removes small segments.

The genome of *B. distachyon* appears to highly dynamic, because it contains many recently inserted transposable elements. The genome has, however, remained compact through the recombinational loss of integrated retrotransposons. Nevertheless, the chromosomes show remarkable differences among them regarding the gain and loss of retrotransposons over time and the relative accumulation of the two superfamilies, *Copia* and *Gypsy*. The dynamic gain and loss of retrotransposons in

B. distachyon, gleaned from examination of a single genome, will provide a promising basis for analyses of multiple genomes from this and other *Brachypodium* species. These will lead to a fuller understanding of the role of retrotransposons in genome dynamics and species diversification.

References

Anca IA, Fromentin J, Bui QT, Mhiri C, Grandbastien MA, Simon-Plas F. Different tobacco retrotransposons are specifically modulated by the elicitor cryptogein and reactive oxygen species. J Plant Physiol. 2014;171:1533–40.

Ansari KI, Walter S, Brennan JM, Lemmens M, Kessans S, McGahern A, et al. Retrotransposon and gene activation in wheat in response to mycotoxigenic and non-mycotoxigenic-associated Fusarium stress. Theor Appl Genet. 2007;114:927–37.

Antonius-Klemola K, Kalendar R, Schulman AH. TRIM retrotransposons occur in apple and are polymorphic between varieties but not sports. Theor Appl Genet. 2006;112:999–1008.

Bachman N, Gelbart ME, Tsukiyama T, Boeke JD. TFIIIB subunit Bdp1p is required for periodic integration of the *Ty1* retrotransposon and targeting of Isw2p to *S. cerevisiae* tDNAs. Genes Dev. 2005;19:955–64.

Beguiristain T, Grandbastien MA, Puigdomènech P, Casacuberta JM. Three Tnt1 subfamilies show different stress-associated patterns of expression in tobacco. Consequences for retrotransposon control and evolution in plants. Plant Physiol. 2001;127:212–21.

Ben-David S, Yaakov B, Kashkush K. Genome-wide analysis of short interspersed nuclear elements SINES revealed high sequence conservation, gene association and retrotranspositional activity in wheat. Plant J. 2013;76:201–10.

Bennett AB, Leitch AR. Nuclear DNA amounts in angiosperms: targets, trends and tomorrow. Ann Bot. 2011;107:467–590.

Bennetzen JL, Wang H. The contributions of transposable elements to the structure, function, and evolution of plant genomes. Annu Rev Plant Biol. 2014;65:505–30.

Biderre C, Pages M, Metenier G, Canning EU, Vivaras CP. Evidence for the smallest nuclear genome (2.9 Mb) in the microsporidium *Encephalitozoon cuniculi*. Mol Biochem Parasitol. 1995;74:229–31.

Böhmdorfer G, Luxa K, Frosch A, Garber K, Tramontano A, Jelenic S, et al. Virus-like particle formation and translational start site choice of the plant retrotransposon *Tto1*. Virology. 2008;372:437–46.

Bossolini E, Wicker T, Knobel PA, Keller B. Comparison of orthologous loci from small grass genomes Brachypodium and rice: implications for wheat genomics and grass genome annotation. Plant J. 2007;49:704–17.

Brady TL, Fuerst PG, Dick RA, Schmidt C, Voytas DF. Retrotransposon target site selection by imitation of a cellular protein. Mol Cell Biol. 2008;28:1230–9.

Butelli E, Licciardello C, Zhang Y, Liu J, Mackay S, Bailey P, et al. Retrotransposons control fruit-specific, cold-dependent accumulation of anthocyanins in blood oranges. Plant Cell. 2012;24:1242–55.

Cavrak VV, Lettner N, Jamge S, Kosarewicz A, Bayer LM, Mittelsten SO. How a retrotransposon exploits the plant's heat stress response for its activation. PLoS Genet. 2014;10:e1004115.

Chalopin D, Naville M, Plard F, Galiana D, Volff JN. Comparative analysis of transposable elements highlights mobilome diversity and evolution in vertebrates. Genome Biol Evol. 2015;7:567–80.

Chang W, Schulman AH. *BARE* retrotransposons produce multiple groups of rarely polyadenylated transcripts from two differentially regulated promoters. Plant J. 2008;56:40–50.

Chang W, Jääskeläinen M, Li S-P, Schulman AH. *BARE* retrotransposons are translated and replicated via distinct RNA pools. PLoS One. 2013;8:e72270.

Chen J, Huang Q, Gao D, Wang J, Lang Y, Liu T, et al. Whole-genome sequencing of Oryza brachyantha reveals mechanisms underlying Oryza genome evolution. Nat Commun. 2013;4:1595.

Choulet F, Wicker T, Rustenholz C, Paux E, Salse J, Leroy P, et al. Megabase level sequencing reveals contrasted organization and evolution patterns of the wheat gene and transposable element spaces. Plant Cell. 2010;22:1686–701.

Ciuffi A, Bushman FD. Retroviral DNA integration: HIV and the role of LEDGF/p75. Trends Genet. 2006;22:388–95.

De La Torre AR, Birol I, Bousquet J, Ingvarsson PK, Jansson S, Jones SJ, et al. Insights into conifer giga-genomes. Plant Physiol. 2014;166:1724–32.

Deragon J, Zhang X. Short Interspersed Elements (SINEs) in plants: origin, classification, and use as phylogenetic markers. Syst Biol. 2006;55:949–56.

Devos KM, Brown JK, Bennetzen JL. Genome size reduction through illegitimate recombination counteracts genome expansion in *Arabidopsis*. Genome Res. 2002;12:1075–9.

Du J, Tian Z, Hans CS, Laten HM, Cannon SB, Jackson SA, et al. Evolutionary conservation, diversity and specificity of LTR-retrotransposons in flowering plants: insights from genome-wide analysis and multi-specific comparison. Plant J. 2010;63:584–98.

El Baidouri M, Panaud O. Comparative genomic paleontology across plant kingdom reveals the dynamics of TE-driven genome evolution. Genome Biol Evol. 2013;5:954–65.

Estep MC, DeBarry JD, Bennetzen JL. The dynamics of LTR retrotransposon accumulation across 25 million years of panicoid grass evolution. Heredity. 2013;110:194–204.

Fattash I, Rooke R, Wong A, Hui C, Luu T, Bhardwaj P, et al. Miniature inverted-repeat transposable elements: discovery, distribution, and activity. Genome. 2013;56:475–86.

Fedoroff N, Wessler S, Shure M. Isolation of the transposable maize controlling elements Ac and Ds. Cell. 1983;35:235–42.

Feschotte C, Mouches C. Evidence that a family of miniature inverted-repeat transposable elements (MITEs) from the *Arabidopsis thaliana* genome has arisen from a pogo-like DNA transposon. Mol Biol Evol. 2000;17:730–7.

Feschotte C, Osterlund MT, Peeler R, Wessler SR. DNA-binding specificity of rice mariner-like transposases and interactions with Stowaway MITEs. Nucleic Acids Res. 2005;33:2153–65.

Fleischmann A, Michael TP, Rivadavia F, Sousa A, Wang W, Temsch EM, et al. Evolution of genome size and chromosome number in the carnivorous plant genus *Genlisea* (Lentibulariaceae), with a new estimate of the minimum genome size in angiosperms. Ann Bot. 2014;114:1651–63.

Fukai E, Umehara Y, Sato S, Endo M, Kouchi H, Hayashi M, et al. Derepression of the plant Chromovirus *LORE1* induces germline transposition in regenerated plants. PLoS Genet. 2010;6:e1000868.

Gao X, Havecker ER, Baranov PV, Atkins JF, Voytas DF. Translational recoding signals between gag and pol in diverse LTR retrotransposons. RNA. 2003;9:1422–30.

Gao X, Hou Y, Ebina H, Levin HL, Voytas DF. Chromodomains direct integration of retrotransposons to heterochromatin. Genome Res. 2008;18:359–69.

Gaut BS, Ross-Ibarra J. Selection on major components of angiosperm genomes. Science. 2008;320:484–6.

Gómez-Orte E, Vicient CM, Martínez-Izquierdo JA. *Grande* retrotransposons contain an accessory gene in the unusually long 3′-internal region that encodes a nuclear protein transcribed from its own promoter. Plant Mol Biol. 2013;81:541–51.

Goodier JL, Kazazian Jr HH. Retrotransposons revisited: the restraint and rehabilitation of parasites. Cell. 2008;135:23–35.

Gorinsek B, Gubensek F, Kordiš D. Phylogenomic analysis of chromoviruses. Cytogenet Genome Res. 2005;110:543–52.

Grandbastien MA. LTR retrotransposons, handy hitchhikers of plant regulation and stress response. Biochim Biophys Acta. 1849;2014:403–16.

Grandbastien MA, Audeon C, Bonnivard E, Casacuberta JM, Chalhoub B, Costa AP, et al. Stress activation and genomic impact of Tnt1 retrotransposons in Solanaceae. Cytogenet Genome Res. 2005;110:229–41.

Gregory TR. Coincidence, coevolution, or causation? DNA content, cell size, and the *C*-value enigma. Biol Rev Camb Philos Soc. 2001;76:65–101.

Gregory TR, Nicol JA, Tamm H, Kullman B, Kullman K, Leitch IJ, et al. Eukaryotic genome size databases. Nucleic Acids Res. 2007;35:D332–8.

Hawkins JS, Kim H, Nason JD, Wing RA, Wendel JF. Differential lineage-specific amplification of transposable elements is responsible for genome size variation in *Gossypium*. Genome Res. 2006;16:1252–61.

Hawkins JS, Proulx SR, Rapp RA, Wendel JF. Rapid DNA loss as a counterbalance to genome expansion through retrotransposon proliferation in plants. Proc Natl Acad Sci U S A. 2009;106:17811–6.

Heitkam T, Holtgrawe D, Dohm JC, Minoche AE, Himmelbauer H, Weisshaar B, et al. Profiling of extensively diversified plant LINEs reveals distinct plant-specific subclades. Plant J. 2014;79:385–97.

Hernández-Pinzón I, Cifuentes M, Hénaff E, Santiago N, Espinás ML, Casacuberta JM. The *Tnt1* retrotransposon escapes silencing in tobacco, its natural host. PLoS One. 2012;7:e33816.

Hickman AB, Chandler M, Dyda F. Integrating prokaryotes and eukaryotes: DNA transposases in light of structure. Crit Rev Biochem Mol Biol. 2010;45:50–69.

Hirsch CN, Foerster JM, Johnson JM, Sekhon RS, Muttoni G, Vaillancourt B, et al. Insights into the maize pan-genome and pan-transcriptome. Plant Cell. 2014;26:121–35.

International Barley Genome Sequencing Consortium, Mayer KF, Waugh R, Brown JW, Schulman A, Langridge P, et al. A physical, genetic and functional sequence assembly of the barley genome. Nature. 2012;491:711–6.

International Brachypodium Initiative. Genome sequencing and analysis of the model grass *Brachypodium distachyon*. Nature. 2010;463:763–8.

International Wheat Genome Sequencing Consortium. A chromosome-based draft sequence of the hexaploid bread wheat (*Triticum aestivum*) genome. Science. 2014;345:1251788.

Jääskeläinen M, Mykkänen A-H, Arna T, Vicient C, Suoniemi A, Kalendar R, et al. Retrotransposon *BARE*-1: expression of encoded proteins and formation of virus-like particles in barley cells. Plant J. 1999;20:413–22.

Jääskeläinen M, Chang W, Moisy C, Schulman AH. Retrotransposon *BARE* displays strong tissue-specific differences in expression. New Phytol. 2013;200:1000–8.

Jiang N, Bao Z, Zhang X, Hirochika H, Eddy SR, McCouch SR, et al. An active DNA transposon family in rice. Nature. 2003;421:163–7.

Jones RN. McClintock's controlling elements: the full story. Cytogenet Genome Res. 2005;109:90–103.

Kalendar R, Tanskanen J, Immonen S, Nevo E, Schulman AH. Genome evolution of wild barley (*Hordeum spontaneum*) by *BARE*-1 retrotransposon dynamics in response to sharp microclimatic divergence. Proc Natl Acad Sci U S A. 2000;97:6603–7.

Kalendar R, Vicient CM, Peleg O, Anamthawat-Jonsson K, Bolshoy A, Schulman AH. LARD retroelements: novel, non-autonomous components of barley and related genomes. Genetics. 2004;166:1437–50.

Kalendar R, Tanskanen JA, Chang W, Antonius K, Sela H, Peleg P, et al. *Cassandra* retrotransposons carry independently transcribed 5S RNA. Proc Natl Acad Sci U S A. 2008;105:5833–8.

Katinka MD, Duprat S, Cornillot E, Metenier G, Thomarat F, Prensier G, et al. Genome sequence and gene compaction of the eukaryote parasite Encephalitozoon cuniculi. Nature. 2001;414:450–3.

Kovach A, Wegrzyn JL, Parra G, Holt C, Bruening GE, Loopstra CA, et al. The *Pinus taeda* genome is characterized by diverse and highly diverged repetitive sequences. BMC Genomics. 2010;11:420.

Kramerov D, Vassetzky N. Short retroposons in eukaryotic genomes. Int Rev Cytol. 2005;247:165–221.

Krishan A, Dandekar P, Nathan N, Hamelik R, Miller C, Shaw J. DNA index, genome size, and electronic nuclear volume of vertebrates from the Miami Metro Zoo. Cytometry A. 2005;65:26–34.

Krishnan L, Engelman A. Retroviral integrase proteins and HIV-1 DNA integration. J Biol Chem. 2012;287:40858–66.

Kronmiller BA, Wise RP. TE nest: automated chronological annotation and visualization of nested plant transposable elements. Plant Physiol. 2007;146:45–59.

Lee SK, Potempa M, Swanstrom R. The choreography of HIV-1 proteolytic processing and virion assembly. J Biol Chem. 2012;287:40867–74.

Li YH, Zhou G, Ma J, Jiang W, Jin LG, Zhang Z, et al. *De novo* assembly of soybean wild relatives for pan-genome analysis of diversity and agronomic traits. Nat Biotechnol. 2014;32:1045–52.

Mager DL, Goodchild NL. Homologous recombination between the LTRs of a human retrovirus-like element causes a 5-kb deletion in two siblings. Am J Hum Genet. 1998;45:848–54.

Maumus F, Quesneville H. Deep investigation of *Arabidopsis thaliana* junk DNA reveals a continuum between repetitive elements and genomic dark matter. PLoS One. 2014;9:e94101.

McClintock B. Mutable loci in maize. Year B Carnegie Inst Wash. 1948;47:155–69.

McCue AD, Nuthikattu S, Reeder SH, Slotkin RK. Gene expression and stress response mediated by the epigenetic regulation of a transposable element small RNA. PLoS Genet. 2012;8: e1002474.

Michael TP, Jackson S. The first 50 plant genomes. Plant Genome. 2013. doi:10.3835/plantgenome2013.03.0001in.

Moisy C, Garrison KE, Meredith CP, Pelsy F. Characterization of ten novel *Ty1/copia*-like retrotransposon families of the grapevine genome. BMC Genomics. 2008;9:469.

Montaño SP, Rice PA. Moving DNA around: DNA transposition and retroviral integration. Curr Opin Struct Biol. 2011;21:370–8.

Montaño SP, Pigli YZ, Rice PA. The μ transpososome structure sheds light on DDE recombinase evolution. Nature. 2012;491:413–7.

Morgante M, De Paoli E, Radovic S. Transposable elements and the plant pan-genomes. Curr Opin Plant Biol. 2007;10:149–55.

Müllers E, Stirnnagel K, Kaulfuss S, Lindemann D. Prototype foamy virus gag nuclear localization: a novel pathway among retroviruses. J Virol. 2011;85:9276–85.

Neumann P, Koblízková A, Navrátilová A, Macas J. Significant expansion of *Vicia pannonica* genome size mediated by amplification of a single type of giant retroelement. Genetics. 2006;173:1047–56.

Neumann P, Navrátilová A, Koblížková A, Kejnovský E, Hřibová E, Hobza R, et al. Plant centromeric retrotransposons: a structural and cytogenetic perspective. Mob DNA. 2011;2:4.

Nystedt B, Street NR, Wetterbom A, Zuccolo A, Lin YC, Scofield DG, et al. The Norway spruce genome sequence and conifer genome evolution. Nature. 2013;497:579–84.

Ohno S. So much 'junk' in our genome. Brookhaven Symp Biol. 1972;23:366–70.

Park M, Jo S, Kwon JK, Park J, Ahn JH, Kim S, et al. Comparative analysis of pepper and tomato reveals euchromatin expansion of pepper genome caused by differential accumulation of *Ty3/Gypsy*-like elements. BMC Genomics. 2011;12:85.

Park M, Park J, Kim S, Kwon JK, Park HM, Bae IH, et al. Evolution of the large genome in Capsicum annuum occurred through accumulation of single-type long terminal repeat retrotransposons and their derivatives. Plant J. 2012;69:1018–29.

Paterson AH, Bowers JE, Bruggmann R, Dubchak I, Grimwood J, Gundlach H, et al. The Sorghum bicolor genome and the diversification of grasses. Nature. 2009;457:551–6.

Pereira V. Insertion bias and purifying selection of retrotransposons in the *Arabidopsis thaliana* genome. Genome Biol. 2004;5:R79.

Piegu B, Guyot R, Picault N, Roulin A, Saniyal A, Kim HI, et al. Doubling genome size without polyploidization: dynamics of retrotransposition-driven genomic expansions in *Oryza australiensis*, a wild relative of rice. Genome Res. 2006;16:1262–9.

Qi LL, Wu JJ, Friebe B, Qian C, Gu YQ, Fu DL, et al. Sequence organization and evolutionary dynamics of *Brachypodium*-specific centromere retrotransposons. Chromosome Res. 2013;21:507–21.

Ramallo E, Kalendar R, Schulman AH, Martínez-Izquierdo JA. *Reme1*, a *Copia* retrotransposon in melon, is transcriptionally induced by UV light. Plant Mol Biol. 2008;66:137–50.

Reinders J, Mirouze M, Nicolet J, Paszkowski J. Parent-of-origin control of transgenerational retrotransposon proliferation in Arabidopsis. EMBO Rep. 2013;14:823–8.

Rosbash M, Ford PJ, Bishop JO. Analysis of the *C*-value paradox by molecular hybridization. Proc Natl Acad Sci U S A. 1974;71:3746–50.

Sabot F, Schulman AH. Parasitism and the retrotransposon life cycle in plants: a hitchhiker's guide to the genome. Heredity. 2006;97:381–8.

Salazar M, Gonzalez E, Casaretto JA, Casacuberta JM, Ruiz-Lara S. The promoter of the TLC1.1 retrotransposon from *Solanum chilense* is activated by multiple stress-related signaling molecules. Plant Cell Rep. 2007;26:1861–8.

Salse J, Bolot S, Throude M, Jouffe V, Piegu B, Quraishi UM, et al. Identification and characterization of shared duplications between rice and wheat provide new insight into grass genome evolution. Plant Cell. 2008;20:11–24.

SanMiguel P, Tikhonov A, Jin YK, Motchoulskaia N, Zakharov D, Melake-Berhan A, et al. Nested retrotransposons in the intergenic regions of the maize genome. Science. 1996;274:765–8.

SanMiguel P, Gaut BS, Tikhoniv A, Nakajima Y, Bennetzen JL. The paleontology of intergene retrotransposons in maize. Nat Genet. 1998;20:43–5.

Schnable PS, Ware D, Fulton RS, Stein JC, Wei F, Pasternak S, et al. The B73 maize genome: complexity, diversity, and dynamics. Science. 2009;326:1112–5.

Schulman AH. Hitching a ride: nonautonomous retrotransposons and parasitism as a lifestyle. In: Grandbastien M-A, Casacuberta JM, editors. Plant transposable elements. Topics in current genetics 24. Berlin: Springer Verlag; 2012. p. 71–88.

Schulman AH. Retrotransposon replication in plants. Curr Opin Virol. 2013;3:604–14.

Schulman AH, Wicker T. A field guide to transposable elements. In: Fedoroff NV, editor. Plant transposons and genome dynamics in evolution. Hoboken, NJ: John Wiley and Sons; 2013. p. 15–40.

Sharma A, Larue RC, Plumb MR, Malani N, Male F, Slaughter A, et al. BET proteins promote efficient murine leukemia virus integration at transcription start sites. Proc Natl Acad Sci U S A. 2013;110:12036–41.

Shirasu K, Schulman AH, Lahaye T, Schulze-Lefert P. A contiguous 66 kb barley DNA sequence provides evidence for reversible genome expansion. Genome Res. 2000;10:908–15.

Slotkin RK, Vaughn M, Borges F, Tanurdzić M, Becker JD, Feijó JA, et al. Epigenetic reprogramming and small RNA silencing of transposable elements in pollen. Cell. 2009;136:461–72.

Soleimani VD, Baum BR, Johnson DA. Quantification of the retrotransposon *BARE*-1 reveals the dynamic nature of the barley genome. Genome. 2006;49:389–96.

Springer NM, Ying K, Fu Y, Ji T, Yeh CT, Jia Y, et al. Maize inbreds exhibit high levels of copy number variation (CNV) and presence/absence variation (PAV) in genome content. PLoS Genet. 2009;5:e1000734.

Staton SE, Bakken BH, Blackman BK, Chapman MA, Kane NC, Tang S, et al. The sunflower (*Helianthus annuus* L.) genome reflects a recent history of biased accumulation of transposable elements. Plant J. 2012;72(1):142–53.

Suzuki Y, Craigie R. The road to chromatin - nuclear entry of retroviruses. Nat Rev Microbiol. 2007;5:187–96.

Swigonova Z, Lai J, Ma J, Ramakrishna W, Llaca V, Bennetzen JL, et al. On the tetraploid origin of the maize genome. Comp Funct Genomics. 2004;5(3):281–4.

Tanskanen JA, Sabot F, Vicient C, Schulman AH. Life without GAG: the *BARE*-2 retrotransposon as a parasite's parasite. Gene. 2007;390:166–74.

Tian Z, Rizzon C, Du J, Zhu L, Bennetzen JL, Jackson SA, et al. Do genetic recombination and gene density shape the pattern of DNA elimination in rice long terminal repeat retrotransposons? Genome Res. 2009;19:2221–30.

Tsukahara S, Kobayashi A, Kawabe A, Mathieu O, Miura A, Kakutani T. Bursts of retrotransposition reproduced in Arabidopsis. Nat Genet. 2009;461:423–6.

Vassetzky NS, Kramerov DA. SINEBase: a database and tool for SINE analysis. Nucleic Acids Res. 2013;41:D83–9.

Vicient CM, Kalendar R, Anamthawat-Jonsson K, Schulman AH. Structure, functionality, and evolution of the *BARE*-1 retrotransposon of barley. Genetica. 1999a;107:53–63.

Vicient CM, Suoniemi A, Anamthawat-Jónsson K, Tanskanen J, Beharav A, Nevo E, et al. Retrotransposon *BARE*-1 and its role in genome evolution in the genus *Hordeum*. Plant Cell. 1999b;11(9):1769–84.

Vicient CM, Kalendar R, Schulman AH. Variability, recombination, and mosaic evolution of the barley *BARE*-1 retrotransposon. J Mol Evol. 2005;61:275–91.

Vitte C, Bennetzen JL. Analysis of retrotransposon structural diversity uncovers properties and propensities in angiosperm genome evolution. Proc Natl Acad Sci U S A. 2006;103:17638–43.

Vitte C, Panaud O. LTR retrotransposons and flowering plant genome size: emergence of the increase/decrease model. Cytogenet Genome Res. 2005;110:91–107.

Vitte C, Estep MC, Leebens-Mack J, Bennetzen JL. Young, intact and nested retrotransposons are abundant in the onion and asparagus genomes. Ann Bot. 2013;112:881–9.

Wang ZN, Huang XQ, Cloutier S. Recruitment of closely linked genes for divergent functions: the seed storage protein (*Glu-3*) and powdery mildew (*Pm3*) genes in wheat (*Triticum aestivum* L.). Funct Integr Genomics. 2010;10:241–51.

Weber B, Heitkam T, Holtgräwe D, Weisshaar B, Minoche AE, Dohm JC, et al. Highly diverse chromoviruses of *Beta vulgaris* are classified by chromodomains and chromosomal integration. Mob DNA. 2013;4:8.

Wegrzyn JL, Lin BY, Zieve JJ, Dougherty WM, Martinez-Garcia PJ, Koriabine M, et al. Insights into the loblolly pine genome: characterization of BAC and fosmid sequences. PLoS One. 2013;8:e72439.

Wei L, Xiao M, An Z, Ma B, Mason AS, Qian W, et al. New insights into nested long terminal repeat retrotransposons in *Brassica* species. Mol Plant. 2013;6:470–82.

Wenke T, Holtgräwe D, Horn AV, Weisshaar B, Schmidt T. An abundant and heavily truncated non-LTR retrotransposon (LINE) family in *Beta vulgaris*. Plant Mol Biol. 2009;71:585–97.

Wessler SR, Bureau TE, White SE. LTR-retrotransposons and MITEs: important players in the evolution of plant genomes. Curr Opin Genet Dev. 1995;5:814–21.

Wicker T, Keller B. Genome-wide comparative analysis of *copia* retrotransposons in Triticeae, rice, and *Arabidopsis* reveals conserved ancient evolutionary lineages and distinct dynamics of individual *copia* families. Genome Res. 2007;17:1072–81.

Wicker T, Stein N, Albar L, Feuillet C, Schlagenhauf E, Keller B. Analysis of a contiguous 211 kb sequence in diploid wheat (*Triticum monococcum*) reveals multiple mechanisms of genome evolution. Plant J. 2001;26(3):307–16.

Wicker T, Sabot F, Hua-Van A, Bennetzen J, Capy P, Chalhoub B, et al. A unified classification system for eukaryotic transposable elements. Nat Rev Genet. 2007;8:973–82.

Wicker T, Taudien S, Houben A, Keller B, Graner A, Platzer M, et al. A whole-genome snapshot of 454 sequences exposes the composition of the barley genome and provides evidence for parallel evolution of genome size in wheat and barley. Plant J. 2009;59:712–22.

Witte CP, Le QH, Bureau T, Kumar A. Terminal-repeat retrotransposons in miniature (TRIM) are involved in restructuring plant genomes. Proc Natl Acad Sci U S A. 2001;98(24):13778–83.

Wu J, Gu YQ, Hu Y, You FM, Dandekar AM, Leslie CA, et al. Characterizing the walnut genome through analyses of BAC end sequences. Plant Mol Biol. 2012;78:95–107.

Yamaguchi K, Kajikawa M, Okada N. Integrated mechanism for the generation of the 5′ junctions of LINE inserts. Nucleic Acids Res. 2014;42:13269–79.

Yieh G, Kassavetis EP, Geiduschek SB, Sandmeyer SB. The Brf and TATA-binding protein subunits of the RNA polymerase III transcription factor IIIB mediate position-specific integration of the gypsy-like element, *Ty3*. J Biol Chem. 2000;275:29800–7.

Yin H, Du J, Li L, Jin C, Fan L, Li M, et al. Comparative genomic analysis reveals multiple long terminal repeats, lineage-specific amplification, and frequent interelement recombination for Cassandra retrotransposon in pear (*Pyrus bretschneideri* Rehd.). Genome Biol Evol. 2014;6:1423–36.

Zhang S, Gu YQ, Singh J, Coleman-Derr D, Brar DS, Jiang N, et al. New insights into Oryza genome evolution: high gene colinearity and differential retrotransposon amplification. Plant Mol Biol. 2007;64:589–600.

Zhou Y, Cahan SH. A novel family of terminal-repeat retrotransposon in miniature (TRIM) in the genome of the red harvester ant, *Pogonomyrmex barbatus*. PLoS One. 2012;7:e53401.

Zonneveld BJM. New record holders for maximum genome size in eudicots and monocots. J Bot. 2010;2010:527357.

Genomic Diversity and Climate Adaptation in *Brachypodium*

Pip Wilson, Jared Streich, and Justin Borevitz

Abstract The *Brachypodium* genus contains the model grasses *B. distachyon*, *B. stacei* and *B. hybridum*, that are useful for molecular and physiological studies relevant to grain, pasture and bioenergy crops, as well as ecology of invasive species and environmental adaptation. In this chapter we discuss the natural variation in climate/geography, genotypic and phenotypic diversity that exists within these species. We describe the utilisation of this diversity via two methods, Genome Wide Association Studies and Landscape Genomics, to examine the interaction between genetic variation, phenotype, and environment. The aim is to identify adaptive loci that control specific traits in specific environments and understand the contribution of background polygenetic variation shaped by demographic processes. With recent developments in high throughput phenotyping, sequencing, and population samples with higher spatial/temporal resolution of climate data, these approaches can exploit the diversity of these *Brachypodium* spp. Experiments using this toolkit will reveal alleles, genes and pathways underlying agriculturally important and environmentally sensitive traits.

Keywords Natural variation • *Brachypodium* • Genome wide association studies • Landscape genomics

Abbreviations

GBS	Genotyping by sequencing
GWAS	Genome wide association studies
MaxEnt	Maximum entropy
QTL	Quantitative trait loci
SNPs	Single nucleotide polymorphisms

P. Wilson • J. Streich • J. Borevitz (✉)
Centre of Excellence in Plant Energy Biology, Australian National University, 134 Linnaeus Way, Canberra, ACT 2601, Australia
e-mail: Justin.Borevitz@anu.edu.au

J. Vogel (ed.), *Genetics and Genomics of Brachypodium*,
Plant Genetics and Genomics: Crops Models, DOI 10.1007/7397_2015_18
Published online: 15 October 2015

Introduction

Brachypodium is part of the Poaceae plant family, one of the five largest. Poaceae are one of two plant families found on all seven continents and grow in most habitats from alpine, to freshwater, and marine environments (Groombridge and Jenkins 2002; Brummitt and Cheek 2007). The Pooideae subfamily of the Poaceae are also widespread, especially in temperate zones, and many genera extend to cold climates in mountainous regions or high latitudes. Similar to other Pooideae genera, the *Brachypodium* genus also has a wide distribution in temperate regions, but is less common in cold climates and more abundant in hot arid climates (Hartley 1973). *Brachypodium* native range is the Mediterranean region with accessions commonly found through southern Europe, northern Africa and Eurasia. However, worldwide herbarium records show *Brachypodium* is now present on all continents except Antarctica (Garvin et al. 2008). In this review we focus on the natural diversity of the *Brachypodium distachyon* complex, now known to contain *Brachypodium distachyon, Brachypodium stacei* and *Brachypodium hybridum* (Catalán et al. 2012). This wide geographic range and climate tolerance in both the native and introduced range, as well as a key evolutionary position, mean the *B. distachyon* complex is a rich resource of allelic diversity for studies of adaptation and comparative genomics.

Diversity in Natural Populations of *Brachypodium*

Climate and Geographic Diversity

A large collection of thousands of *Brachypodium distachyon, Brachypodium stacei and Brachypodium hybridum* (referred hereon collectively as the *B. distachyon* complex) accessions are now available (Filiz et al. 2009; López-Alvarez et al. 2012; Mur et al. 2011; Vogel et al. 2006a, b, 2009). They come from a range of seasonal temperature profiles that span from alpine to hot arid deserts (Fig. 1). The majority of accessions have been collected from temperate grasslands with hot dry summers and wet winters (Mediterranean). These include accessions from southern Europe and South-Western USA. Many accessions come from temperate regions that also get summer rainfall, such as France and South-Eastern Australia. A few accessions come from regions with colder climates, such as mountainous regions in Turkey and on the Spanish/French border. Lastly, several accessions come from arid regions in the Middle East. These different climate types favour the development of different life history strategies allowing optimization of reproductive success, as well as abiotic stress tolerance mechanisms to deal with heat, cold and drought.

Like all plant species, the geographic distribution of *Brachypodium* species is dependent on climate and the history of its dispersal. *Brachypodium* species have had a long association with human civilisation. There is evidence of use of

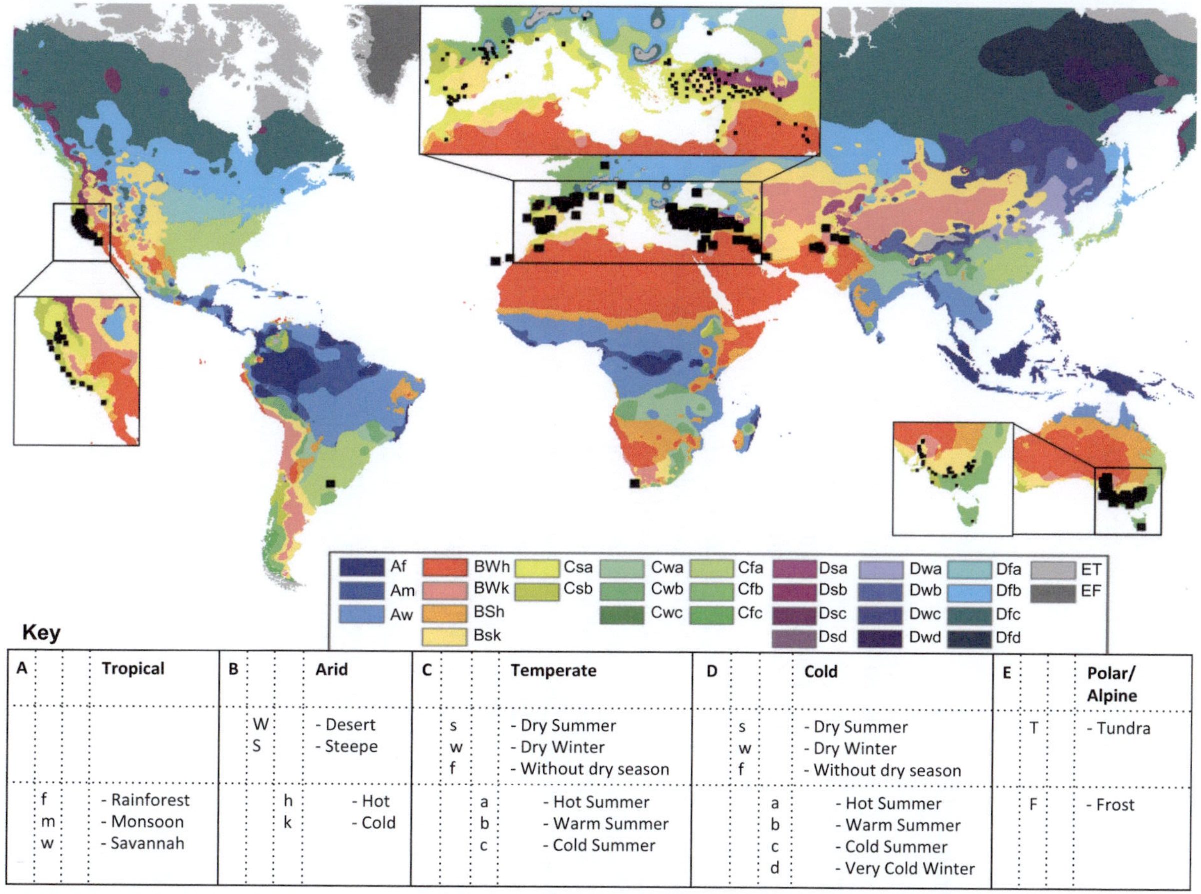

Fig. 1 Geographic and climatic distribution of *Brachypodium* accessions. Distribution of 350 collection sites, >1400 accessions, of *Brachypodium* overlaid on the Köppen-Geiger climatic regions (Peel et al. 2007). Many other private collections exist

Brachypodium species as food in Palaeolithic societies in the region of Italy as long as 30,000 years ago (Revedin et al. 2010), which may have aided its dispersal through its native range. The native range of the *B. distachyon* complex is southern Europe, northern Africa and south-western Asia through to India (Global Biodiversity Information Facility, 2013). In more modern times, the co-localisation of *Brachypodium hybridum* with cereal crops may have aided its dispersal through joint harvest and planting, especially from its native range to the "New World" of the Americas and Australia.

Genomic Diversity

There is a large amount of genomic diversity in the *Brachypodium distachyon* complex, with populations from Spain and Turkey most well described. The complex was originally described as one species with a polyploid series (Robertson 1981), however, there are now three species described: *B. distachyon* ($2n = 2x = 10$), *B. stacei* ($2n = 2x = 20$) and *B. hybridum* ($2n = 4x = 30$) (Hasterok et al. 2004; Catalán et al. 2012). *B. hybridum* is thought to be allotetraploids, resulting from a recent hybridization events fusing *B. distachyon* and *B. stacei* (Betekhtin et al. 2014; Catalán et al. 2012).

Within each species there is significant population genomic structure, as shown by a range of marker based techniques (Catalán et al. 2012; Filiz et al. 2009; Hammami et al. 2014; Jaroszewicz et al. 2012; Mur et al. 2011; Vogel et al. 2009). Two distinct groups of *B. distachyon* were shown by SSR repeat analysis of 187 Turkish and Iraqi lines, one group similar to the reference line Bd21 and the other like Bd1-1 (Vogel et al. 2009). We refer to these groups as the A (Bd21-like) and B (Bd1-1-like) subgroups. While the two subgroups can be intercrossed in the laboratory (Garvin et al. 2008), deep population structure in the wild indicates that gene flow has been restricted between the two groups. A further third *B. distachyon* subgroup, known as C, may exist in Sicily, Italy (Tyler et al. 2015). Strong population structure has also been found in *B. hybridum* populations in California with distinct genotypes (Bakker et al. 2009). Though currently under represented in germplasm collections, *B. stacei* appears to have one prominent haplotype, and many rare haplotypes (Catalán et al. 2012). Strong population structure can be attributed to the autogamous, selfing, nature of *B. distachyon* complex species.

Interestingly, AFLP analysis of 116 *B. distachyon* accessions from 45 sites in Turkey showed significant genotypic variation within a site but very similar genotypes were located at multiple sites across Turkey (Filiz et al. 2009). The authors suggested that animal based seed dispersal over long distance was likely common. There are also some indications of significant genetic diversity within a population/site in Spain with 24 % variance explained by differences within population of the same species (Hammami et al. 2014).

The recent development of Genotyping by Sequencing (GBS) provides an efficient way to characterize large population genomic variation within and between locations and (sub)species within a complex (Morris et al. 2011). As multiple foundering genotypes have been seen within individual sites, it is important to independently collect and sequence multiple individuals from a site to identify different genotypes. Multiplexing samples for genotyping reduces the individual sample cost of sequencing allowing many more samples to be sequenced, up to 384 in a single sequencing lane. GBS uses restriction digestion, barcoded adaptor ligation, pooling, second generation sequencing, and de-multiplexing to identify SNPs and call genotypes at >10,000 loci (Elshire et al. 2011). The large number of SNPs dispersed across the whole genome allows greater resolution of genetic distance between individuals than traditional individual genes or markers based methods. This method has extended our understanding of the *B. distachyon* complex phylogenetic and population genetic structure.

Whole genome sequencing has now been completed for seven *B. distachyon* accessions: Bd21, Bd21-3, Bd1-1, Bd3-1, Bd30-1, BdTR12c and Koz-3 (http://phytozome.jgi.doe.gov/pz/portal.html; Gordon et al. 2014; The International *Brachypodium* Initiative 2009). Gordon and colleagues (2014) identified close to 2.5 million SNPs and over 1 million InDels between the original reference genome, Bd21, and the six further accessions sequenced. A further 113 accessions of *Brachypodium* are currently being sequenced, including *B. stacei* and *B. hybridum* accessions, with 48 accessions almost ready for publication (J. Vogel pers. comm.; http://jgi.doe.gov/our-science/science-programs/plant-genomics/brachypodium/). Whole genome sequencing of multiple accessions will identify candidate functional variation and define haplotype boundaries but care needs to be taken to determine which accessions are used for genome wide association studies to avoid excessive population and family structure.

All three *Brachypodium* species are morphologically similar to each other, but express different traits. *B. stacei* and *B. hybridum* have higher seed yield, more biomass and quicker growth rate than *B. distachyon* (Catalán et al. 2012; Vogel et al. 2009). They also have different ranges with the majority of accessions found in North America and Australia being *B. hybridum* while *B. stacei* has been found throughout Mediterranean regions. *B. hybridum* was found to have a higher frequency in drier environments in Spain and showed greater drought tolerance in greenhouse trials (Manzaneda et al. 2012). These species provide ideal models for studies of allopolyploidy with parallels in particular to wheat, *Triticum aestivum*.

Adaptive Phenotypic Variation

The genetic makeup of a population is the result of neutral historical process of non random migration and non random mating. It is also shaped by natural selection. Selective pressures can include biotic and abiotic stresses, the prevalence of which can vary at different times of year and from year to year. These pressures can select for favourable combinations of existing alleles in a mixed population and/or favour new advantageous mutations. Grime (1977) defined classic strategies for plant survival in selective environments including competition (C), stress tolerant (S) and tolerant to frequently-disturbed environments (R) the CSR model. Plants that are competitive tend to have rapid growth with extensive lateral spread above and below ground and low seed yield. Stress tolerant plants tend to have slow growth with small leaves and low seed yields. Plants that colonize disturbed sites (ruderal plants) tend to have rapid growth, with small stature and high yield. Most *B. distachyon complex* accessions seem to favour a ruderal strategy, however the diversity of life strategies, growth rates, and plant architectural traits seen suggest that competition and stress tolerance are also important selective pressures in certain environments and seasons.

Isolation by distance can cause populations to diverge over time and may have had a major impact on *B. distachyon complex* diversity. This differentiation may occur via physical barriers, such as a mountain range, ice sheet or seas. We see evidence of divergence due to geographic separation in our preliminary GBS data (unpublished). However, genetic isolation may also occur from a difference in reproductive timing as different life strategies initiate flowering at different times of year, eliminating the chance of cross-pollination which is already low. Spatial and temporal factors can explain the division of the A and B subgroups in *B. distachyon.* Post reproductive barriers could also be involved, such as genetic incompatibilities and hybrid breakdown.

To date, a variety of natural variation phenotyping studies have been undertaken in *B. distachyon*, and to a lesser extent, *B. hybridum* on small to medium numbers of accessions. These studies have investigated a number of traits including morphology, seed protein make-up, weediness traits, dormancy, flowering time, biomass, root architecture, tolerance to abiotic and biotic stresses (e.g. Filiz et al. 2009; Wang et al. 2010; Bakker et al. 2009; Barrero et al. 2012; Schwartz et al. 2010; Tyler et al. 2014; Pacheco-Villalobos and Hardtke 2012; Li et al. 2012; Luo et al. 2011; Azhaguvel et al. 2008; Cui et al. 2012). These studies have demonstrated significant phenotypic variation between natural accessions that is being dissected through traditional bi-parental mapping populations, preliminary genome wide association studies and the mixed approach of nested association mapping studies. Environmental association studies have correlated phenotypes obtained in a common growing condition with the climatic variables at the site of collection. Luo and colleagues (2011) found significant variation in drought tolerance of 57 *B. distachyon* accessions but found no correlation to environmental variables obtained from the geographic locations at the site of collection. Incorporating genetic variation into environmental association, or landscape genomic studies, can identify factors shaping the range of particular genotypes and also potentially the identify the particular genetic variants underlying adaptation (Bragg et al. 2015). Landscape genomics (discussed further below) requires large, diverse, targeted collections and with comprehensive genomic analysis to directly get to the genetic basis underlying climate adaptation. This knowledge could predict and invasion success in *Brachypodium* spp. and be a powerful model for other invasive plants.

Case Study: Flowering Time

The timing of germination and reproduction, also know as life strategy, are good examples of phenological variation that are strongly shaped by natural selection in local climates. *B. distachyon* accessions collected from high altitude and high latitude sites are generally strong 'vernalisation requiring' ecotypes (Schwartz et al. 2010). This allows extended vegetative growth and the synchronisation of flowering with the more favourable conditions of late spring rather than risking damage to cold-sensitive floral structures in winter. In Mediterranean regions, winter rainfall determines the growing season for 'rapid cycling' ecotypes to avoid the hot dry summers. Accessions from hot arid climates with intermittent rainfall as well as disturbed agricultural sites also tend to harbour 'rapid cycling' ecotypes with fast generation times, low dormancy and low vernalisation requirements, e.g. Bd21 (Barrero et al. 2012). This fast generation time allows a full life cycle to occur when moisture is available while enduring dry periods as seed. *Brachypodium* accessions from warmer, wet summer regions that are later flowering allowing more vegetative growth and increased summer seed set. Some natural accessions may also have the ability to produce different phenotypes in different environmental conditions because of phenotypic plasticity. For example, accessions from variable climates may have some vernalisation requirements, but also a strong photoperiod response such that if there is a cold winter, flowering can be promoted, but in a mild winter flowering is initiated by a strong photoperiod response.

An experiment to investigate the genotypic and environmental basis for the diversity of flowering time of *B. distachyon* was undertaken at The Australian National University (ANU) in 2014. Plants were grown in specially modified growth chambers to simulate a temperate region with a cold winter and hot summer. Chambers were set for either a Winter or Spring germination. The modified chamber controls allowed us to program down to 1 min changes in light intensity, light spectrum, temperature and humidity; mimicking diurnal and seasonal changes in climate. In this experiment 256 accessions were grown in each chamber. Ear emergence was chosen as an indicator of transition to reproduction as many accessions flowered within the ear and hence were difficult to score for flowering. The Winter germination chamber had cold (5 °C) nights, resulting in the vernalisation requirements of all lines being met and the majority of lines reaching ear emergence between 100 and 160 days post emergence (Fig. 2a). However, in the Spring germination treatment the transition to reproduction was more variable. The majority of lines in the Spring germination reached ear emergence fairly quickly between 60 and 120 days post emergence. However, although there were still a few weeks with cold nights, the accessions requiring stronger vernalisation did not reach ear emergence by the end of the experiment at 180 days (e.g. Tek1, ABR4, UKR-99-137). Interestingly, while the life strategy of the majority of lines remained fairly unchanged between environments (early flowering), those accessions from Western Europe tended to show more phenotypic plasticity between

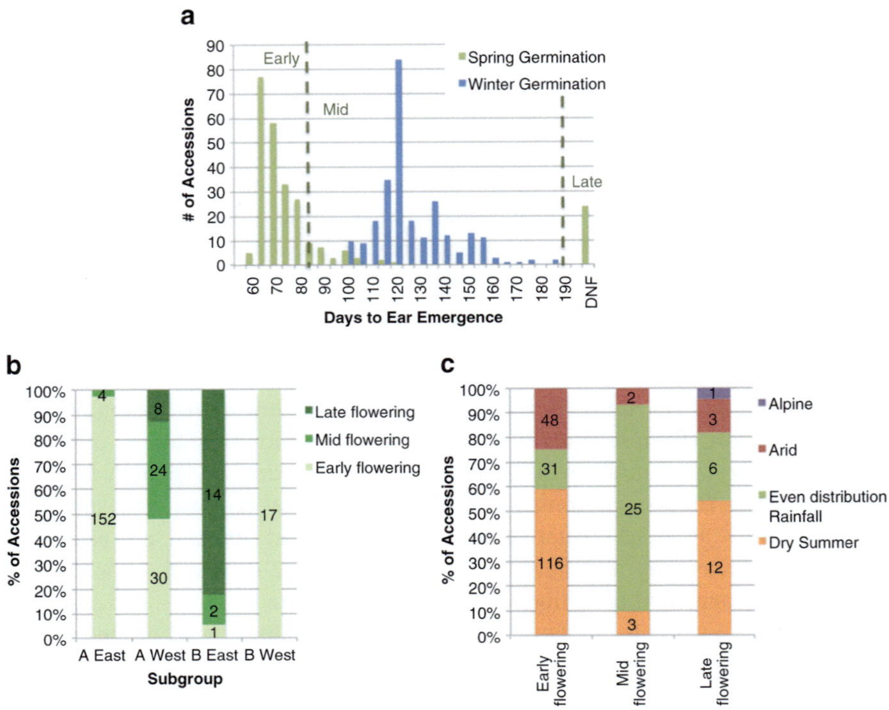

Fig. 2 Life strategy of *B. distachyon* accessions is associated with subgroup, region and climate type at the collection site. Two hundred fifty-six *Brachypodium* accessions were grown in simulated growth conditions to mimic a Winter and Spring sowing date. Days from seedling emergence to ear emergence was monitored as an indicator of transition to the reproductive stage. (**a**) Comparison of life strategy in the two environments; *dashed lines* indicate the cutoffs for classification of accessions as early, mid or late flowering based on their flowering time under the Spring germination. (**b**) Subgroup and geographic origin show varying distributions of early, mid and late flowering accessions; *numbers* in columns indicate number of accessions in each category and (**c**) Early, middle and late flowering groups of accessions are sourced from different proportions of climate types

environments with a wider range of delay in ear emergence with the Winter sowing than the Spring sowing. (Supplementary phenotypic data can be found at http://borevitzlab.anu.edu.au/resources/populations.)

The accessions were then classified as early, mid or later flowering based on the days to ear emergence in the Spring sowing (Fig. 2a). When these groupings are compared to the subgroups classifications we can see there is no firm trend for flowering time between the subgroups, as the patterns differ significantly in the Eastern European and Western European subsets (Fig. 2b). For example, while accessions from the A subgroup tend to be early flowering in Eastern Europe, a mix of life strategies are present in Western Europe. This mix would indicate that different flowering time alleles may have evolved in the different subgroups. Crosses between accessions from different geographic regions and also between subgroups within a region are needed to determine the true number of alleles controlling flowering time. Just as flowering time is variable within subgroups and geographic regions, it also varies with climatic type at the site of collection. When we categorise our accessions by climatic region, the life strategy patterns are not fixed based one on selective pressures in these climates alone. Accessions from climates with dry summers can be early or late flowering (Fig. 2c). A similar trend is seen for accessions from arid regions, where a wide range of flowering times are seen. Mid flowering accessions tend to be from climates with rainfall evenly distributed through the year, indicating that the accessions with long growing seasons and no vernalisation requirements have a fitness advantage in this climate type. Alpine accessions were late flowering, suggesting that vernalisation is needed for survival in this environment. We must keep in mind sampling bias in any collection. The observed trends can be influenced by when and where the collections are made as well as genetic relatedness among accessions. Netherless, there are trends with exceptions, illustrating the impact of the genotype and climate on the geographic distribution and phenotypic expression of accessions.

Understanding fitness of an individual accession in a particular climate is complex, complicating studies aimed at elucidating the genetic architecture of adaptive traits. Luckily, new approaches have been developed that control for genetic and spatial structure, which can result in outcomes that have applications in crops species breeding programs, weed management and ecosystem restoration. Next we outline two of these approaches: Genome Wide Association Studies (GWAS) and Landscape Genomics.

Using Local Adaptation to Identify Causative Alleles

Genome Wide Association Studies

Natural populations provide a valuable genetic resource when the diversity of genotypic variation is found in a range of extreme environments. Historically these populations have been studied for phenotypic diversity in small numbers of accessions using SSR and microsatellites. The development of comprehensive next generation sequencing has meant that thousands of accessions can now be screened and powerful subsets selected for Genome Wide Association Studies.

Genome wide association studies (GWAS) were originally developed for studies of human disease to find causative mutations (Klein et al. 2005). The aim of GWAS is to identify the genetic architecture (number of alleles, loci and effect sizes) of phenotypic traits of interest. By phenotyping across different environments we can also identify which alleles are sensitive and which are robust across environments. GWAS have been performed in a number of plant species including Arabidopsis, maize, rice, barley and sorghum (e.g. Atwell et al. 2010; Wen et al. 2014; Huang et al. 2010; Pasam et al. 2012; Morris et al. 2013). These studies have successfully identified causative QTLs for many traits including disease resistance, proline accumulation, flowering time, starch profile, plant height and grain weight.

Genotype Data

To perform GWAS you must genotype a large number of SNPs tagging the many haplotype blocks that are differentially assorted in the sample being studied. The density of SNPs needed depends on ancestral recombination in the genome and will affect the resolution of the QTLs identified. As previously described, GBS can be used to get high density SNP genotyping. In addition the, small genomes of *B. distachyon, B. stacei* and *B. hybridum* size allows the use of low pass whole genome re-sequencing as an high density genotyping method. Samples can be pooled for efficient sequencing by ligating barcoded adapters to randomly sheared DNA. Sequencing to $1\times$ or less genome coverage is sufficient as missing genotypes at common SNPs can be imputed based on linkage disequilibrium (Halperin and Stephan 2009). GWAS must be performed on a sizable number of appropriately diverse accessions, at least 100 but optimally 300 or more, to maximise statistical power to detect common (>10 % allele frequency) moderate effect (10 % phenotypic variation) loci. SNPs are usually filtered at 5 % minor allele frequency. A genome scan is then performed testing each SNP for association with the quantitative trait while controlling for ancestral relatedness among samples using the kinship matrix as a random effect. Once a QTL is identified to perhaps 100 kb resolution, candidate genes can be selected. The resolution of QTLs is dependent on the rate of decay of linkage disequilibrium in the test set and at the particular genomic region. Ultimately, the causative alleles can be confirmed by transgenic complementation of knockout lines or transformation with different functional alleles.

GWAS can have some advantages over traditional bi-parental mapping (e.g. RILs), nested association mapping (NAM) or multi-parent advanced generation intercrossed (MAGIC) populations in that the source of genetic variation is greater. Furthermore, the resolution of mapping of the causative allele can be higher due to the higher frequency of recombination events in a large ancestral population evolved over centuries compared to a hand-crossed population with only a handful of generations of crossing. However, rare alleles of large effect contribute noise to GWAS. In addition, population structure from non-random mating can cause false-positives at loci associated with the causative one. Even when appropriately controlled for statistically, population structure can cause false-negative results by downplaying the effect at the truly causative locus.

There are a couple approaches to mitigate the effects of population structure. The first is to balance the structure. GBS can be performed on a very large number of diverse lines from across the world. Once the structure of the global population is known, a core GWAS set can be selected balancing membership roughly equally from across different lineages (Fig. 3a). Whole genome sequencing is performed on this GWAS set, to increase resolution of mapping causative alleles. A kinship matrix is also made from the SNP data to include in the statistical analysis to reduce the rate of false positives. The second approach is to perform the same initial GBS screen of a wide range of accessions, but then to select a region within a population structure group or a hybrid zone where the groups have become well mixed. The resulting population sample is less confounded such that segregation at the causative alleles will be largely uncoupled from the background genetic variation. These approaches are further described in Brachi et al. (2011) and an example of its success in Arabidopsis is given by Li et al. (2010), Baxter et al. (2010), Horton et al. (2012).

In consideration to the first approach outlined above, GWAS has limited success when performed across the A and B subgroups of *B. distachyon* due to a high level of population structure. Hence we suggest that GWAS sets are developed for each subgroup separately. Furthermore, the current *B. distachyon* public collections contain a lot of groups of very closely related accessions, e.g. BdTR1_lines (Vogel et al. 2009). Selection of a GWAS set must also limit this family structure. Including closely related accessions excludes ancestral recombination and introduces bias towards the common alleles in that family. The second approach of finding a well-admixed population may be challenging due to the selfing nature of *Brachypodium*. However, it is our experience that several field sites and even some maternal lines are segregating both genotypes and phenotypes, indicating that this strategy of identifying naturally admixed sites as regional mapping populations (Horton et al. 2012) is a possibility.

Phenotyping Data

There are a number of GWAS requirements for phenotype data. The traits must be due to common genetic variation rather than multiple rare variants. It is helpful to measure heritability, or the proportion of total phenotypic variation (Vp) due to genetic variation (Vg) among lines. A simple way to estimate heritability (H^2) is to subtract environmental variation (Ve) measured within replicate inbred lines, $H^2 = Vg/Vp = (Vp - Ve)/Vp$. If the variability within an inbred line (Ve) is low, then more of the variability seen in the total phenotypic variation will be due to genetic factors (Fig. 3b). Alternatively the pseudo-heritability is the amount of phenotypic variation explained by the kinship, or pairwise genetic diversity among lines. This can be calculated prior to GWAS.

Due to the large number of accessions to be phenotyped for GWAS, the phenotypes must also be quantifiable in a high throughput manner. For some measures, such as plant height or ear emergence this can be done manually with relative ease. However, for traits that require continuous measurement, such as growth rates, or those where measurements and analysis are very time consuming, such as photosynthetic measures, high throughput phenotyping systems can be very beneficial. One such system is the PlantScreen phenotyping platform (Brown et al. 2014). This platform allows up to 300 plants to be screened at a time; including individual pot weighing and watering, thermal imaging, RGB stereo imaging and chlorophyll fluorescence. We have used this system to monitor abiotic stress tolerance of accessions by measuring photosynthetic efficiency and photoinhibition; water use; stomatal conductance/transpiration cooling; and photoprotective mechanisms such as pigment accumulation and non-photochemical quenching (NPQ).

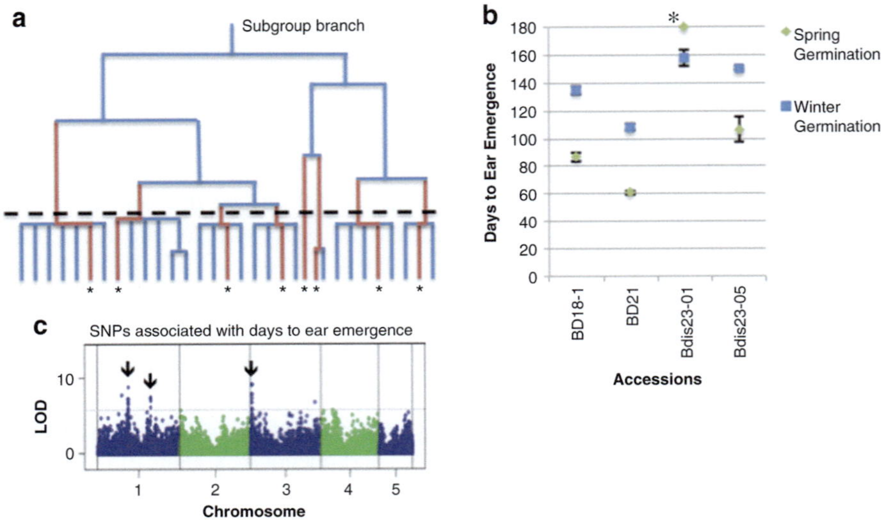

Fig. 3 Considerations when performing GWAS in *Brachypodium*. (**a**) Selection of a balanced set of accessions is key for successful GWAS. Initial GBS of all available accessions facilitates selection of a subset of genotypes (*asterisks*) for intense phenotyping and whole genome sequencing (**b**). Replication of a few key lines allow for the heritability of a phenotypic trait to be estimated. While variance for flowering time is low, traits with large environmental variance will return poor GWAS results. Points are the average time to ear emergence of 3–6 biological replicates ± S.D.; *did not flower in the Spring germination (**c**) A genome wide screen of SNPs identifies those associated with a quantitative trait. Manhattan plot with *arrows* showing SNPs associated with days to ear emergence with a Spring germination. A balanced set of 95 accessions from the A subgroup was used in analysis. The *dotted line* indicates the 5 % empirical genome-wide significance threshold

Genotype by Environment Interactions

An important extension of GWAS is to understand how the growth environment affects the genetic architecture of traits. Certain loci may be more important to the expression of a trait when grown in a particular environment. A simple example is where a vernalisation-sensitive allele may be important in controlling flowering time in an alpine climate, but a photoperiod-sensitive allele may be more important in a lowland climate. The interaction between genetic loci may also vary between environments. Understanding the genetic architecture of phenotypic plasticity is also of great interest, especially in a world where climate variability and change is accelerating. Hence it is important to consider how the growth environment will influence the phenotypes seen in GWAS experiments and how this will relate to extrapolation of results to applications of plant breeding or ecological work.

Using dynamic growth chambers, such as the SpectralPhenoClimatron (Brown et al. 2014; Li et al. 2006), in combination with climate modelling software such as SolarCalc, GWAS can be used to simultaneously compare contrasts such as life strategies at different times of year (Fig. 2, Li et al. 2010); growth and yield in current and future climates (e.g. Li et al. 2014); and the effect of fluctuating light, as would be seen under a forest canopy, on growth and initiation of photoprotective mechanisms.

Applications

GWAS allows us to understand the genetic architecture of interesting phenotypic traits, whether it is flowering time, biomass production, components crop yield or abiotic stress tolerance. With this approach we can identify segregating alleles in natural populations causing advantageous phenotypes (Fig. 3c). This understanding can then be applied to breeding better crops by selecting genotypes predicted to have optimal phenotypes, aka crop design (Huang and Han 2014). Understanding how the genetic architecture of traits is affected by environment allows us to select accessions with the appropriate genetic makeup to thrive in specific environments, current or future. This can apply to both agriculture and foundation species of natural systems as climate change is occurring more rapidly than traditional breeding by phenotypic selection can keep up with (Hoffmann et al. 2015).

Landscape Genomics

GWAS is a powerful method to dissect the genetic basis of complex traits and the dependence on the growth environment. It is of interest to know if these are adaptive traits and contribute to yield and fitness in particular environments as this can help predict invasive potential and resilience under environmental change. Common gardens or reciprocal transplant studies can identify where the home genotype performs superior to a geographically distinct genotype (e.g. Clausen et al. 1940). Provenance trials do this on a larger scale, evaluating phenotypes from a broader range of genotypes across many locations. However, these studies are massive and can be hampered by starting conditions and weather variation and don't get at the genetic basis of fitness.

Even highly selfing species such as *B. distachyon* (Vogel et al. 2009) will eventually outcross birthing new allelic-combinations to the local environment. Genes controlling adaptive traits may show a signature of selection on the landscape if one allele provided an advantage in a particular environment and/or the other allele was detrimental in an opposing environment. Indeed direct associations between environmental variables at geographic locations and alleles at quantitative trait loci have been found, confirming their role in adaptation (Li et al. 2010 and others). This opens the door to a genome scan to naively identify adaptive loci associated with environmental variables. This could work without knowledge of what traits are under selection. This so called Landscape Genomics approach combines spatial/temporal climate data from collection sites with second generation sequencing technology to explore the differences in climatic niche breadth among genotypes to find adaptive alleles. Of the most successful Landscape Genomics studies, many use model species (Hancock et al. 2011; Fournier-Level et al. 2011).

Though promising, Landscape Genomics has caveats including needing to control for background variation due to population structure. Geographic location is confounded with many environmental variables limiting our ability to separate selection at adaptive loci from isolation by distance that shapes the entire genome. An initial study showed the complexity of *B. distachyon* across landscapes (Dell'Acqua et al. 2014). Strong family and population structure highlight the need for large, diverse sampling across climatic and geographic space.

By understanding the genetic basis of local adaptation, we may find allelic variants that allow some groups to be specialists, while other alleles aid generalists (Storz 2005). The *B. distachyon* complex range distribution has a similar breadth to *Arabidopsis thaliana*. How genetic diversity is associated with geographic and climatic space depends on differences in migration and mating patterns as well as selection on particular adaptive loci (Platt et al. 2010; Horton et al. 2012; Banta et al. 2012). Each *B. distachyon* subgroup may show associations with different adaptive alleles across geographic and climatic space when ancestral variation has not been broadly distributed. This adaptive scale must be kept in mind when designing landscape genomic studies (Bragg et al. 2015).

Species or even genotype level distributions can be inferred and predicted using Maximum Entropy (MaxEnt) software and BioClim (BioClim, 2015), a global grid of biologically relevant climate layers (Phillips et al. 2006; Phillips and Dudík 2008; WorldClim 2014). MaxEnt couples climate data of training locations as model input and compares known locations to a geographic background and outputs heat maps of similar ecological niches and response curves of each climate variable contribution. A recent study about environmental niche variation showed the variation of climate and species distribution of all three *B. distachyon* complex species (López-Alvarez et al. 2015). They show the overlapping range of the three species in paleo and current environments, which will aid the diagnoses of how and where species in each group evolved and where species specific collecting could occur. Figure 4a provides an example of MaxEnt modelling, showing the niche breadth of *B. hybridum* in Australia using ~80 locations. Though *B. distachyon* is in Australia, the current number of genetically distinct populations is too few for precise subgroup modelling or identification of adaptive alleles. Targeting particular admixed locations is an attractive approach once they are identified.

Investigating climate association with genomic variants can diagnose which particular climate variables drive natural selection. These climate variables may be different for different groups of accessions. For example, an ordered heat-map of 80 Australian collection locations by their climate variables results in two large clades of environment types with five sub-groups (Fig. 4b). While there is similarity among climate at a local scale extending East and West, there is a large split North to South. Three regions are highlighted: Mediterranean regions (green) cluster near desert regions (grey) while temperate regions (red) are climate separated, but Mediterranean regions are geographically separated from deserts.

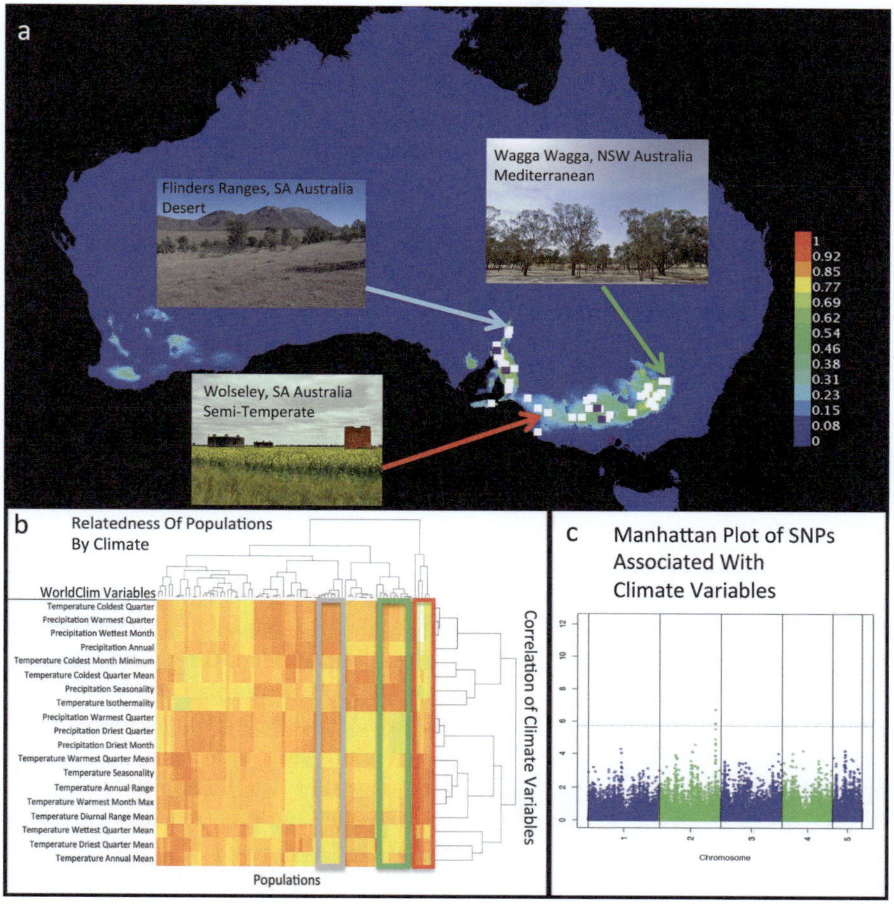

Fig. 4 Predicting species distribution and detecting segregating SNPs by climate input. (**a**) MaxEnt produced probable distribution of *Brachypodium* in Australia using 80 geographic locations as presence only data correlated with WorldClim/BioClim variables. (**b**) Eighty Australian populations collected by the Borevitz lab of *B. hybridum* show the environmental relatedness of each population (no genetic data used). The *vertical* dendrogram describes environmental variable correlation across all populations. The *horizontal* dendrogram describes individual population clustering by their environmental relatedness to inform researchers about where certain genotypes/haplotypes might grow. If specific haplotypes are found in multiple locations one can dissect which strains are more likely to be pervasive. Likewise the variation and possible selection pressure one or more environmental variable(s) may have across many populations. Using a custom R script, we normalize 19 WorldClim environmental variables point sampled across locations. Variables with the smallest normalised distribution from the mean are likely to be a strong selection pressure species-wide. Variables with larger distribution could be random, but also could be associated with specific alleles. (**c**) A cartoon representation of SNPs associated with the environmental variable of Precipitation in the Warmest Quarter, calculated much the same way as a LOD with phenotype data

One can also use environment data as a fitness phenotype to test if certain alleles are associated to climate. Measuring fitness directly is difficult and is the result of selection at many different life stages and loci. However, if a large set of individuals spanning a suitable range of space is sequenced, presence data can be used as an indirect observation that a particular combination of alleles can survive there. A GWAS scan can then use climate data from collection sites rather than phenotype data to identify adaptive loci (Fig. 4c).

Landscape genomics seeks to find allelic associations with climate and environmental space to better understand evolutionary patterns and adaptation. It aims to understand how genotypes and phenotypes segregate across geographic space, which can aid crop breeding, control of invasive species, and preservation of environments.

Future Collections

Extensive public collections of *Brachypodium* exist with over 1060 accessions published (e.g. Catalán et al. 2012; Dell'Acqua et al. 2014; Draper et al. 2001; Filiz et al. 2009; Hammami et al. 2014; Mur et al. 2011; Vogel et al. 2006a) and at least 181 accessions available from the USDA National Plant Germplasm System, including 141 *B. distachyon* (ars-grin.gov/npgs/). At least a further 3000 accessions are estimated to be present in private collections and available by request or through collaborations. However many geographic regions and climate niche, remain unrepresented or under sampled. Having those spaces filled would be beneficial for the whole *Brachypodium* community. As mentioned above there is a strong division in the current public collection between Eastern and Western European accessions. Central Southern Europe might provide a source of admixture populations between the Eastern and Western genotypes or completely new genotype groups. Furthermore, it would be advantageous to have collections across the Middle East and North Africa as it is highly possible that *B. distachyon* complex species were pushed south during the last ice age and extant lineages could be sources of new maternal lines for research. With genotype data, we can focus further collections on hybrid zones and polymorphic sites to increase the number of recombinant genotypes. These genotypes provide unique opportunities to study natural selection on adaptive traits. They provide natural genetic mapping resources to dissect complex adaptive traits. Admixed populations also inform the evolutionary history of *Brachypodium* and provide an opportunity to test natural selection on segregating variation.

To study the natural variation of stress tolerance in *B. distachyon,* it would be good to have further collections from more extreme environments. This would include higher altitude/latitude environments such as Northern Europe and more arid environments in the Middle East and North Africa. Areas with multiple, recent introductions are also of interest as they may provide examples of strong selection on standing variation or even new mutations.

Conclusion

The *B. distachyon* complex has a large and structured genetic diversity with a distribution spanning a wide range of climatic regions. Through Landscape Genomics we can use this diversity to predict the niche breadth and potential range size of a particular background genotype. We can also start to understand the genetic basis of adaptation to a particular niche. Genome Wide Association Studies will utilise *Brachypodium* genetic diversity in combination with its broad phenotypic diversity to elucidate the genetic architecture of adaptive traits and which loci are varying among different climate types. Both techniques rely on a thorough understanding of the population and family structure of available collections with particular attention to species and subspecies divisions. Future developments in these techniques will rely heavily on the expansion of the collection of publicly available accessions and development of bioinformatic and statistical methods to increase the power and accuracy of both the Landscape Genomics and GWAS approaches.

Acknowledgments We would like to acknowledge our collaborators: Location data and seed—Shuangshuang Liu of the Kent Bradford lab at UC Davis United states, Pilar Catalan from the University of Zaragoza in Huesca Spain, Luis Mur at Aberystwyth University at Aberystwyth Wales, Dave Garvin from University of Minnesota/USDA United States and John Vogel from JGI/UC Berkeley.

Glossary

Accession A collection of seeds from one location. This includes bulk collections and maternal descent lines.

Ecotype An individual or group whose genetic distinction is strongly associated to an environment or type.

Genotype This general term is used either to describe the genotype at a locus such as a SNP (AA, Aa, aa) or a background whole genome genotype which can have levels of species, subgroup, population genetic structure group, family, individual maternal line.

Phenotype (qualitative and quantitative) Measurable traits expressed by plants.

Population Non-random mating between groups within a specified geographic space.

Subgroup In this paper, subgroup is a major hierarchical cluster of genotype groups and their respective families and/or genotypes. Subgroups could interbreed but don't in natural environments due to some sort of natural barrier.

References

Atwell S, Huang YS, Villhjálmsson BJ, Willems G, Horton M, Li Y, et al. Genome-wide association study of 107 phenotypes in *Arabidopsis thaliana* inbred lines. Nat Lett. 2010;465:627–31.

Azhaguvel P, Li W, Rudd JC, Gill BS, Michels Jr GJ, Weng Y. Aphid feeding response and microsatellite genetic diversity among diploid *Brachypodium distachyon* (L.) Beauv accessions. Plant Genet Resour. 2008;7:72–9.

Bakker EG, Montgomery B, Nguyen T, Eide K, Chang J, Mockler TC, et al. Strong population structure characterises weediness gene evolution in the invasive grass species *Brachypodium distachyon*. Mol Ecol. 2009;18:2588–601.

Banta JA, Ehrenreich IM, Gerard S, Chou L, Wilczek A, Schmitt J, et al. Climate envelope modeling reveals intraspecific relationships among phenology, niche breadth and potential range size in *Arabidopsis thaliana*. Ecol Lett. 2012;15:769–77.

Barrero JM, Jacobsen JV, Talbot MJ, White RG, Swain SM, Garvin DF, et al. Grain dormancy and light quality effects on germination in the model grass *Brachypodium distachyon*. New Phytol. 2012;193:376–86.

Baxter I, Brazelton JN, Yu D, Huang YS, Lahner B, Yukubova E, et al. A coastal cline in sodium accumulation in *Arabidopsis thaliana* is driven by natural variation of the sodium transporter AtHKT1;1. PLoS Genet. 2010. doi:10.1371/journal.pgen.1001193.

Betekhtin A, Jenkins G, Hasterok R. Reconstructing the evolution of Brachypodium genomes using comparative chromosome painting. PLoS One. 2014. doi:10.1371/journal.pone.0115108.

BioClim [Internet]. WorldClim—Global Climate Data, free climate data for ecology. [cited 5 Jan 2015]. http://www.bioclim.org/.

Brachi B, Morris GP, Borevitz JO. Genome-wide association studies in plants: the missing heritability in the field. Genome Biol. 2011;12:232.

Bragg JG, Supple MA, Andrew RL, Borevitz JO. Tansley review: genomic variation across landscapes: insights and applications. New Phytol. 2015;208:953.

Brown TB, Cheng R, Sirault XRR, Rungrat T, Murray KD, Trtilek M, et al. TraitCapture: genomic and environmental modelling of plant phenomic data. Curr Opin Plant Biol. 2014;18:73–9.

Brummitt RK, Cheek MR. Hoplestigmataceae. In: Heywood VH, Brummitt RK, Culham A, Seberg O, editors. Flowering plant families of the world. Kew: Royal Botanic Gardens; 2007. p. 167.

Clausen J, Keck DD, Hiesey WM. Experimental studies on the nature of species. Washington, DC: Carnegie Institution of Washington; 1940.

Catalán P, Müller J, Hasterok R, Jenkins G, Mur LAJ, Langdon T, et al. Evolution and taxonomic split of the model grass *Brachypodium distachyon*. Ann Bot. 2012;109(2):385–405.

Cui Y, Lee MY, Huo N, Bragg J, Yan L, Yuan C, et al. Fine mapping of the *Bsr1* barley stripe mosaic virus resistance gene in the model grass *Brachypodium distachyon*. PLoS One. 2012;7: e38333.

Dell'Acqua M, Zuccolo A, Tuna M, Gianfranceschi L, Enrico PM. Targeting environmental adaptation in the monocot model *Brachypodium distachyon:* a multi-faceted approach. BMC Genomics. 2014;15:801.

Draper J, Mur LA, Jenkins G, Ghosh-Biswas GC, Bablak P, Hasterok R, et al. *Brachypodium distachyon*. A new model system for functional genomics in grasses. Plant Physiol. 2001;127:1539–55.

Elshire RJ, Glaubitz JC, Sun Q, Poland JA, Kawamoto K, Buckler ES, et al. A robust simple genotyping-by-sequencing (GBS) approach for high diversity species. PLoS One. 2011. doi:10.1371/journal.pone.0019379.

Filiz E, Ozdemir BS, Budak F, Vogel JP, Tuna M, Budak H. Molecular, morphological, and cytological analysis of diverse *Brachypodium distachyon* inbred lines. Genome. 2009;52: 876–90.

Fournier-Level A, Korte A, Cooper MD, Nordborg M, Schmitt J, Wilczek AM. A map of local adaptation in Arabidopsis thaliana. Science. 2011;334(6052):86–9.

Garvin DF, Gu YQ, Hasterok R, Hazen SP, Jenkins G, Mockler TC, et al. Development of genetic and genomic research resources for *Brachypodium distachyon*, a new model system for grass crop research. Crop Sci. 2008;48:S69–84.

Gordon SP, Priest H, Des Marais DL, Schackwitz W, Figueroa M, Martin J, et al. Genome diversity in *Brachypodium distachyon*: deep sequencing of highly diverse inbred lines. Plant J. 2014;79:361–74.

The Global Biodiversity Information Facility: GBIF Backbone Taxonomy, 2013-07-01. http://www.gbif.org/species/5290143. Accessed 2015-05-08.

Grime JP. Evidence for the existence of three primary strategies in plants and its relevance to ecological and evolutionary theory. Am Nat. 1977;111(982):1169–94.

Groombridge B, Jenkins M. World atlas of biodiversity: Earth's living resources in the 21st century. University of California Press: Berkeley, CA; 2002.

Halperin E, Stephan DA. SNP imputation in association studies. Nat Biotechnol. 2009;27:349–51.

Hammami R, Jouve N, Soler C, Frieiro E, González JM. Genetic diversity of SSR and ISSR markers in wild population of *Brachypodium distachyon* and its close relatives *B. stacei* and *B. hybridum* (Poaceae). Plant Syst Evol. 2014;300:2029–40.

Hancock AM, Brachi B, Faure N, Horton MW, Jarymowycz LB, Sperone FG, et al. Adaptation to climate across the Arabidopsis thaliana genome. Science. 2011;334(6052):83–6.

Hartley W. Studies on the origin, evolution and distribution of the *Gramineae*. V. The subfamily *Festucoideae*. Aust J Bot. 1973;21:201–34.

Hasterok R, Draper J, Jenkins G. Laying the cytotaxonomic foundations of a new model grass, *Brachypodium distachyon* (L.) Beauv. Chromosome Res. 2004;12:397–403.

Hoffmann A, Griffin P, Dillon S, Catullo R, Rane R, Byrne M, et al. A framework for incorporating evolutionary genomics into biodiversity conservation and management. Clim Chang Responses. 2015;2:1. doi:10.1186/s40665-014-0009-x.

Horton MW, Hancock AM, Huang YS, Toomajian C, Atwell S, Auton A, et al. Genome-wide patterns of genetic variation in worldwide *Arabidopsis thaliana* accessions from the RegMap panel. Nat Genet. 2012;44(2):212–7.

Huang X, Han B. Natural variations and genome-wide association studies in crop plants. Annu Rev Plant Biol. 2014;65:531–51.

Huang X, Wei X, Sang T, Zhao T, Zhao Q, Feng Q, et al. Genome-wide association studies of 14 agronomic traits in rice landraces. Nat Genet. 2010;42:961–7.

Jaroszewicz AM, Kosina R, Stankiewicz PR. RAPD, karyology and selected morphological variation in a model grass, Brachypodium distachyon. Weed Res. 2012;52:204–16.

Klein RJ, Zeiss C, Chew EY, Tsai JY, Sackler RS, Haynes C, et al. Complement Factor H polymorphism in age-related macular degeneration. Science. 2005;308(5720):385–9.

Li C, Rudi H, Stockinger EJ, Cheng H, Cao M, Fox SE, et al. Comparative analyses reveal potential uses of *Brachypodium distachyon* as a model for cold stress responses in temperate grasses. BMC Plant Biol. 2012;12:65.

Li Y, Roycewicz P, Smith E, Borevitz JO. Genetics of local adaptation in the laboratory: flowering time quantitative trait loci under geographic and seasonal conditions in Arabidopsis. PLoS One. 2006;1(1):e105.

Li Y, Huang Y, Bergelson J, Nordborg M, Borevitz JO. Association mapping of local climate-sensitive quantitative trait loci in *Arabidopsis thaliana*. Proc Natl Acad Sci U S A. 2010;107 (49):21199–204.

Li Y, Cheng R, Spokas KA. Genetic variation for life history sensitivity to seasonal warming in *Arabidopsis thaliana*. Genetics. 2014;196:569–77.

López-Alvarez D, Manzaneda AJ, Rey PJ, Giraldo P, Benavente E, Allainguillaume J, et al. Environmental niche variation and evolutionary diversification of the *Brachypodium distachyon* grass complex species in their native circu-Mediterranean range. Am J Bot. 2015;102:1073.

López-Alvarez D, López-Herranz ML, Betekhtin A, Catalán P. A DNA barcoding method to discriminate between the model plant *Brachypodium distachyon* and Its close relatives *B. stacei* and *B. hybridum* (Poaceae). PLoS One. 2012;7:e51058.

Luo N, Liu J, Yu X, Jiang Y. Natural variation of drought response in *Brachypodium distachyon*. Physiol Plant. 2011;141:19–29.

Manzaneda AJ, Rey PJ, Bastida JM, Weiss-Lehman C, Raskin E, Mitchell-Olds T. Environmental aridity is associated with cytotype segregation and polyploidy occurrence in *Brachypodium distachyon* (Poaceae). New Phytol. 2012;193:797–805.

Morris GP, Grabowski PP, Borevitz JO. Genomic diversity in switchgrass (Panicum virgatum): from the continental scale to a dune landscape. Mol Ecol. 2011;20(23):4938–52.

Morris GP, Ramu P, Deshpande SP, Hash CT, Shah T, Upadhyaya HD, et al. Population genomics and genome-wide association studies of agroclimatic traits in sorghum. Proc Natl Acad Sci U S A. 2013;110(2):453–8.

Mur LAJ, Allainguillaume J, Catalan P, Hasterok R, Jenkins G, Lesniewska K, et al. Exploiting the *Brachypodium* tool box in cereal and grass research. New Phytol. 2011;191:334–47.

Pacheco-Villalobos D, Hardtke CS. Natural genetic variation of root system architecture from *Arabidopsis* to *Brachypodium*: towards adaptive value. Philos Trans R Soc B. 2012;367:1552–8.

Pasam RK, Sharma R, Malosetti M, van Eeuwijk FA, Haseneyer G, Kilian B, et al. Genome-wide association studies for agronomical traits in a world wide spring barley collection. BMC Plant Biol. 2012;12:16.

Peel MC, Finlayson BL, McMahon TA. Updated world map of the Köppen-Geiger climate classification. Hydrol Earth Syst Sci. 2007;11:1633–44.

Phillips SJ, Anderson RP, Schapire RE. Maximum entropy modeling of species geographic distributions. Ecol Model. 2006;190(3-4):231–59.

Phillips SJ, Dudík M. Modeling of species distributions with Maxent: new extensions and a comprehensive evaluation. Ecography. 2008;31:161–75.

Platt A, Horton M, Huang YS, Li Y, Anastasio AE, Mulyati NW, et al. The scale of population structure in *Arabidopsis thaliana*. PLoS Genet. 2010;6(2):e1000843.

Revedin A, Aranguren B, Becattini R, Longo L, Marconi E, Lippi MM, et al. Thirty thousand-year-old evidence of plant food processing. Proc Natl Acad Sci U S A. 2010;107(44):18815–9.

Robertson JH. Chromosome numbers in *Brachypodium* Beauv. (Gramineae). Genetica. 1981;56:55–60.

Schwartz CJ, Doyle MR, Manzaneda AJ, Rey PJ, Mitchell-Olds T, Amasino RM. Natural variation of flowering time and vernalization responsiveness in *Brachypodium distachyon*. Bioenerg Res. 2010;3:38–46.

Storz JF. Using genome scans of DNA polymorphism to infer adaptive population divergence. Mol Ecol. 2005;14:671–88.

The International Brachypodium Initiative. Genome sequencing and analysis of the model grass *Brachypodium distachyon*. Nature. 2009;463:763–8.

Tyler L, Fangel JU, Fagerstrom AD, Steinwand MA, Raab TK, Willats WGT, et al. Selection and phenotypic characterisation of a core collection of *Brachypodium distachyon* inbred lines. BMC Plant Biol. 2014;14:25.

Tyler L, Lee SJ, Young ND, DeIulio GA, Benavente E, Reagon M, et al. Building tools for genome-wide association studies: flowering time helps explain population structure in Brachypodium. In: The 3rd international Brachypodium conference, University of Massachusetts, Amherst, 16–19 June 2015.

Vogel JP, Garvin DF, Leong OM, Hayden DM. *Agrobacterium*-mediated transformation and inbred line development in the model grass *Brachypodium distachyon*. Plant Cell Tiss Org Cult. 2006a;84:199–211.

Vogel JP, Gu YQ, Twigg P, Lazo GR, Laudencia-Chingcuanco D, Hayden DM, et al. EST sequencing and phylogenetic analysis of the model grass *Brachypodium distachyon*. Theor Appl Genet. 2006b;113:186–95.

Vogel JP, Tuna M, Budak H, Huo N, Gu YQ, Steinwand MA. Development of SSR markers and analysis of diversity in Turkish populations of *Brachypodium distachyon*. BMC Plant Biol. 2009;9:88.

Wang K, Han X, Dong K, Gao L, Li H, Ma H, et al. Characterisation of seed proteome in Brachypodium distachyon. J Cereal Sci. 2010;52:177–86.

Wen W, Li D, Li W, Gao Y, Li W, Li H, et al. Metabolome-based genome-wide association study of maize kernel leads to novel biochemical insights. Nat Commun. 2014;5(3438):1–10.

WorldClim [Internet]. WorldClim—Global Climate Data, free climate data for ecology. [Cited 5 Jan 2014]. http://www.worldclim.org/.

The microRNAs of *Brachypodium*

Karl R. Franke and Pamela J. Green

Abstract Since the proposal of *Brachypodium distachyon* as a model organism, a substantial amount of research has been done investigating the miRNAs which aid in the post-transcriptional gene regulation of this grass. Techniques such as the sequencing of small RNA libraries have allowed for the identification and annotation of over 500 miRNAs present in the *Brachypodium* genome. Furthermore, experimental evidence for the cleavage of a large number of target mRNAs has also been found, including a few instances of differential target cleavage by distinct miRNA family members. This research has not only provided the community with a wealth of publicly available sequencing data, but it has also led to many insights regarding how these small 21–24 nt RNA molecules may play critical roles in different tissues, stress responses, and developmental processes of *Brachypodium*.

Keywords *Brachypodium* • Grass • miRNA • miRNA variants • Small RNA • PARE • miRNA target cleavage • Post-transcriptional gene regulation • Pooideae

Abbreviations

AGO1	Agonaute
ESTs	Expressed Sequence Tags
GEO	Gene Expression Omnibus
GSSs	Genome Survey Sequences
PARE	Parallel Analysis of RNA Ends
RISC	RNA-induced silencing complex

K.R. Franke
Delaware Biotechnology Institute, University of Delaware, Newark, DE, USA

Department of Biological Sciences, University of Delaware, Newark, DE, USA

P.J. Green (✉)
Delaware Biotechnology Institute, University of Delaware, Newark, DE, USA

Department of Plant and Soil Sciences, University of Delaware, Newark, DE, USA
e-mail: green@dbi.udel.edu

J. Vogel (ed.), *Genetics and Genomics of Brachypodium*,
Plant Genetics and Genomics: Crops Models, DOI 10.1007/7397_2015_9
Published online: 06 October 2015

Introduction

A new mechanism of post-transcriptional gene regulation was brought to light in 1993 when microRNAs (miRNAs) were discovered (Lee et al. 1993; Wightman et al. 1993). miRNA genes are transcribed by RNA polymerase II into single-stranded precursors forming stem-loop secondary structures. In plants, these are processed in the nucleus by DCL1 to generate a miRNA-miRNA* duplex (Kurihara & Watanabe 2004). These small 21–24 nt RNA molecules primarily help direct plant RNA decay by association with Agonaute (AGO1)/RNA-induced silencing complex (RISC) and guiding the endoribonucleolytic cleavage of target mRNAs by means of complementary base pairing (Bohmert et al. 1998; Llave et al. 2002). miRNAs can also impact gene expression through means other than decay such as translational repression of the target transcript, the main mechanism in animal systems, but also evident in plants (Brodersen et al. 2008). Here, we cover the advances made in the discovery of miRNAs encoded by the *Brachypodium* genome, the prediction and validation of the targets of those miRNAs, and what insights have been uncovered regarding the roles miRNAs play in the regulation of *Brachypodium* development and stress responses.

Brachypodium miRNA Discovery: A Historical Perspective

The complexity of plant small RNA populations can make accurate identification and annotation of miRNAs a challenging process. The community has established a set of criteria (Meyers et al. 2008) to avoid misannotation of siRNAs as miRNAs whereby the most critical factor is the precise excision of the miRNA-miRNA* duplex from the stem of a single stranded stem-loop precursor. Therefore, many studies with a focus on large scale miRNA discovery employ bioinformatic pipelines which include miRNA precursor identification based on stem-loop structure prediction. The use of a set of miRNAs from other species as a basis for searching for potential precursors allows for the identification of conserved miRNAs; alternatively, next generation sequencing of small RNA libraries can also be leveraged to search for novel miRNAs as well. It is important to keep in mind that differences in the structure of these pipelines and/or variations in cutoff stringency can lead to different results. With that said, one trend is clear: As *Brachypodium* research has progressed, these analyses have yielded greater results as more resources have become available, such as the reference genome, and through the use of more advanced techniques, specifically deep sequencing of small RNA libraries. Many of the small RNA libraries sequenced during these studies have been made public (Table 1) and provide a powerful resource to researchers studying miRNAs in *Brachypodium*.

One of the first studies investigating miRNAs in *Brachypodium* was done by Unver and Budak in 2009 (Unver & Budak 2009). The authors utilized a homology

Table 1 Publicly available *Brachypodium* small RNA (smRNA) and PARE libraries

GEO accession	Name/description	Type	References
GSM406302	Young spike	smRNA	Wei et al. (Wei et al. 2009), Baev et al. (Baev et al. 2011), Lucas et al. (Lucas et al. 2014)
GSM406303	Root, shoot, leaf	smRNA	Wei et al. (Wei et al. 2009), Baev et al. (Baev et al. 2011), Lucas et al. (Lucas et al. 2014)
GSM506621	BDI05 panicle 1	smRNA	Vogel et al. (Vogel et al. 2010), Baev et al. (Baev et al. 2011), Jeong et al. (Jeong et al. 2013), Lucas et al. (Lucas et al. 2014)
GSM506620	OBD01 panicle 2	smRNA	Vogel et al. (Vogel et al. 2010), Baev et al. (Baev et al. 2011), Jeong et al. (Jeong et al. 2013), Lucas et al. (Lucas et al. 2014)
GSM1266839	BDI08 root	smRNA	Jeong et al. (Jeong et al. 2013)
GSM1266840	OBD03 seedling	smRNA	Jeong et al. (Jeong et al. 2013)
GSM1266841	BDI04 leaf 1	smRNA	Jeong et al. (Jeong et al. 2013)
GSM1266842	BDI09 leaf 2	smRNA	Jeong et al. (Jeong et al. 2013)
GSM1266843	BDI06 stem	smRNA	Jeong et al. (Jeong et al. 2013)
GSM1266844	OBD02 leaf and stem	smRNA	Jeong et al. (Jeong et al. 2013)
GSM1266845	BDI02 shoot control for stress	smRNA	Jeong et al. (Jeong et al. 2013)
GSM1266846	BDI03 drought-stressed shoot	smRNA	Jeong et al. (Jeong et al. 2013)
GSM1266847	BDI17 salt-stressed shoot	smRNA	Jeong et al. (Jeong et al. 2013)
GSM1266848	BDI01 cold-stressed shoot	smRNA	Jeong et al. (Jeong et al. 2013)
GSM1266849	BDI18 heat-stressed shoot	smRNA	Jeong et al. (Jeong et al. 2013)
GSM1266850	BDI19 submergence-stressed shoot	smRNA	Jeong et al. (Jeong et al. 2013)
GSM1266851	BDI15 shoot control for low phosphate	smRNA	Jeong et al. (Jeong et al. 2013)
GSM1266852	BDI16 phosphate-starved shoot	smRNA	Jeong et al. (Jeong et al. 2013)
GSM1266853	OBD04 pooled stresses	smRNA	Jeong et al. (Jeong et al. 2013)
GSM1331683	Bd21 control	smRNA	Wang et al. (2015)
GSM1331684	Salt 48 h	smRNA	Wang et al. (2015)
GSM1331685	Cold 6 h	smRNA	Wang et al. (2015)
GSM1331686	Cold 24 h	smRNA	Wang et al. (2015)
GSM1331687	Heat air 1 h	smRNA	Wang et al. (2015)
GSM1331688	Heat air 3 h	smRNA	Wang et al. (2015)
GSM1331689	Heat water 1 h	smRNA	Wang et al. (2015)

(continued)

Table 1 (continued)

GEO accession	Name/description	Type	References
GSM1331690	Heat water 3 h	smRNA	Wang et al. (2015)
GSM1266854	BDI20 root	PARE	Jeong et al. (Jeong et al. 2013)
GSM1266855	BDI21 leaf	PARE	Jeong et al. (Jeong et al. 2013)
GSM1266856	BDI23 stem	PARE	Jeong et al. (Jeong et al. 2013)
GSM1266857	BDI25 panicle	PARE	Jeong et al. (Jeong et al. 2013)
GSM1173198	DW—control	PARE	Zhang et al. (Zhang et al. 2013)
GSM1173199	DC—cold	PARE	Zhang et al. (Zhang et al. 2013)
GSM1173200	DW—control replicate	PARE	Zhang et al. (Zhang et al. 2013)
GSM1173201	DC—cold replicate	PARE	Zhang et al. (Zhang et al. 2013)

These libraries have been made available to the community for further analysis and can be downloaded from GEO (http://www.ncbi.nlm.nih.gov/geo/)

based approach and searched for sequences within the *Brachypodium* Expressed Sequence Tags (ESTs) and Genome Survey Sequences (GSSs) available at the time which matched known plant miRNAs. After performing stem-loop structure prediction, they predicted 26 miRNA and pre-miRNAs belonging to 19 families, 6 of which they validated successfully using qRT-PCR. Due to the method chosen by the authors, their predictions were limited to only conserved miRNA sequences.

Another study published that year by Wei et al. took advantage of small RNA sequencing with Solexa technology which allowed for the identification of conserved as well as *Brachypodium* specific miRNAs (Wei et al. 2009). Two small RNA libraries were generated from BD21-3 plants, one made from young spikes and the other from root, stem, and leaf tissue. The authors sequenced the libraries to a depth of over 5 million reads and identified 94 conserved miRNAs from 28 families by blasting to known plant miRNA sequences from miRBase. After mapping the small RNA sequences to BAC ends and ESTs they utilized stem-loop structure prediction to identify 12 putative novel miRNAs; however, only 15 of the 94 conserved miRNAs had identifiable precursors. Three of the novel miRNAs were validated successfully using both Northern blotting for precursor detection and qRT-PCR for small RNA detection. The discovery of these novel miRNAs may not have been possible without the use of small RNA sequencing and these libraries were the first to be made publicly available to the community.

The availability of the 4X draft genome would shortly allow for the identification of many more precursors as seen in the study by Zhang et al. in late 2009 (Zhang et al. 2009). Similar methods were used in this study (small RNA sequencing, bioinformatic precursor prediction); however, searching the genome instead of ESTs and BACs for potential precursors led to the prediction of 129 novel miRNAs with matching precursors in addition to 27 conserved miRNAs from 21 families.

Since the 8X genome was released in 2010 (Vogel et al. 2010), more studies have been published which drastically increased our knowledge of *Brachypodium* miRNAs. The first, conducted by researchers Baev et al., (Baev et al. 2011) used a bioinformatic pipeline to analyze four publicly available small RNA libraries (Wei et al. 2009; Vogel et al. 2010) to identify 56 conserved miRNAs and miRNA precursors from 21 families and 46 novel miRNAs. These results included 11 conserved miRNAs and 21 precursors which had not previously been identified. Eight more precursors were later identified from a reanalysis of these four small RNA libraries (Lucas et al. 2014), illustrating how new results can be obtained with a different analysis and the importance of making these datasets available to the community.

Two recent studies created *Brachypodium* small RNA datasets which were much larger and more diverse than any previously discussed. In 2013 Bertolini et al. published a study in which eight small RNA libraries were generated from proliferating and expanding leaf zones under both control and drought stress conditions (Bertolini et al. 2013). The authors' subsequent analysis of these libraries increased the number of known conserved miRNA genes in *Brachypodium* by 28 and the number of *Brachypodium* specific miRNA genes by 94. Later that year the amount of small RNA library data available to the community more than tripled when researchers Jeong et al. published the sequencing and analysis of 17 small RNA libraries representing a number of different Bd21 tissues and abiotic stress conditions (Jeong et al. 2013). The authors showed evidence for 81 new miRNA precursors which included the confirmation of 16 precursors reported in the previous study (Bertolini et al. 2013).

Annotated miRNA Families in miRBase: Current Status

In total, the *Brachypodium* community has annotated 317 miRNA precursors and 525 mature miRNAs from 182 families in miRBase 21 (full table available at http://brachymirnas.appspot.com) which even surpasses Arabidopsis (Kozomara & Griffiths-Jones 2014). Of the monocots included in miRBase, *Brachypodium* is only eclipsed by rice with regards to number of annotations. The vast majority of *Brachypodium* miRNA sequences have experimental evidence for expression as 475 of them are present in at least one publicly available *Brachypodium* library (data not shown). Many of the miRNA families present in *Brachypodium* give credence to the use of this model plant studying the regulatory mechanisms of miRNAs in monocots, but also miRNAs in general. All of the highly conserved miRNA families (Table 2) found to be expressed in angiosperms (Axtell & Bowman 2008) such as miR156, miR159/319, miR396, and miR168, as well as few monocot specific families miR444, miR528, miR1432, and miR2275 are found in *Brachypodium*. miRNAs thus far unique to *Brachypodium* are also present, such as miR5163b-3p which has been shown to initiate the production of phased siRNAs from its target, an NB-ARC domain-containing R protein (Jeong et al. 2013). The miR444 family possesses some rather distinctive features; not only is the miR444 pre-miRNA overlapping and antisense to the mature miRNA target, but it also

Table 2 Conserved miRNA families in *Brachypodium*

Family/ species miRNA	Rice	Wheat	Barley	*Sorghum*	Maize	*Arabidopsis*	*Medicago*	Soybean
miR156	*	*	*	*	*	*	*	*
miR529	*			*	*			
miR159	*	*	*	*	*	*	*	*
miR319	*	*		*	*	*	*	*
miR160	*	*		*	*	*	*	*
miR162	*			*	*	*	*	*
miR164	*	*		*	*	*	*	*
miR166	*		*	*	*	*	*	*
miR167	*	*		*	*	*	*	*
miR168	*		*	*	*	*	*	*
miR169	*	*	*	*	*	*	*	*
miR171	*	*	*	*	*	*	*	*
miR172	*			*	*	*	*	*
miR390[a]	*			*	*	*	*	*
miR393	*			*	*	*	*	*
miR394	*			*	*	*		*
miR395	*	*		*	*	*	*	*
miR396	*	*		*	*	*	*	*
miR397	*	*	*	*	*	*	*	*
miR398	*	*		*	*	*	*	*
miR399	*	*	*	*	*	*	*	*
miR408	*	*		*	*	*	*	*
miR437	*			*				
miR444	*	*	*		*			
miR528	*			*	*			
miR530	*	*					*	*
miR531	*	*						
miR827	*				*	*		
miR845						*		
miR1122		*						
miR1127		*						
miR1135		*						
miR1139		*						
miR1432	*			*	*			
miR1878	*							
miR2118[a]	*			*	*		*	*
miR2275[a]	*	*			*			
miR5049		*	*					
miR5062		*						
miR5175		*						
miR5179	*							

(continued)

Table 2 (continued)

Family/ species miRNA	Rice	Wheat	Barley	*Sorghum*	Maize	*Arabidopsis*	*Medicago*	Soybean
miR5200[b]		*						
miR5281							*	
miR7757		*						

An asterisk indicates the miRNA family is conserved between Brachypodium and the specified species according to the miRBase 21 annotations. Underlined families are monocot specific. **Bolded** families are Pooideae specific

[a]Capable of initiating phased loci

[b]Evidence for miR5200 in barley has been shown via RNA gel blot (Wu et al. 2013) although it is not yet annotated. miR5200 has also been found and annotated in *Aegilops tauschii* (Jia et al. 2013), a member of the Pooideae not shown in this table

contains introns which must be spliced before the pre-miRNA stem loop can form (Lu et al. 2008b). The miR390, miR2118, and miR2275 families are known to initiate the production of phased siRNAs from their target transcripts in other plants (Axtell et al. 2006; Johnson et al. 2009) as well as *Brachypodium* (Jeong et al. 2013). Additionally the *Pooideae* specific miRNA miR5200 is also present, which has been shown to play a role in flowering time (Wu et al. 2013) and the submergence stress response (Jeong et al. 2013).

It is important to note that while a miR162 precursor is present in the genome, multiple findings of very low expression make further study necessary. Not only is this miRNA widely conserved across flowering plants (Axtell & Bowman 2008; Xie et al. 2003), but it also plays a very important role in the regulation of miRNA biogenesis as its target is *DCL1*. Given this and the known regulation of AGO1 via miR168 (Vaucheret et al. 2006), it appears that some miRNAs play a critical role in regulating their own abundances and the miRNA machinery post-transcriptionally. The analysis of expression data from small RNA libraries of two separate studies (Bertolini et al. 2013; Jeong et al. 2013), however, showed only five reads present in over 20 small RNA libraries which could be attributed to miR162. Furthermore, no decay intermediates were found which matched the predicted *DCL1* target site in Parallel Analysis of RNA Ends (PARE) data and a change in one nucleotide in the precursor leading to decreased stability between the miRNA:miRNA* pairing and impaired *DCL1* processing was proposed as a possible explanation (Jeong et al. 2013). Considering similar low levels of miR162 expression were observed in barley, sorghum, and wheat (Wei et al. 2009; Schreiber et al. 2011; Zhang et al. 2011; Yao et al. 2007) and the potential importance of this negative feedback loop (Jeong et al. 2013), additional research is necessary to better understand the differences in *DCL1* regulation present in these monocots.

Confirmation of miRNA Guided Target Cleavage

Knowing which genes a miRNA targets greatly aids in developing an understanding of the role the miRNA plays in the plant. Many computational tools are available for target prediction which search for regions within transcripts with levels of complementarity to a miRNA based on known criteria (Rhoades et al. 2002; Wang et al. 2004). A number of papers have been published which include numerous predicted targets of *Brachypodium* miRNAs (Unver & Budak 2009; Wei et al. 2009; Zhang et al. 2009; Baev et al. 2011; Lucas et al. 2014; Bertolini et al. 2013; Jeong et al. 2013; Budak & Akpinar 2011; Zhang et al. 2013). Experimental validation of predicted targets on small scale is often accomplished via RNA ligase-mediated 5′ rapid amplification of cDNA ends (RLM 5′ RACE) to clone the decay intermediate resulting from miRNA mediated cleavage. This method has been used to validate a handful of targets in *Brachypodium* (Wei et al. 2009; Budak & Akpinar 2011) including *ARF6*, a target of miR167 which is highly conserved across many plant species.

Experimental validation of predicted targets on a genomic scale became possible with the advent the library construction technique known as PARE (German et al. 2008; German et al. 2009) involving a modified 5′-RACE before high throughput sequencing. The resulting PARE data contain the 3′ decay intermediates of mRNAs undergoing decay through deadenylation independent pathways such as miRNA guided cleavage; computational analysis can be performed to match these intermediates with the corresponding predicted targets. This method was first utilized by the *Brachypodium* community in 2013 when the sequencing of four PARE libraries made from root, leaf, stem, and panicle tissues provided experimental evidence for the miRNA-mediated cleavage of 264 predicted target sites (Jeong et al. 2013). Another study investigating the cold stress of *Brachypodium* published an analysis of four PARE libraries made which showed evidence for the cleavage of 46 miRNA targets (Zhang et al. 2013). All eight of these PARE libraries have been uploaded to the publicly available database, Gene Expression Omnibus (GEO; Table 1) and, alongside the small RNA libraries produced in previously discussed studies, represent a wealth of information available to the community. The rich set of annotated miRNAs, numerous miRNA targets with experimental evidence of miRNA guided cleavage, miRNA expression levels in various tissues and stress conditions, sequence variation data, and abundances of miRNA target decay intermediates resulting from these studies have given and will continue to give deeper insights into the important gene regulatory roles of miRNAs in *Brachypodium* and related plant species.

Insights into miRNA Function in Tissues and During Development

miRNAs play a pivotal role in many aspects of plant development such as the formation of root, stem, leaf, and floral organs, control of cell division, regulation of hormone responses, and patterning. (Achard et al. 2004; Mallory et al. 2004; Meng et al. 2010; Palatnik et al. 2003; Vaucheret et al. 2004; Jones-Rhoades et al. 2006) This is often made possible by means of preferential expression of different miRNAs within various tissues and organs during the development process. Insights into the expression patterns of *Brachypodium* miRNAs began with one of the earliest studies by Wei et al. (Wei et al. 2009) when one small RNA library was constructed from panicles and the other from the remaining plant tissue. Of the miRNAs identified in this study, 16 of them showed a twofold or greater difference in expression between the libraries including Bdi-miR172 which was found in greater abundances in the panicle library and Bdi-miR319 which was observed to have greater expression in the library made from vegetative tissues. The targets of miR172 include transcription factors such as AP2 which have been shown to play a role in flowering time and floral organ development in Arabidopsis (Allen et al. 2005; Aukerman & Sakai 2003; Chen 2004). miR319 has also been shown to play a role in Arabidopsis floral organ development (Nag et al. 2009) as well as the development of leaf margins which takes place during vegetative growth in tomato (Ori et al. 2007). Since then the amount of miRNA expression data available has increased significantly and the developmental roles of other miRNAs have been explored.

miRNA Family Members Exhibiting Differential Expression and Target Cleavage

Differences in expression and sequence of members of the same miRNA family can allow for a higher level of specification in subsequent gene regulation. This can be true for miRNAs regulating different targets or in some cases the same target under different circumstances. An example of the first scenario was reported in Arabidopsis (Palatnik et al. 2007) where mature miRNA sequences originating from the MIR159 loci guide cleavage of *MYB* transcripts whereas those miRNAs from MIR319 loci guide the cleavage of transcripts coding for members of the *TCP* transcription factor family. The *MYB* transcripts are not regulated via miR319 due to the expression pattern of the miRNA, and the *TCP* transcripts are not regulated by miR159 due to a slight variation in sequence. The second scenario can be seen in rice (Jeong et al. 2011) with members of the miR171 family regulating multiple *GRAS* transcripts. It was shown that miRNA guided cleavage of *GRAS13* takes place via miR171g.2 and miR171h.2 in root tissue at one target site but by miR171i in panicle at another target site. Another target, *GRAS8*, is regulated not only by the previously mentioned miR171 family members but also by miR171b-f in both root and panicle at a third target site.

In *Brachypodium*, two examples of the second scenario have been reported (Jeong et al. 2013). The first involves multiple members of the miR166 family which were predicted to target an HD-ZIP III transcription factor at two distinct sites. One of these sites was unique to Bdi-miR166f. The authors observed not only varying tissue preferential expression patterns among the miRNA family members, but the unique Bdi-miR166f target site allowed for a correlation to be seen between increased Bdi-miR166f expression in root tissue and increased abundance of PARE sequences mapping to the predicted target site in libraries made from root tissue as compared to panicle. It has been proposed that miRNA guided regulation of this family of transcription factors may take place through spatial and temporal mechanisms which is supported by tissue preferential expression of miR166 family members observed in Arabidopsis (Jung & Park 2007). These results indicate a similar mechanism may exist in *Brachypodium* by demonstrating not only differential expression of the miRNAs but also differential target cleavage. Given that members of the Class III HD-ZIP transcription factor family have been shown to regulate meristem initiation, vascular development, and leaf development in Arabidopsis (Prigge et al. 2005), it is likely that future research will uncover unique roles for the miR166 family during *Brachypodium* development.

The second example reported by Jeong et al. (Jeong et al. 2013) investigates the targeting of *BdiSPL14L* by miR156 and miR529. Decreased expression of miR156 has been observed in panicle compared to other tissues in two different studies (Wei et al. 2009; Jeong et al. 2013) whereas the inverse was reported for miR529 (Jeong et al. 2013). In rice it has been proposed that the miRNA guided regulation of *OsSPL14* is controlled predominantly by miR156 during juvenile development and by miR529 during reproductive development and flowering; these relationships have been shown to play a role in tillering and panicle-branching where a single nucleotide polymorphism in *OsSPL14* resulted in the disruption of the miRNA target site, increased panicle branching, fewer tillers, and higher yields (Jeong et al. 2011; Jiao et al. 2010; Miura et al. 2010). Further analysis of PARE data by Jeong et al. (Jeong et al. 2013) showed a higher abundance of sequences mapping to the predicted miR156 target site in *BdiSPL14L* as compared to the miR529 target site in libraries made from leaf tissue with the opposite being true for libraries made from panicle. These results indicate *Brachypodium* shares an interesting miRNA guided mechanism with rice where differential expression of two members of the same family, miR156 and miR529, along with slight differences in sequence result in the majority of juvenile SPL14 regulation taking place via miR156 at one target site and reproductive SPL14 regulation via miR529 at another target site (Jeong et al. 2011). This regulatory mechanism which utilizes multiple family members for distinct roles is not only of interest to the miRNA community but it also has implications for biofuel crop development. In contrast to rice, increased tillering is more advantageous for these crops where total biomass production is of primary concern. However, while overexpression of miR156 has been shown to cause the development of more tillers in *Brachypodium*, rice, switchgrass and maize, it can also lead to other morphological phenotypes including dwarfism (Xie et al. 2006; Chuck et al. 2007; Chuck et al. 2011; Fu et al. 2012). Overexpression of miR529, on the other hand, has not been investigated and could lead to a similar outcome with fewer negative side effects.

Regulation of Flowering Time by Pooideae *Specific miR5200*

The regulatory mechanisms involved in the initiation of flowering are complex, rely on multiple signals including various hormones, day length, temperature, and have been explored much more in depth in Arabidopsis and other eudicots as compared to monocots (Amasino 2010). Interpretation of these signals results in the activation of FLOWERING LOCUS T (FT) and its orthologs. This is followed by production of florigen in the leaves which translocates to the shoot apex and initiates the flowering process (Bäurle & Dean 2006). A recent study by Wu et al. (Wu et al. 2013) investigated the role of Bdi-miR5200, a miRNA only shown to be conserved in *Pooideae*, in this process. The authors identified two predicted targets of Bdi-miR5200, (BdiFTL1 and BdiFTL2) verified miRNA guided cleavage using 5′ RACE, and found that miR5200 localizes to leaf tissue where the repression of FT mRNA was observed. The expression of Bdi-miR5200 was found to be photoperiod dependent with increases in miRNA abundance during short-day conditions as compared to long-day. In contrast, vernalization did not affect miR5200 expression. Overexpression of miR5200 resulted in late flowering under long day conditions whereas transgenic miR5200 target mimic plants flowered earlier under short-day conditions indicating repression of FTL1 and FTL2 via miR5200 takes places during short-day conditions to prevent flowering. The authors determined one aspect of the regulation of miR5200 expression involves the repressive H3K27me3 histone modification during long-day conditions to allow for flowering initiation.

These findings mark a significant advancement in the understanding of monocot flowering initiation, especially in *Pooideae*. The limited conservation observed for miR5200 gives evidence to this mechanism being more recently evolved. These findings are likely conserved in wheat (Table 2), and have implications for the development of other crops as well with a new potential transgenic strategy for altering flowering time.

Brachypodium miRNAs Regulate the Stress Responses

The roles of miRNAs in plants are not limited to development; these small RNA molecules have also been implicated in the sophisticated regulatory networks plants exhibit during their adaptation to environmental stresses. These mechanisms are of particular interest when it comes to the development of transgenic crops which can be grown under harsher conditions.

Cold Stress

A study by Zhang et al. (Zhang et al. 2009) found miR397, miR169e, and miR172ab to be upregulated in cold stressed *Brachypodium* plants based on small

RNA library analysis. These three miRNAs are not only conserved across many plants species (Axtell & Bowman 2008) but their association with cold stress is conserved as well. All three have been found to exhibit changes in expression in Arabidopsis plants under cold stress (Sunkar & Zhu 2004; Zhou et al. 2008; Liu et al. 2008) while miR169 and miR397 were also shown to be cold-associated in *Populus* (Lu et al. 2008a). A later study published by the same group (Zhang et al. 2013) utilized PARE libraries to show evidence for the cleavage of the predicted targets of these miRNAs. While these miRNA had previously been shown to be cold associated in other dicots, the results of this study provide evidence for their involvement in the cold stress response of monocots.

Drought Stress

Two studies have investigated the roles of *Brachypodium* miRNAs in the drought stress response. The first study reported a miRNA microarray analysis which indicated seven miRNAs showed differential regulation under drought conditions (Budak & Akpinar 2011). The expression levels of five miRNAs were upregulated: miR1881, miR1450, and miR171c-3p showed increases in RNA samples made from root tissue, miR390 was upregulated in leaf and root samples, and miR1850 was upregulated in leaf samples. Two miRNAs observed as downregulated under drought conditions, miR528 and miR406 in leaf and root samples respectively. It is important to note that miR1850, miR1450, miR1881 and miR406 have not been annotated in *Brachypodium*, but given the nature of microarray experiments it is possible another miRNAs or siRNAs with sequence similarity to these probes was the source of the signals.

The second study investigated the differences in miRNA expression in proliferating versus expanding leaf cells under drought conditions using small RNA libraries (Bertolini et al. 2013). Under drought stress they found that miR319b showed increased expression in proliferating cells whereas miR164a, miR164c, miR396b, miR396e, miR156, and miR528 were upregulated in expanding cells. Some of these findings are consistent with the known functions of theses miRNAs and targets in other plants. miR319 is known to target *TCP* genes which regulate cell proliferation (Palatnik et al. 2003) and has been shown to play a role in leaf development in Arabidopsis along with miR164 and miR396 (Rodriguez et al. 2010). The increase in miR528 expression seen here, however, conflicts with the previous study (Budak & Akpinar 2011) where decreased levels of the miRNA were observed which could indicate a temporal mechanism of expression causing different results in stress treatments of varying design.

Phosphate Starvation

Analysis of small RNA libraries by Jeong et al (Jeong et al. 2013) made from phosphate starved and phosphate replete *Brachypodium* plants showed increases in both miR399 and miR827 under stress conditions and these results were further confirmed using by splint ligation-mediated miRNA detection. The role of these miRNAs in response to phosphate starvation is well conserved in multiple plants including Arabidopsis (Fujii et al. 2005), rice (Bari et al. 2006), and barley (Hackenberg et al. 2013) where increased expression of miR399 leads to decreased expression of its target *PHO2*, a negative regulator of phosphate uptake; regulation via miR827 is more complex as this miRNA targets *SPX-MSF1* under starvation conditions whereas normal conditions lead to *SPX-MSF2* targeting (Hackenberg et al. 2013).

Submergence

Interestingly the *Pooideae* specific miR5200 discussed previously was found to play a role in the *Brachypodium* submergence response (Jeong et al. 2013). Data from small RNA libraries showed increased miR5200 expression under stress conditions and which was confirmed by splint ligation-mediated miRNA detection. Timecourse experiments revealed a coherent relationship between miR5200 expression and the expression of its target *FTL1* where the initiation of miR5200 after the induction of stress was followed by a decrease in target mRNA levels. RNA samples from plants subsequently removed from stress conditions showed decreases in miR5200 during the recovery phase and increases in *FTL1* mRNA. These results are in line with the previously discussed study (Wu et al. 2013) which show that flowering time in *Brachypodium* is regulated by miR5200; additionally the repression of flowering time under submergence conditions seen here has also been proposed in rice and Arabidopsis (Peña-Castro et al. 2011).

Summary and Conclusions

Considering that *Brachypodium* is a relatively new model organism, the advances which have been made in the research of miRNAs are quite significant. Over 500 miRNAs have been annotated including members of all of the highly conserved plant miRNA families. Researchers wishing to study miRNAs in *Brachypodium* are well equipped with 35 publicly available small RNA and PARE libraries representing a number of different stress conditions and tissue samples in addition to a list of over 260 target mRNAs which have experimental evidence for miRNA-guided cleavage. Multiple mechanisms of miRNA regulation of *Brachypodium* development have been revealed, some of which are conserved with other plant species such as the targeting of *SPL14* by miR156/miR529 while others are more specific such as the regulation of flowering time via miR5200 which has only been observed in the *Pooideae*. The stress responses of *Brachypodium* have shown to be of particular interest to researchers in that miRNAs have already been associated with cold, drought, phosphate starvation, and submergence stresses. The impact of these findings could extend to the development of future crops capable of withstanding harsher environments. Outside of the stress responses, mechanisms such as those seen in the regulation of *SPL14*, have implications in grain and biomass yields. As we gain a deeper understanding of these regulatory networks through further functional analysis of miRNAs in *Brachypodium*, and begin to leverage more advance techniques, such as artificial miRNAs and target mimics, we will realize how the study of small RNA molecules in a small grass leads to big implications for larger crops used throughout the world.

Acknowledgements We thank Dr. Dong-Hoon Jeong, Hallym University, for helpful comments on the manuscript. *Brachypodium* research in our lab was supported by the Office of Biological and Environmental Research of the Department of Energy under Plant Feedstock Genomics for Bioenergy award number DE-FG02-07ER64450 to P.J.G. K.R.F. was supported, in part, by NIH/NIGMS CBI Training Grant 5T32GM008550-21.

References

Achard P, Herr A, Baulcombe DC, Harberd NP. Modulation of floral development by a gibberellin-regulated microRNA. Development. 2004;131(14):3357–65.

Allen E, Xie Z, Gustafson AM, Carrington JC. microRNA-directed phasing during trans-acting siRNA biogenesis in plants. Cell. 2005;121(2):207–21.

Amasino R. Seasonal and developmental timing of flowering. Plant J. 2010;61(6):1001–13.

Aukerman MJ, Sakai H. Regulation of flowering time and floral organ identity by a MicroRNA and its APETALA2-like target genes. Plant Cell. 2003;15(11):2730–41.

Axtell MJ, Bowman JL. Evolution of plant microRNAs and their targets. Trends Plant Sci. 2008;13(7):343–9.

Axtell MJ, Jan C, Rajagopalan R, Bartel DP. A two-hit trigger for siRNA biogenesis in plants. Cell. 2006;127(3):565–77.

Baev V, Milev I, Naydenov M, Apostolova E, Minkov G, Minkov I, et al. Implementation of a de novo genome-wide computational approach for updating Brachypodium miRNAs. Genomics. 2011;97(5):282–93.

Bari R, Datt Pant B, Stitt M, Scheible W-R. PHO2, microRNA399, and PHR1 define a phosphate-signaling pathway in plants. Plant Physiol. 2006;141(3):988–99.

Bäurle I, Dean C. The timing of developmental transitions in plants. Cell. 2006;125(4):655–64.

Bertolini E, Verelst W, Horner DS, Gianfranceschi L, Piccolo V, Inzé D, et al. Addressing the role of microRNAs in reprogramming leaf growth during drought stress in Brachypodium distachyon. Mol Plant. 2013;6(2):423–43.

Bohmert K, Camus I, Bellini C, Bouchez D, Caboche M, Benning C. AGO1 defines a novel locus of Arabidopsis controlling leaf development. EMBO J. 1998;17(1):170–80.

Brodersen P, Sakvarelidze-Achard L, Bruun-Rasmussen M, Dunoyer P, Yamamoto YY, Sieburth L, et al. Widespread translational inhibition by plant miRNAs and siRNAs. Science. 2008;320(5880):1185–90.

Budak H, Akpinar A. Dehydration stress-responsive miRNA in Brachypodium distachyon: evident by genome-wide screening of microRNAs expression. OMICS. 2011;15(11):791–9.

Chen X. A microRNA as a translational repressor of APETALA2 in Arabidopsis flower development. Science. 2004;303(5666):2022–5.

Chuck G, Cigan AM, Saeteurn K, Hake S. The heterochronic maize mutant Corngrass1 results from overexpression of a tandem microRNA. Nat Genet. 2007;39(4):544–9.

Chuck GS, Tobias C, Sun L, Kraemer F, Li C, Dibble D, et al. Overexpression of the maize Corngrass1 microRNA prevents flowering, improves digestibility, and increases starch content of switchgrass. Proc Natl Acad Sci U S A. 2011;108(42):17550–5.

Fu C, Sunkar R, Zhou C, Shen H, Zhang J-Y, Matts J, et al. Overexpression of miR156 in switchgrass (Panicum virgatum L.) results in various morphological alterations and leads to improved biomass production. Plant Biotechnol J. 2012;10(4):443–52.

Fujii H, Chiou T-J, Lin S-I, Aung K, Zhu J-K. A miRNA involved in phosphate-starvation response in Arabidopsis. Curr Biol. 2005;15(22):2038–43.

German MA, Pillay M, Jeong D-H, Hetawal A, Luo S, Janardhanan P, et al. Global identification of microRNA-target RNA pairs by parallel analysis of RNA ends. Nat Biotechnol. 2008;26 (8):941–6.

German MA, Luo S, Schroth G, Meyers BC, Green PJ. Construction of Parallel Analysis of RNA Ends (PARE) libraries for the study of cleaved miRNA targets and the RNA degradome. Nat Protoc. 2009;4(3):356–62.

Hackenberg M, Shi B-J, Gustafson P, Langridge P. Characterization of phosphorus-regulated miR399 and miR827 and their isomirs in barley under phosphorus-sufficient and phosphorus-deficient conditions. BMC Plant Biol. 2013;13:214.

Jeong D-H, Park S, Zhai J, Gurazada SGR, De Paoli E, Meyers BC, et al. Massive analysis of rice small RNAs: mechanistic implications of regulated microRNAs and variants for differential target RNA cleavage. Plant Cell. 2011;23(12):4185–207.

Jeong D-H, Schmidt SA, Rymarquis LA, Park S, Ganssmann M, German MA, et al. Parallel analysis of RNA ends enhances global investigation of microRNAs and target RNAs of Brachypodium distachyon. Genome Biol. 2013;14(12):R145.

Jia J, Zhao S, Kong X, Li Y, Zhao G, He W, et al. Aegilops tauschii draft genome sequence reveals a gene repertoire for wheat adaptation. Nature. 2013;496(7443):91–5.

Jiao Y, Wang Y, Xue D, Wang J, Yan M, Liu G, et al. Regulation of OsSPL14 by OsmiR156 defines ideal plant architecture in rice. Nat Genet. 2010;42(6):541–4.

Johnson C, Kasprzewska A, Tennessen K, Fernandes J, Nan G-L, Walbot V, et al. Clusters and superclusters of phased small RNAs in the developing inflorescence of rice. Genome Res. 2009;19(8):1429–40.

Jones-Rhoades MW, Bartel DP, Bartel B. MicroRNAS and their regulatory roles in plants. Annu Rev Plant Biol. 2006;57:19–53.

Jung J-H, Park C-M. MIR166/165 genes exhibit dynamic expression patterns in regulating shoot apical meristem and floral development in Arabidopsis. Planta. 2007;225(6):1327–38.

Kozomara A, Griffiths-Jones S. miRBase: annotating high confidence microRNAs using deep sequencing data. Nucleic Acids Res. 2014;42(Database issue):D68–73.

Kurihara Y, Watanabe Y. Arabidopsis micro-RNA biogenesis through Dicer-like 1 protein functions. Proc Natl Acad Sci U S A. 2004;101(34):12753–8.

Lee RC, Feinbaum RL, Ambros V. The C. elegans heterochronic gene lin-4 encodes small RNAs with antisense complementarity to lin-14. Cell. 1993;75(5):843–54.

Liu H-H, Tian X, Li Y-J, Wu C-A, Zheng C-C. Microarray-based analysis of stress-regulated microRNAs in Arabidopsis thaliana. RNA. 2008;14(5):836–43.

Llave C, Xie Z, Kasschau KD, Carrington JC. Cleavage of Scarecrow-like mRNA targets directed by a class of Arabidopsis miRNA. Science. 2002;297(5589):2053–6.

Lu S, Sun Y-H, Chiang VL. Stress-responsive microRNAs in Populus. Plant J. 2008a;55 (1):131–51.

Lu C, Jeong D-H, Kulkarni K, Pillay M, Nobuta K, German R, et al. Genome-wide analysis for discovery of rice microRNAs reveals natural antisense microRNAs (nat-miRNAs). Proc Natl Acad Sci U S A. 2008b;105(12):4951–6.

Lucas SJ, Baştaş K, Budak H. Exploring the interaction between small RNAs and R genes during Brachypodium response to Fusarium culmorum infection. Gene. 2014;536(2):254–64.

Mallory AC, Reinhart BJ, Jones-Rhoades MW, Tang G, Zamore PD, Barton MK, et al. MicroRNA control of PHABULOSA in leaf development: importance of pairing to the microRNA 5′ region. EMBO J. 2004;23(16):3356–64.

Meng Y, Ma X, Chen D, Wu P, Chen M. MicroRNA-mediated signaling involved in plant root development. Biochem Biophys Res Commun. 2010;393(3):345–9.

Meyers BC, Axtell MJ, Bartel B, Bartel DP, Baulcombe D, Bowman JL, et al. Criteria for annotation of plant MicroRNAs. Plant Cell. 2008;20(12):3186–90.

Miura K, Ikeda M, Matsubara A, Song X-J, Ito M, Asano K, et al. OsSPL14 promotes panicle branching and higher grain productivity in rice. Nat Genet. 2010;42(6):545–9.

Nag A, King S, Jack T. miR319a targeting of TCP4 is critical for petal growth and development in Arabidopsis. Proc Natl Acad Sci U S A. 2009;106(52):22534–9.

Ori N, Cohen AR, Etzioni A, Brand A, Yanai O, Shleizer S, et al. Regulation of LANCEOLATE by miR319 is required for compound-leaf development in tomato. Nat Genet. 2007;39 (6):787–91.

Palatnik JF, Allen E, Wu X, Schommer C, Schwab R, Carrington JC, et al. Control of leaf morphogenesis by microRNAs. Nature. 2003;425(6955):257–63.

Palatnik JF, Wollmann H, Schommer C, Schwab R, Boisbouvier J, Rodriguez R, et al. Sequence and expression differences underlie functional specialization of Arabidopsis microRNAs miR159 and miR319. Dev Cell. 2007;13(1):115–25.

Peña-Castro JM, van Zanten M, Lee SC, Patel MR, Voesenek LAJC, Fukao T, et al. Expression of rice SUB1A and SUB1C transcription factors in Arabidopsis uncovers flowering inhibition as a submergence tolerance mechanism. Plant J. 2011;67(3):434–46.

Prigge MJ, Otsuga D, Alonso JM, Ecker JR, Drews GN, Clark SE. Class III homeodomain-leucine zipper gene family members have overlapping, antagonistic, and distinct roles in Arabidopsis development. Plant Cell. 2005;17(1):61–76.

Rhoades MW, Reinhart BJ, Lim LP, Burge CB, Bartel B, Bartel DP. Prediction of plant microRNA targets. Cell. 2002;110(4):513–20.

Rodriguez RE, Mecchia MA, Debernardi JM, Schommer C, Weigel D, Palatnik JF. Control of cell proliferation in Arabidopsis thaliana by microRNA miR396. Development. 2010;137 (1):103–12.

Schreiber AW, Shi B-J, Huang C-Y, Langridge P, Baumann U. Discovery of barley miRNAs through deep sequencing of short reads. BMC Genomics. 2011;12:129.

Sunkar R, Zhu J-K. Novel and stress-regulated microRNAs and other small RNAs from Arabidopsis. Plant Cell. 2004;16(8):2001–19.

Unver T, Budak H. Conserved microRNAs and their targets in model grass species Brachypodium Distachyon. Planta. 2009;230:659–69.

Vaucheret H, Vazquez F, Crété P, Bartel DP. The action of ARGONAUTE1 in the miRNA pathway and its regulation by the miRNA pathway are crucial for plant development. Genes Dev. 2004;18(10):1187–97.

Vaucheret H, Mallory AC, Bartel DP. AGO1 homeostasis entails coexpression of MIR168 and AGO1 and preferential stabilization of miR168 by AGO1. Mol Cell. 2006;22(1):129–36.

Vogel JP, Garvin DF, Mockler TC, Schmutz J, Rokhsar D, Bevan MW, et al. Genome sequencing and analysis of the model grass Brachypodium distachyon. Nature. 2010;463(7282):763–8.

Wang H-LV, Dinwiddie BL, Lee H, Chekanova JA. Stress-induced endogenous siRNAs targeting regulatory intron sequences in Brachypodium. RNA. 2015;21(2):145–63.

Wang X-J, Reyes JL, Chua N-H, Gaasterland T. Prediction and identification of Arabidopsis thaliana microRNAs and their mRNA targets. Genome Biol. 2004;5(9):R65.

Wei B, Cai T, Zhang R, Li A, Huo N, Li S, et al. Novel microRNAs uncovered by deep sequencing of small RNA transcriptomes in bread wheat (Triticum aestivum L.) and Brachypodium distachyon (L.) Beauv. Funct Integr Genomics. 2009;9(4):499–511.

Wightman B, Ha I, Ruvkun G. Posttranscriptional regulation of the heterochronic gene lin-14 by lin-4 mediates temporal pattern formation in C. elegans. Cell. 1993;75(5):855–62.

Wu L, Liu D, Wu J, Zhang R, Qin Z, Liu D, et al. Regulation of FLOWERING LOCUS T by a microRNA in Brachypodium distachyon. Plant Cell. 2013;25(11):4363–77.

Xie Z, Kasschau KD, Carrington JC. Negative feedback regulation of Dicer-Like1 in Arabidopsis by microRNA-guided mRNA degradation. Curr Biol. 2003;13(9):784–9.

Xie K, Wu C, Xiong L. Genomic organization, differential expression, and interaction of SQUAMOSA promoter-binding-like transcription factors and microRNA156 in rice. Plant Physiol. 2006;142(1):280–93.

Yao Y, Guo G, Ni Z, Sunkar R, Du J, Zhu J-K, et al. Cloning and characterization of microRNAs from wheat (Triticum aestivum L.). Genome Biol. 2007;8(6):R96.

Zhang J, Xu Y, Huan Q, Chong K. Deep sequencing of Brachypodium small RNAs at the global genome level identifies microRNAs involved in cold stress response. BMC Genomics. 2009;10:449.

Zhang L, Zheng Y, Jagadeeswaran G, Li Y, Gowdu K, Sunkar R. Identification and temporal expression analysis of conserved and novel microRNAs in Sorghum. Genomics. 2011;98 (6):460–8.

Zhang J, Mao Z, Chong K. A global profiling of uncapped mRNAs under cold stress reveals specific decay patterns and endonucleolytic cleavages in Brachypodium distachyon. Genome Biol. 2013;14(8):R92.

Zhou X, Wang G, Sutoh K, Zhu J-K, Zhang W. Identification of cold-inducible microRNAs in plants by transcriptome analysis. Biochim Biophys Acta. 2008;1779(11):780–8.

Transformation and T-DNA Mutagenesis

Mon Mandy Hsia and John P. Vogel

Abstract An efficient method to stably insert DNA into a genome is absolutely essential for a modern model system. In general, grasses are notoriously difficult to transform. Fortunately, *Brachypodium distachyon* has proven to be highly amenable to tissue culture and transformation. Indeed, current transformation methods make *B. distachyon* one of the easiest and most efficiently transformed grasses. This amenability to transformation is one of the key features that led to the widespread adoption of *B. distachyon* as a model for the grasses. In addition to being used for a host of functional experiments aimed at specific genes, high-efficiency transformation has enabled the creation of a large collection of sequence indexed insertional mutants. This T-DNA collection is a valuable experimental resource that has been utilized by hundreds of researchers around the world.

Keywords *Brachypodium* • Agrobacterium • Model system • Grass • Bioenergy

Transformation Methods

The stable insertion of foreign DNA into the genome, transformation, is a foundational tool that allows researchers to perform an array of experiments aimed at elucidating the function of genes and regulatory elements. For example, researchers can over-express genes, complement mutations with wild-type genes, knock out genes, test promoters, etc. In addition, transformation enables the engineering of novel properties into plants either one gene at a time or in more comprehensive ways that are being developed for emerging synthetic biology approaches. Thus, efficient transformation is an absolute requirement for a modern model system. Unfortunately, grasses are generally recalcitrant to transformation and those that can be transformed can often only be transformed at low

M.M. Hsia
USDA ARS Western Regional Research Center, Albany, CA, USA

J.P. Vogel (✉)
DOE Joint Genome Institute, Walnut Creek, CA, USA
e-mail: jpvogel@lbl.gov

J. Vogel (ed.), *Genetics and Genomics of Brachypodium*,
Plant Genetics and Genomics: Crops Models, DOI 10.1007/7397_2015_16
Published online: 02 October 2015

efficiency. Fortunately, *Brachypodium distachyon* has proven highly amenable to transformation and high efficiency transformation methods have been developed

The creation of stable transgenic grasses typically requires a tissue culture step where the plant is grown as an undifferentiated or partially differentiated callus. This callus is then transformed, typically by particle bombardment or by co-cultivation with *Agrobacterium*, and encouraged to regenerate plantlets. The first reported tissue culture of *B. distachyon* was published in 1995 (Bablak et al. 1995). The authors demonstrated that an embryogenic callus could be formed and fertile plants regenerated. The success of this early work set the stage for the development of *B. distachyon* as a model system. Building upon this early success, researchers at the same institution published the first report of *Brachypodium* transformation using a biolistic method in 2001 (Draper et al. 2001). However, the transformed line was polyploid and is now known as *B. hybridum* (Catalán et al. 2012). Biolistic transformation was further refined in 2005. In this case, the authors transformed the true *B. distachyon* (Christiansen et al. 2005). However, biolistic transformation often results in complex DNA integration events that contain multiple insertions and/or rearrangements of both inserted and endogenous DNA (Svitashev and Somers 2002). Thus for many applications that benefit from simple insertions (e.g. T-DNA tagging) biolistic transformation is not ideal.

Agrobacterium-mediated transformation typically produces much simpler insertions events than biolistic transformation and is thus the preferred method of plant transformation (2015). The first report of *Agrobacterium*-mediated transformation of *B. distachyon* was in 2006 (Vogel et al. 2006). In this paper the authors transformed both *B. distachyon* and *B. hybridum*, though at the time *B. hybridum* was considered a polyploidy form of *B. distachyon*. They achieved a reasonably high efficiency (up to 15 %) for *B. hybridum* transformation. While, the average transformation efficiency of the true *B. distachyon* was lower (3.2 %), it was clearly transformable by *Agrobacterium*. Efforts to improve *Agrobacterium*-mediated transformation were undertaken in several labs. A method based on the transformation of immature embryos was very efficient (55 %) (Pǎcurar et al. 2008). However, this method is very labor intensive because it requires the dissection of an immature embryo for every transgenic line produced. Two similar methods that transform embryogenic callus generated from immature embryos were published in 2008 (Vain et al. 2008; Vogel and Hill 2008). Both of these methods expand the initial callus through sub-culturing. Thus, they are less labor intensive because many transgenic lines can be created from each dissected embryo. The lines used in these studies also differed, one used Bd21 (Vain et al. 2008) and one used Bd21-3 (Vogel and Hill 2008). This somewhat confusing situation arose because the original work on *Agrobacterium*-mediated transformation used seeds from early in the inbreeding of line Bd21 (Vogel et al. 2006). Unfortunately, those seeds were bulked from several individual plants grown from seed from the original collection PI 254867 obtained from the National Plant Germplasm System (http://www.ars-grin.gov/npgs/). Since this collection was presumably made from many plants at one location there are multiple genetically distinct individuals in accession PI 254867. Thus, the transformation optimization was done on multiple genotypes. After optimization of the protocol an attempt was made to transform the final inbred line, Bd21, developed from accession PI 254867. This line was called Bd21 and not Bd21-1 because it was the only line developed from PI 254867 at that time (Vogel et al. 2006). Unfortunately, the high efficiency transformation method did not work well on Bd21 (Vogel and Hill 2008). Realizing that it was probably another genotype from the original seed lot that was being transformed, several inbred lines were developed from the original seed lot provided to the transformation group. The third line tested from that set could be efficiently transformed and was designated Bd21-3. Thus, Bd21 and Bd21-3 are two independently generated inbred lines that were originally collected from the same location in Iraq (Vogel and Hill 2008). Subsequently, line Bd21 was sequenced to produce the reference genome for *B. distachyon* and Bd21-3 was resequenced (IBI 2010; Gordon et al. 2014). A comparison of the two sequences revealed that while the lines were more closely related than most other accessions, they were quite distinct as evidenced by the presence of >400,000 SNPs (Gordon et al. 2014). The second high-efficiency transformation method used Bd21 (Vain et al. 2008). The inclusion of copper sulfate seems to be the key to creating embryogenic callus from Bd21 and is the major difference between the two methods. Thus, at this stage there were two methods and two lines with high transformation efficiencies.

In the course of creating the JGI T-DNA collection (formerly the WRRC T-DNA collection) the transformation method was further optimized by shortening incubation times, adding copper sulfate and eliminating some subcultures (Bragg et al. 2012, 2015). Currently, *B. distachyon* is one of if not the most efficiently transformed grass(es). Indeed, during the peak of T-DNA mutant production each person created 100 T-DNA lines per week, including transformation and plant care. Future improvements in transformation techniques will undoubtedly occur and updated protocols can be found at the JGI *Brachypodium* resource page http://jgi.doe.gov/our-science/science-programs/plant-genomics/brachypodium/. Plant genotype has a huge effect on transformation efficiency and in many systems it is necessary to examine many genotypes to identify one that can be efficiently transformed. The initial *B. distachyon Agrobacterium* transformation work was conducted on only three *B. distachyon* lines and only one line was transformable (Vogel et al. 2006). Thus, exploration of additional *B. distachyon* germplasm may reveal superior lines for transformation. The main limitation to *B. distachyon* transformation seems to be regeneration from callus and not transformation per-se. An examination of additional lines from Turkey did reveal large variation in regeneration potential with the best line, Adi9 greatly outperforming Bd21 and Bd21-3, the lines currently transformed (Fig. 1). Unfortunately, Adi9 requires longer vernalization making it less desirable for use as a laboratory standard.

Fig. 1 Variation in tissue culture performance. In an effort to find a more efficient line for transformation, several inbred lines were evaluated for callus formation and regeneration, the limiting steps in transformation. Note that the line on the right, Adi9, has a much higher regeneration rate than the line on the left, Adi3. The regeneration rate of Adi3 is similar to the regeneration rate of standard lines Bd21 and Bd21-3. Unfortunately, Adi9 requires a long vernalization period and thus is not as desirable as Bd21 or Bd21-3

T-DNA Mutant Resources

With an efficient transformation protocol in hand, the next obvious step was the creation of sequence-indexed mutant collections. The Arabidopsis community has amassed an enormous T-DNA collection that has revolutionized plant research by allowing researchers to easily order multiple mutants in almost any gene. A collection of 5000 *B. distachyon* T-DNA lines, of which 1000 were sequenced, was created by researchers at the John Innes Centre (Thole et al. 2010). Unfortunately, this collection is no longer publicly available. The creation of a much larger collection was funded by grants from the U.S. Department of Energy. Currently, this JGI *Brachypodium* T-DNA collection consists of 23,649 lines all of which have been sequenced. The first 7145 mutants were sequenced using Sanger sequencing via an inverse PCR approach (Bragg et al. 2012). This identified insertions at 5285 loci. The remaining 16,504 lines were sequenced using an Illumina-based approach (unpublished). In total, 23,496 insertion sites were identified. A table of insertion sites and instructions for ordering seeds are available at the *Brachypodium* T-DNA Collection website (http://jgi.doe.gov/our-science/science-programs/plant-genomics/brachypodium/brachypodium-t-dna-collection/). In addition, the insertion sites are available as a jbrowse track in the Phytozome database (http://phytozome.jgi.doe.gov/pz/portal.html). This collection has been heavily utilized by the research community and >12,000 lines have been distributed to over 100 laboratories around the world (Fig. 2).

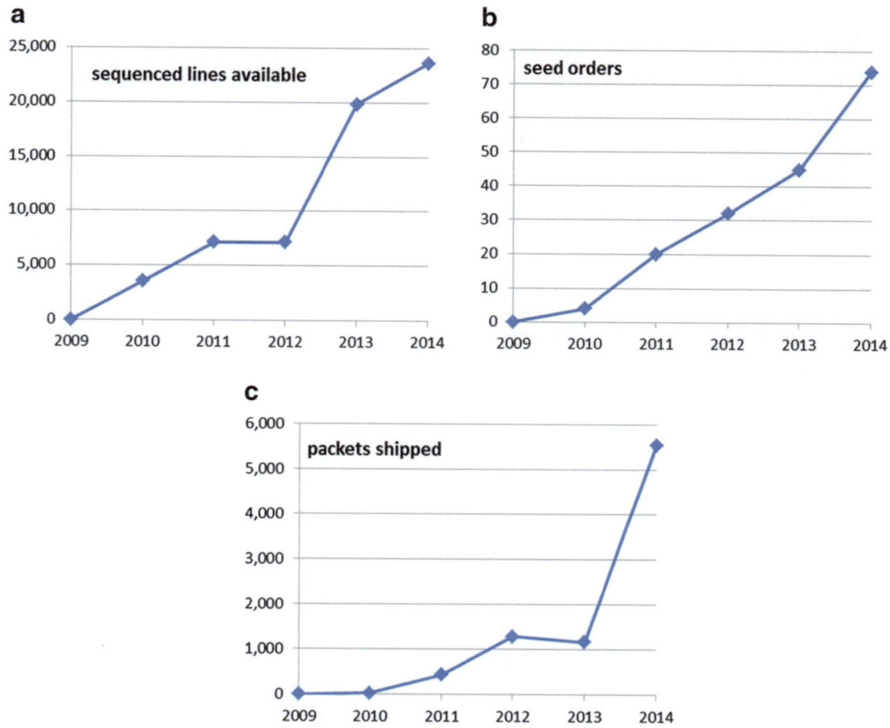

Fig. 2 Creation and utilization of the JGI T-DNA collection. Note that as the number of freely available lines has increased (**a**) the number of requests for seed (**b**) and the number of seed packets shipped (**c**) has increased dramatically

Directed Mutagenesis

While T-DNA mutagenesis is a powerful tool for reverse genetics it suffers from limitations including the inability to target specific DNA sequences and the cost of creating and maintaining the hundreds of thousands of lines necessary to reach saturation. In recent years the technology to target specific DNA sequences for cleavage and subsequent mutagenesis by endogenous DNA damage repair machinery has exploded. The most recent advance in targeting specific DNA sequences is based on an adaptive bacterial immune system that uses DNA sequence information stored in tandem arrays, called clustered regularly interspaced short palindromic repeats or CRISPRs, to target homologous invading DNA for destruction (reviewed in Belhaj et al. 2013). CRISPRs, were first described in 1987 (Ishino et al. 1987), but their function was not understood until 2007 when it was shown that phage resistance could be conferred by sequences in the CRISPR and that some of the CRISPR associated proteins were required for this resistance (Barrangou et al. 2007). In 2012 it was shown that the Cas9 protein used the base pairing of the guide RNA transcribed from the CRISPR to specify the location of DNA cleavage by the nuclease domain of Cas9 (Jinek et al. 2012). In 2013 several reports of CRISPR/Cas9 targeted mutagenesis in plants clearly demonstrated the huge potential of the system (Jiang et al. 2013; Nekrasov et al. 2013; Shan et al. 2013; Upadhyay et al. 2013). What makes CRISPR/Cas-9 system so powerful is the fact that DNA target specificity is conferred by a small guide RNA. Thus, researchers need only engineer the 23 bp DNA that is transcribed into the sgRNA to target any desired location in the genome. Furthermore, it is possible to target multiple loci by simply adding more sgRNA sequences (e.g. Li et al. 2013). In addition to creating small indels, CRISPR/Cas-9 technology can also be used to delete larger pieces of DNA or replace sections of DNA with alternate sequence (reviewed in Lozano-Juste and Cutler 2014). The efficient *B. distachyon* transformation system makes it particularly amenable to CRISPR/Cas-9 genome modification and at least two groups have successfully used the system to efficiently create targeted mutations (D. O'Connor, E, Abrush pers. comm.).

Conclusions

With the low cost of DNA sequencing and increasingly sophisticated methods for genome assembly the number of species with high quality genome sequences is rising rapidly. By contrast, the biological factors that make an organism useful as a model system (e.g. rapid generation time, self fertility, etc.) and the effort required to develop methods and create infrastructure (e.g. efficient transformation systems and characterized mutant collections) has not changed. Thus, the amenability of *B. distachyon* to high efficiency transformation has been and will continue to be crucial for its utility as a model for the grasses.

Acknowledgement The T-DNA project was supported by Office of Biological and Environmental Research, Office of Science, US Department of Energy, interagency agreements DE-SC0001526 and DE-AI02-07ER64452. JV is supported by the U.S. Department of Energy Joint Genome Institute, a DOE Office of Science User Facility, that is supported under Contract No. DE-AC02-05CH11231.

References

Bablak P, Draper J, Davey MR, Lynch PT. Plant regeneration and micropropagation of *Brachypodium distachyon*. Plant Cell Tiss Org Cult. 1995;42(1):97–107.

Bragg NB, Anderton A, Nieu R, Vogel JP. *Brachypodium distachyon. Agrobacterium* protocols: volume 1. In: Wang K, editor. Methods in molecular biology, vol. 1223. New York: Springer; 2015. p. 17–33.

Barrangou R, Fremaux C, Deveau H, Richards M, Boyaval P, Moineau S, et al. CRISPR provides acquired resistance against viruses in prokaryotes. Science. 2007;315(5819):1709–12.

Belhaj K, Chaparro-Garcia A, Kamoun S, Nekrasov V. Plant genome editing made easy: targeted mutagenesis in model and crop plants using the CRISPR/Cas system. Plant Methods. 2013;9(1):39.

Bragg JN, Anderton A, Nieu R, Vogel JP. *Brachypodium distachyon*. Methods Mol Biol. 2015;1223:17–33.

Bragg JN, Wu J, Gordon SP, Guttman ME, Thilmony R, Lazo GR, et al. Generation and characterization of the Western Regional Research Center *Brachypodium* T-DNA insertional mutant collection. PLoS One. 2012;7(9):e41916.

Catalán P, Müller J, Hasterok R, Jenkins G, Mur LAJ, Langdon T, et al. Evolution and taxonomic split of the model grass *Brachypodium distachyon*. Ann Bot. 2012;109(2):385–405.

Christiansen P, Didion T, Andersen CH, Folling M, Nielsen KK. A rapid and efficient transformation protocol for the grass *Brachypodium distachyon*. Plant Cell Rep. 2005;23 (10–11):751–8.

Draper J, Mur LAJ, Jenkins G, Ghosh-Biswas GC, Bablak P, Hasterok R, et al. *Brachypodium distachyon*. A new model system for functional genomics in grasses. Plant Physiol. 2001;127 (4):1539–55.

Gordon SP, Priest H, Des Marais DL, Schackwitz W, Figueroa M, Martin J, et al. Genome diversity in *Brachypodium distachyon*: deep sequencing of highly diverse inbred lines. Plant J. 2014;79(3):361–74.

IBI. Genome sequencing and analysis of the model grass *Brachypodium distachyon*. Nature. 2010;463:763–8.

Ishino Y, Shinagawa H, Makino K, Amemura M, Nakatura A. Nucleotide sequence of the iap gene, responsible for alkaline phosphatase isoenzyme conversion in *Escherichia coli*, and identification of the gene product. J Bacteriol. 1987;169(12):5429–33.

Jiang W, Zhou H, Bi H, Fromm M, Yang B, Weeks DP. Demonstration of CRISPR/Cas9/sgRNA-mediated targeted gene modification in Arabidopsis, tobacco, sorghum and rice. Nucleic Acids Res. 2013;41(20):e188.

Jinek M, Chylinski K, Fonfara I, Hauer M, Doudna JA, Charpentier E. A programmable dual-RNA-guided DNA endonuclease in adaptive bacterial immunity. Science. 2012;337 (6096):816–21.

Li JF, Norville JE, Aach J, McCormack M, Zhang D, Bush J, et al. Multiplex and homologous recombination-mediated genome editing in Arabidopsis and *Nicotiana benthamiana* using guide RNA and Cas9. Nat Biotechnol. 2013;31(8):688–91.

Lozano-Juste J, Cutler SR. Plant genome engineering in full bloom. Trends Plant Sci. 2014;19 (5):284–7.

Nekrasov V, Staskawicz B, Weigel D, Jones JDG, Kamoun S. Targeted mutagenesis in the model plant *Nicotiana benthamiana* using Cas9 RNA-guided endonuclease. Nat Biotechnol. 2013;31 (8):691–3.

Păcurar DI, Thordal-Christensen H, Nielsen KK, Lenk I. A high-throughput *Agrobacterium*-mediated transformation system for the grass model species *Brachypodium distachyon* L. Transgenic Res. 2008;17(5):965–75.

Shan Q, Wang Y, Li J, Zhang Y, Chen K, Liang Z, et al. Targeted genome modification of crop plants using a CRISPR-Cas system. Nat Biotechnol. 2013;31(8):686–8.

Svitashev SK, Somers DA. Characterization of transgene loci in plants using FISH: a picture is worth a thousand words. Plant Cell Tiss Org Cult. 2002;69(3):205–14.

Thole V, Worland B, Wright J, Bevan MW, Vain P. Distribution and characterization of more than 1000T-DNA tags in the genome of *Brachypodium distachyon* community standard line Bd21. Plant Biotechnol J. 2010;8(6):734–47.

Upadhyay SK, Kumar J, Alok A, Tuli R. RNA-guided genome editing for target gene mutations in wheat. G3 (Bethesda). 2013;3(12):2233–8.

Vain P, Worland B, Thole V, McKenzie N, Alves SC, Opanowicz M, et al. *Agrobacterium*-mediated transformation of the temperate grass *Brachypodium distachyon* (genotype Bd21) for T-DNA insertional mutagenesis. Plant Biotechnol J. 2008;6:236–45.

Vogel J, Hill T. High-efficiency *Agrobacterium*-mediated transformation of *Brachypodium distachyon* inbred line Bd21-3. Plant Cell Rep. 2008;27(3):471–8.

Vogel JP, Garvin DF, Leong OM, Hayden DM. *Agrobacterium*-mediated transformation and inbred line development in the model grass *Brachypodium distachyon*. Plant Cell Tiss Org Cult. 2006;85:199–211.

Chemical and Radiation Mutagenesis: Induction and Detection by Whole Genome Sequencing

Fabienne Granier, Aurélie Lemaire, Yin Wang, Philippe LeBris, Sébastien Antelme, John Vogel, Debbie Laudencia-Chingcuanco, and Richard Sibout

Abstract *Brachypodium distachyon* has emerged as an effective model system to address fundamental questions in grass biology. With its small sequenced genome, short generation time and rapidly expanding array of genetic tools, *B. distachyon* is an ideal system to elucidate the molecular basis of important traits in crops and bioenergy feedstocks. Induced mutations are one of the pillars of modern molecular genetics and are particularly useful for assigning function to individual genes. Due to their ease of use and low cost, mutagenic chemicals and ionizing radiation have been widely used to create mutant populations of many different organisms. The major limitations for these mutagens are the difficulty of identifying the specific mutation responsible for an observed phenotype and the difficulty of identifying mutations in a gene of interest. As a step toward addressing these limitations, Targeting Induced Local Lesions in Genomes (TILLING) has been developed as an efficient method to rapidly identify mutations in genes of interest. Recently, the decreasing cost of DNA sequencing has made it feasible to detect mutations throughout the genome using whole genome sequencing. This promises to revolutionize the use of chemical and radiation mutants in research. In this chapter we describe the status of *B. distachyon* mutagenesis including the methods, mutagens, TILLING populations and initial results using whole genome sequencing to identify induced genetic variation.

Keywords Mutagenesis • EMS • Sodium azide • Fast neutron • Whole genome sequence • TILLING

F. Granier • A. Lemaire • Y. Wang • P. LeBris • S. Antelme • R. Sibout (✉)
Institut Jean-Pierre Bourgin, INRA, AgroParisTech, CNRS, Université Paris-Saclay, RD10, 78026 Versailles Cedex, France
e-mail: richard.sibout@versailles.inra.fr

J. Vogel
US-DOE, Joint Genome Institute, 2800 Mitchell Drive, Walnut Creek, CA 94598, USA

D. Laudencia-Chingcuanco
USDA-ARS, Western Regional Research Center, 800 Buchanan Street, Albany, CA 94710, USA

J. Vogel (ed.), *Genetics and Genomics of Brachypodium*,
Plant Genetics and Genomics: Crops Models, DOI 10.1007/7397_2015_20
Published online: 24 November 2015

Introduction

Mutagenesis has been widely used to introduce genetic variation in plants for both commercial and research purposes. The introduction of new marketable traits in crops through mutagenesis is often called mutation breeding and has been used for nearly a century (Muller 1927; Stadler 1928a, b; Ahloowalia et al. 2004). As a low-cost method to create genetic variation, mutation breeding has been encouraged by the Food and Agriculture Organization (FAO) and the International Atomic Energy Agency (IAEA) (http://www-pub.iaea.org/books/IAEABooks/1220/Manual-on-Mutation-Breeding-2nd-Edition; https://www.iaea.org/). Such introduced variation is particularly useful for crops that are difficult to breed through traditional methods due to poor seed production, crossing difficulties, low recombination rates and limited natural diversity (Pathirana 2011). When breeders want to disrupt a specific gene in a crop that is difficult to improve with biotechnology, because of recalcitrance to genetic transformation for instance, chemical and radiation mutagenesis are also very useful. In this context, improved varieties developed through mutation breeding have an advantage over transgenic varieties because chemical and radiation mutants are not subject to government regulation. However, the spectrum of improvements that can be introduced by mutation breeding is more limited than those that can be introduced by transgenesis so these breeding methods are highly complementary. Since the mid-twentieth century, more than 3200 mutant plant varieties (produced by radiation and chemical mutagenesis or by somaclonal variation) have been commercialized (Pathirana 2011). Asia is now the most active in this field. Indeed, China and Japan combined released 60 % of the new mutagenized varieties since 2000 (data from Joint FAO/IAEA, http://mvgs.iaea.org/Search.aspx). Grasses make up an important part of the 150 species improved by mutation breeding because China produces mainly rice and wheat varieties.

Researchers routinely use chemical or radiation mutagenesis to assign function to specific genes. This is best exemplified by the great successes using ethyl methanesulfonate (EMS) mutagenesis in the model plant Arabidopsis (Kim et al. 2006; Maple and Møller 2007; Austin et al. 2011). Many questions in diverse biological fields, including development, metabolism, signaling and epigenetics were resolved through characterization of Arabidopsis EMS mutants (Hou et al. 2010; Qu and Qin 2014). A limitation of chemical and radiation mutants is the difficulty in identifying the causal allele. For chemical mutants the mutation is typically a single nucleotide polymorphism (SNP) that can be identified using a positional cloning approach. Similarly, radiation mutants are typically cloned by positional methods, but larger deletions can be easier to identify. Of course, if the deletion is so large that it knocks out more than one gene it becomes impossible to assign function to a single gene. Thus, for researchers interested in specific genes and for breeders that need to identify the exact mutation/gene to protect intellectual property rights, the time and effort required to identify causal mutations is a significant bottleneck (Yadav et al. 2012).

Insertional mutagenesis allows more efficient identification of causal alleles because the known sequence of the inserted DNA (T-DNA or transposon) can be used as a marker to identify the exact insertion site. The most powerful application of insertional mutagenesis is the large collection of sequence-indexed Arabidopsis T-DNA lines. This resource has transformed plant biology and today Arabidopsis researchers can order lines containing mutations in almost every gene in the genome (https://www.arabidopsis.org/). Despite the great utility of insertional mutants, chemical and radiation mutants are still extremely useful, even for Arabidopsis researchers. Why is this? First, the huge cost and long time required to create and sequence a T-DNA population make the creation of mutant populations for specific purposes prohibitively expensive. Thus, researchers wishing to do a suppressor screen in a specific mutant background or working with a species without an established T-DNA population routinely turn to chemical or radiation mutagenesis. Second, the higher mutation rate of chemical and radiation mutagens is an advantage for forward genetic screens that require significant labor. Simply put, if you have a labor-intensive screen it is less expensive to screen an EMS population with ~1000 mutations per line than a T-DNA collection with 1–2 insertions per line. Third, the SNPs caused by chemical mutagenesis can lead to hypomorphic mutations, conditional mutations and dominant mutations that are only rarely created by T-DNA mutagenesis (Till et al. 2003; Triques et al. 2007). This wider spectrum of mutations is especially useful for studying essential or functionally redundant genes.

In order to harness the wide spectrum of chemical mutations for reverse genetic studies it was necessary to develop a high-throughput method to identify mutations in genes of interest from large collections of chemical mutants. The most widely used method is Targeting Induced Local Lesions in Genomes (TILLING) (McCallum et al. 2000a). Briefly, PCR (using DNA from pools of mutants as a template) is used to amplify DNA from the gene of interest. If one of the mutants contains a mutation (SNP or small deletion) in the amplified sequence then there will be a mismatch in some of the amplicons after denaturation and reannealing. These mismatches can be then be detected by several approaches (e.g. gel electrophoresis following incubation with an enzyme that cleaves mismatch sites, gel electrophoresis under conditions that detect conformational changes, conformation sensitive capillary electrophoresis and high resolution melt analysis, Gady et al. 2009; Lochlainn et al. 2011). Depending on the pooling strategy used, additional round(s) of PCRs on individual mutants may be required to identify the line containing the mutation. Detection of mutations in the PCR products by pooling and sequencing many PCR products at once is now also possible (Perez-de-Castro et al. 2012; Wang et al. 2012). However, even with the use of high throughput sequencing to detect mutations, TILLING is still labor and time intensive when compared with ordering T-DNA mutants from a stock center.

Due to the rapidly decreasing cost of DNA sequencing it is now feasible to sequence thousands of genomes from species with small genomes and exomes for species with larger genomes. Thus, using Whole Genome Sequencing (WGS) to identify the mutations in populations of chemical and/or radiation mutants is becoming a possibility (Schneeberger et al. 2009; Fekih et al. 2013; Henry et al. 2014). This offers a new paradigm that truly combines the main advantage of T-DNA mutants (known insertion sites) with the advantages of chemical mutagenesis (wide spectrum of mutations, high mutation rate, ease of creating a population) (Till et al. 2003; Triques et al. 2007). Thus, the creation of sequence indexed chemical and radiation mutants will undoubtedly become commonplace for both model plants and crops.

Brachypodium distachyon is an excellent candidate for mutagenesis for several reasons. First, as described by the other chapters in this book, it is an excellent model system with many resources and a large community of researchers that would benefit from mutant populations. Second, simple diploid genetics, inbreeding nature and small genome size simplify the creation and genetic analysis of *B. distachyon* mutations. Third the excellent reference genome sequence greatly facilitates the identification of mutations using a whole genome sequencing approach. In this chapter, we describe the current status of *B. distachyon* mutagenesis techniques, describe two mutant populations, explore the feasibility of using WGS to identify mutations and describe a recently initiated project to sequence EMS, sodium azide and fast neutron mutants.

Chemical and Radiation Mutagenesis of *B. distachyon*

B. distachyon has been mutagenized by several means including EMS (http://jgi.doe.gov/our-science/science-programs/plant-genomics/brachypodium/; http://www.brutnelllab.org/#!brachypodium/c1svt; Brutnell et al. 2015), sodium azide (NaN_3) (Engvild 2005; Dalmais et al. 2013), fast-neutron radiation (FNR) (Derbyshire and Byrne 2013) heavy ion beam (https://pag.confex.com/pag/xxi/webprogram/Paper6496.html) and gamma irradiation (Lee et al. 2013).

EMS is an alkylating agent (like diethyl sulfate and dimethyl sulfate) that is commonly used as mutagen on a wide range of organisms (animals, plants, microbes). EMS causes the alkylation of guanine (G) by transfer of an ethyl group to create O6–ethylguanine. This modified nucleotide preferentially pairs to thymine (T) instead of cytosine (C) and when DNA replicates or when this mispairing causes a single strand break the alkylated guanine is most often replaced by adenine (A). In some cases, chromosome breakages may happen (Hoffmann 1980). The most frequent mutations induced by EMS are called transitions: the replacement of a purine with a purine (e.g. G to A) or the replacement of a pyrimidine with a pyrimidine (e.g. C to T). Transversions, the substitution of purines and pyrimidines, are rarely observed after EMS treatment (McCallum et al. 2000a). Interestingly, the proportion of transitions to transversions can be modulated by the nature and the dose of the alkylating agent (Krieg 1963; Kovalchuk et al. 2000; Greene et al. 2003).

In contrast to EMS, the exact mechanism of mutagenesis by NaN_3 is not completely understood and several possible mechanisms have been proposed.

NaN_3 can induce mutations through the creation of an organic metabolite, L-azidoalanine. At very low pH, the acid form of NaN_3 penetrates into the tissues and it is metabolized by O-acetylserine (thiol)-lyase. This enzyme catalyzes the addition of azide (N_3^-) to O-acetylserine which leads to the synthesis of azidoalanine (Owais et al. 1986; Owais and Kleinhofs 1988). L-azidoalanine is a highly mutagenic amino acid analogue (reviewed in Al-Qurainy and Khan 2009; Gruszka et al. 2012). Azide itself may also play a role in mutagenesis (LaVelle and Mangold 1987). However, no direct interaction between NaN_3 and DNA was observed by Owais (1993). Some studies report that the substitution of the thiol groups of methionine and cysteine by azide during DNA replication may cause mutations (Vattem et al. 2006). Finally, oxygen free radicals from azide-derived peroxide accumulation might also be responsible for the mutagenic activity of NaN_3 (reviewed in Gruszka et al. 2012). While the mutation rate following NaN_3 is often reported to be low (reviewed in Gruszka et al. 2012), some studies report rates comparable to alkylating agents (Sugihara et al. 2013). An advantage of NaN_3 is that it is easier to handle than EMS because it is not volatile and it is less toxic to mammalian cells (Owais and Kleinhofs 1988).

A method for NaN_3 mutagenesis of barley seeds (Nilan 1973; Olsen et al. 1993) has been adapted for the mutagenesis of several grasses: maize, barley, oat, wheat and rice (reviewed in Al-Qurainy and Khan 2009 and in Pathirana 2011). Due to the success of NaN_3 mutagenesis in grasses, we selected NaN_3 to create a *B. distachyon* mutant population (Dalmais et al. 2013). The main INRA (French National Institute for Agricultural Research) mutant population was made from line Bd21-3 (Fig. 1a). In addition, mutagenized populations of Bd3-1 and of *Brachypodium stacei* were also produced at INRA (unpublished). The mutagenesis protocol is similar to the protocol developed for oat by Rines (1985) and is described in Dalmais et al. (2013). Briefly, the seeds were pre-soaked in distilled water for 2 h and then imbibed in different concentrations of fresh NaN_3 solutions (0.5–10 mM) prepared in a phosphate buffer (pH 3). Prior to sowing, the seeds were washed four times in water. Kill curves were used to estimate the appropriate concentration of mutagen.

Fast Neutron Mutagenesis

Fast-neutron radiation (FNR) has been shown to be an effective mutagen to knockout genes in various plants e.g. Arabidopsis (Belfield et al. 2012), soybean (Bolon et al. 2011; O'Rourke et al. 2013) *Medicago* (Rogers et al. 2009) and rice (Wu et al. 2005; Bruce et al. 2009). Early work in Arabidopsis showed that FNR can be as potent in generating genetic variation as chemical mutagens like EMS (Koornneef et al. 1982). Traditionally, FNR has been thought to primarily induce medium to large deletions and chromosomal structural variations (e.g. inversions, duplications, translocations). This view has probably been reinforced by the approaches used to detect the mutations: genomic hybridization (Bolon et al. 2011), fragment amplification (Li et al. 2002) or southern blots (Bruce et al. 2009; Bruggemann et al. 1998). However, a recent study using

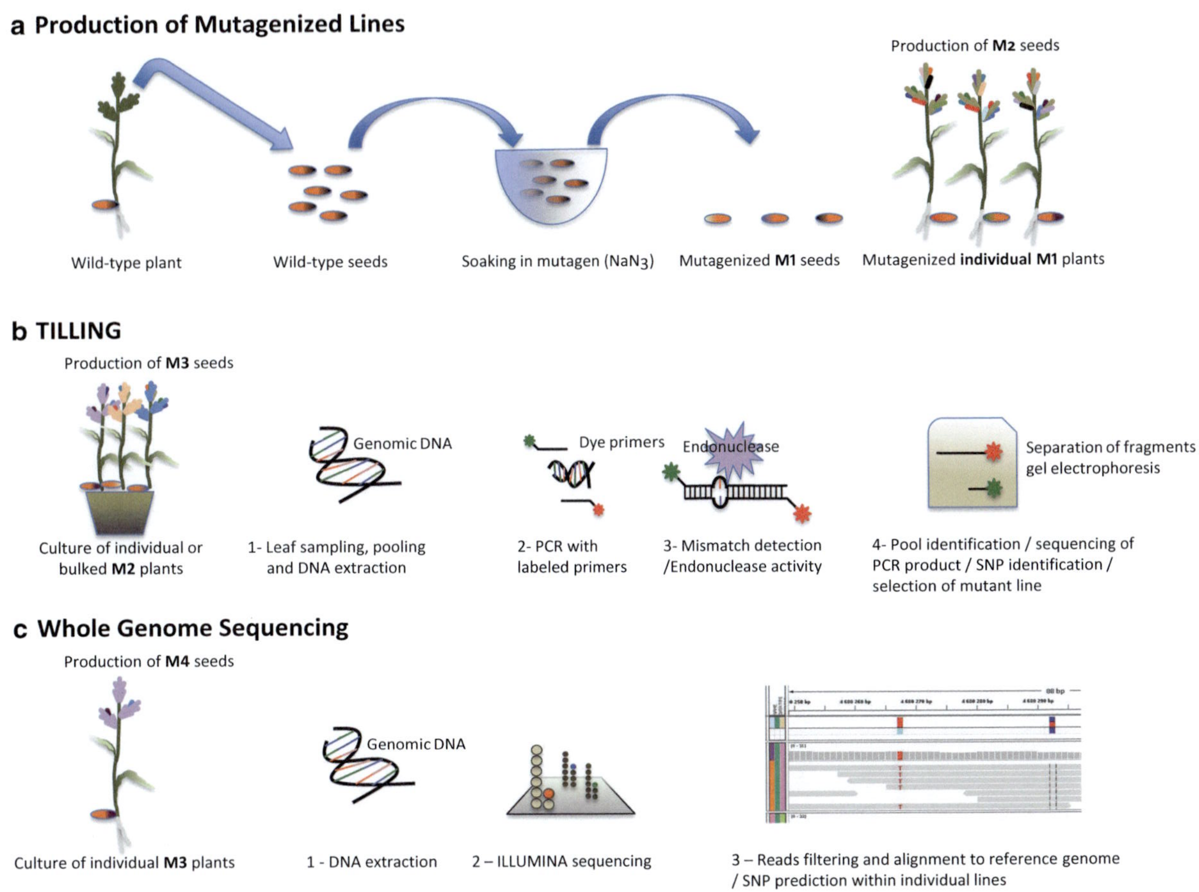

Fig. 1 Illustration of mutagenesis, TILLING and whole genome sequencing. (**a**) Mature M_1 seeds are treated by sodium azide (NaN_3). M_2 seeds are produced from M_1 plants. (**b**) DNA isolated from bulked plants is used for labeled-PCR amplicon production. SNPs are detected by the activity of an endonuclease. (**c**) DNA from individual plants is used for whole genome sequencing (ILLUMINA)

Table 1 Bd21-3 dose response to fast neutron radiation

Dose (Gray)	% M_1 germination	% M_1 survival	% M_1 fertile	% M_2 albino
0*	100	100	98	0
6	106	100	100	0
12	101	93	98	0.25
18	104	94	92	1.12
24	102	95	57	1.68
36	74	0	0	nd

*untreated seeds; Germination reported as % relative to untreated seeds; % M_2 Albino was determined from 500-1,000 M_2 seeds planted per dose; Survival, indicates plants that reached flowering stage; nd – not determined
1 Gray = 100 rads

Next-Generation Sequencing (NGS) to detect FNR induced mutations in Arabidopsis (Belfield et al. 2012) revealed that FNR also induces a significant number of SNPs.

To create a *B. distachyon* FNR mutagenized population, dry seeds of Bd21-3 were exposed to five different doses of FNR: 6, 12, 18, 24 and 36 Gray (1 Gray = 100 rads). A pool of 10,000 seeds was exposed for each dose. A small sample of M_1 seeds from each dose was planted to determine germination rate, plant survival and fertility (Table 1). The resulting M_2 seeds were used to determine the frequency of albino plants for each treatment. Based on the dose response, the 18 and 24 Gray doses are optimal for maximizing the mutation rate while maintaining reasonable fertility. This result is consistent with the findings for Bd21 mutagenesis using FNR (Derbyshire and Byrne 2013).

About 20 M_1 seeds were planted per pot to generate M_2 seeds. M_2 seeds from a single head per M_1 plant was harvested and planted to produce M_3 seeds. M_2 seeds from the remaining M_1 plants in each pot were also pooled separately as an additional resource. We currently have M_3 seeds from more than 1100 independent M_2 lines and phenotypes have been observed (Fig. 2). Genomic DNA from each M_2 line was extracted for whole genome sequencing to establish a sequence-indexed library of mutations as described below (http://jgi.doe.gov/our-projects/csp-plans/fy-2015-csp-plans/).

TILLING for *B. distachyon* Mutants

TILLING has been used in several plants species to identify mutations in genes of interest (reviewed by Henikoff et al. 2004, Fig. 1b). Thus, it is not surprising that several groups have created *B. distachyon* TILLING collections. The Brutnell lab at the Danforth Center has established a TILLING population that can be utilized by the community for a fee (http://www.brutnelllab.org/#!brachypodium/c1svt). Similarly, the Rasmussen lab at the University of Copenhagen has established a

Fig. 2 Phenotypes observed in *B. distachyon* FNR and NaN_3 mutant populations. (**a** and **b**) colored stems, (**c**) curved leaves, (**d**) lesions, (**e**) abnormal-shaped leaves with excrescence, (**f**) yellow-midrib leaves, (**g**) no tiller production, (**h**) virescent plant, (**i**) twisted plant with abnormal spikelets, (**j**) curly leaves, (**k** and **l**) dwarf plants. Photographs **f**, **g**, **h** courtesy of Oliver Darack

TILLING collection than can be accessed through collaborations (http://plen.ku.dk/english/research/plant_soil/breeding/quality/brachylife/). In 2013, we collaborated with A. Bendahmane's group to publish an article (Dalmais et al. 2013) describing the establishment of an INRA TILLING platform that can be used by the community for a fee (located at Evry, France http://www-urgv.versailles.inra.fr/tilling/brachypodium.htm).

The INRA TILLING collection was extensively validated and characterized by looking for mutations in seven genes in the lignin biosynthetic pathway. The identification of mutations in individual genes was carried out using the endonuclase ENDO1 as described in (Triques et al. 2007). Briefly, each gene of interest was amplified by PCR with dye labelled primers from pooled genomic DNAs of mutagenized lines. The amplicons will contain heteroduplexes if at least one line possesses a SNP in the targeted gene. To identify the line of interest, the amplified fragment was subjected to an endonuclease that specifically cleaves mismatches in heteroduplexes (McCallum et al. 2000a, b). Using this approach we identified 128 mutations (out of a population of 5580 mutagenized lines) in the seven genes we targeted. An examination of these mutations revealed that 90 % of the mutations were transitions. Primarily, Guanine (G) was substituted by Adenine (A) and Cytosine (C) was substituted by Thymine (T). Interestingly, >60 % of the mutations were non-synonymous (including 5 % that created premature stop codons). A single nucleotide deletion adjacent to a transition was also detected. Only one mutation was predicted to alter a splice site (Dalmais et al. 2013). The average mutation rate was estimated to be approximately one substitution per 400 kb or ~700 mutations per genome. This mutation frequency is similar to previous reports of EMS-mutagenized grasses (Dalmais et al. 2013). The predominance of transitions is in accordance with the mutation type and frequency observed in the NaN_3-mutagenized barley population by (Olsen et al. 1993) and in a population of rice mutagenized with a combination of NaN_3 plus methyl-nitrosourea by (Till et al. 2007). Nevertheless, Olsen et al. (1993) detected three times higher AT to GC changes than GA to CT while we observed the reverse as did Till et al. (2007).

The Genetically Effective Cell Number (GECN) is the number of germline cells that ultimately go on to form gametes. The GECN is important for calculating the expected diversity in a mutant population because the effective number of mutagenized individuals is really the number of mutagenized germline cells. We were able to estimate the GECN for *B. distachyon* seeds because M_1 plants (the plants that grew from the mutagenized seeds) were grown individually and we sequenced several M_2 plants from each tilled family. The estimated GECN for *B. distachyon* seeds is at least four (Dalmais et al. 2013). This lies between the estimated GECN for oat (6) (Hodgdon et al. 1981) and the GECN for Arabidopsis (2) (Page and Grossniklaus 2002).

Mutation Detection by Whole Genome Sequencing (WGS)

While TILLING is effective in identifying mutations in specific genes, only a short stretch of DNA is interrogated by each PCR reaction. Thus, even when mutations are detected by high-throughput sequencing, it is still time and labor intensive. The low cost of DNA sequencing has now made the identification of mutations by WGS of individual mutagenized plants feasible. Thus, mutant detection by WGS is a natural evolution from TILLING (Fig. 1c). By sequencing the entire genome instead of analyzing individual PCR products, the throughput of mutant detection is increased by several orders of magnitude. In order to determine the feasibility of large scale WGS for creating a comprehensive collection of sequence indexed *B. distachyon* mutants, we sequenced 18 mutants from the INRA TILLING collection using Illumina HiSeq technology (2×100 bp paired end reads, average depth $22\times$).

We used the "mutDetect" pipeline developed by the bioinformatics team of Institute Jean-Pierre Bourgin (IJPB) to identify mutations (unpublished). One complication for mutation detection is that the mutant lines were produced in a different genetic background (Bd21-3) than the line that was sequenced for the reference genome (Bd21). Despite the fact that these lines were collected from the same location, their genomes differ by over 500,000 SNPs (Gordon et al. 2014). Thus, it was necessary to filter out the natural variation from the induced mutations. First, we filtered out any known SNPs (Gordon et al. 2014). Next, we filtered out any SNP shared by more than two of the 18 mutant lines because these are probably undetected natural variants. After filtering, 36,681 high and low quality SNP remained as putative mutations induced by NaN_3. While this is perhaps an overestimate of the number of mutations, a higher filtering stringency would miss some real mutations, especially in lines with low sequence coverage (three lines had less than $10\times$ sequence coverage). As an example, we confirmed genuine SNPs (by genomic PCR) that were covered by only one or two reads. With high stringency filters these SNPs would have been lost. Thus, for constructing a sequence-indexed mutant population (see below) we think it is desirable to call SNPs with a lower stringency and accept that some percentage of putative mutants will be false positives.

The WGS of the 18 mutants can be used to describe the genome-wide spectrum of putative mutations induced by NaN_3 (Table 2). Seventy percent of the mutations were transitions (highlighted in grey in Table 2). The most common mutations were G-A and C-T substitutions similar to those induced by EMS (Till et al. 2003). The reverse transitions A-G to T-C (found at high frequency in barley by Olsen et al. 1993) occurred at a much lower rate.

To evaluate the functional consequence of NaN_3 mutagenesis we predicted the impact of the 36,681 SNPs on protein function. Among the 18 mutants, 330 mutations were predicted to have a high impact on protein function (Table 3). High impact mutations are those that will undoubtedly have a major negative impact on protein function (e.g. frameshifts, splice site mutations, premature stops, loss of the stop codon or the start codon). Among these mutations we confirmed a previously

Table 2 Spectrum of mutations induced by sodium azide

Mutation type[a]	Average mutations per line[b]	Percent of all mutations
A–C	50	2.5
A–G	136	6.7
A–T	55	2.7
C–A	141	7.0
C–G	62	3.0
C–T	529	26.1
G–A	613	30.3
G–C	60	2.9
G–T	141	7.0
T–A	56	2.8
T–C	135	6.7
T–G	46	2.3
Total transitions	1413	69.9
Total transversions	610	30.2
Total EMS-type mutations (G–A or C–T)	1142	56.4

[a]Transitions are highlighted gray
[b]Based on 18 mutant lines

identified mutation (G/A at position 65,504,763 on chromosome 1 of line *Bd4442*) in *BdLAC5* (Dalmais et al. 2013; Wang et al. 2015). This mutation alters the splice site for the last intron which is predicted to result in a truncated protein with an aberrant C-terminal. *BdLAC5* encodes a laccase involved in lignification and, interestingly, the mutant releases almost 40 % more glucose in saccharification assays when compared to wild type (Wang et al. 2015). Since *BdLAC5* mutants have no obvious deleterious phenotypes, *LACCASE* genes are excellent targets for creating improved biomass crops (Berthet et al. 2011; Wang et al. 2015). In addition to high impact mutations, we identified 2587 non-synonymous mutations. These mutations can have a major impact on protein function if they change functionally important amino acids. As an example, we detected a previously described mutation in line *Bd4179* that alters the *BdCAD1* gene encoding a Cinnamyl Alcohol Dehydrogenase (CAD) (Bouvier d'Yvoire et al. 2013). The mutation changes glycine to aspartic acid in the highly conserved glycine-rich NADPH binding site GLG**G**VG(188-195) (Bouvier d'Yvoire et al. 2013). Reduced CAD activity can result in altered lignin composition and sometimes also reduced lignin content. Phenotypically grasses with mutations in CAD often have a reddish midrib due to the incorporation of atypical lignin monomers into their lignin (Saballos et al. 2009; Sattler et al. 2010; Zhang et al. 2006). Not surprisingly, individuals from line *Bd4179* that were homozygous for the mutation in *BdCAD1* had reddish midribs and altered lignin composition (Bouvier d'Yvoire et al. 2013). We used this easily scored phenotype to select a heterozygous individual from *Bd4179* to include as one of the 18 sequenced mutants. We specifically wanted to test whether we could detect a heterozygous mutation with low coverage sequencing (<30 fold).

Table 3 Predicted functional consequence of sodium azide-induced mutations

Predicted effect or location	*Average number per line*
Frameshift[a]	8.9
Splice site acceptor[a]	1.9
Splice site donor[a]	1.9
Start lost[a]	0.2
Stop gained[a]	5.4
Total high impact mutations	18.3
Codon change plus codon deletion[b]	0.3
Codon change plus codon insertion[b]	2.6
Codon deletion[b]	0.3
Codon insertion[b]	1.7
Non synonymous coding[b]	138.9
Total moderate impact mutations	143.9
Start gained	0.7
Stop lost	0.2
Synonymous coding	68.4
3′ UTR	50.3
5′ UTR	15.6
Others	135.1
Effect by regions	*Average number*
Exon	227.0
Intron	375.6
Downstream	1337.2
Upstream	1394.4
Intergenic	2147.3

[a]High impact mutation
[b]Moderate impact mutation

The G575A transition in the sequenced line was found in only three reads among seven that covered the region of interest. While this result illustrates that heterozygous mutations are detectable with our pipeline, it also suggests that sequence coverage should not go below 15 fold and that filtering should be applied carefully to low coverage regions as discussed above. Using a CIRCOS plot (Krzywinski et al. 2009) we mapped the location and predicted effect of each mutation along the five chromosomes (Fig. 3). Overall, the mutations were evenly distributed indicating that there is no major bias in mutation frequency.

Scaling Up

Based on our preliminary results there are no significant hurdles to applying this approach to a larger population. This is exactly the goal of a recently initiated project at the DOE Joint Genome Institute through the Community Science

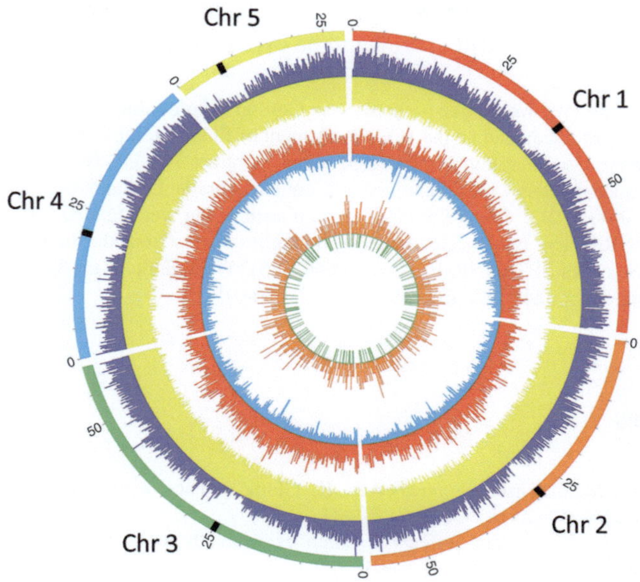

Fig. 3 Circos plot illustrating the position of NaN_3-induced SNPs along the chromosomes. Outside colored bars represents *B. distachyon* chromosomes with centromere position (*black bands*). From the center to the edge: the mutations with high impact (*green*), the non-synonymous mutations (*orange*), all detected SNPs (*blue*), the EMS-type SNPs (*red*), the GC content percentage (*yellow*), gene density (*purple*) are represented. Circos software was used for plotting (Krzywinski et al. 2009)

Program (http://jgi.doe.gov/our-projects/csp-plans/fy-2015-csp-plans/). The initial goal of this project is to sequence up to 2000 mutants from three different populations treated with EMS, NaN_3 or FNR. The first 92 mutants have been sequenced and are currently being analyzed to select the best mix of mutagen, dose and sequencing depth for the remaining sequencing (unpublished). When completed, the sequenced mutants will be an unprecedented resource for forward and reverse genetics.

Conclusion

Physical and chemical mutagenesis have been used for nearly a century to generate useful genetic variation in both crops and model organisms. Methods for the mutagenesis of *B. distachyon* have been developed and several mutant populations created. TILLING has been successfully used by at least three laboratories and TILLING services are available to all researchers. A pilot project to evaluate the feasibility of WGS for the genome wide identification of mutations identified up to 1700 SNPs per NaN_3 mutagenized line. This set the stage for a recently initiated project to sequence up to 2000 mutants. Thus, chemical and radiation mutagenesis of *B. distachyon* are and will continue to be extremely important research tools.

Acknowledgments We would like to thank Sean Gordon for providing high and low confidence SNP files. The work conducted by the U.S. Department of Energy Joint Genome Institute, a DOE Office of Science User Facility, is supported under Contract No. DE-AC02-05CH11231. Yin Wang was funded by the China Scholarship Council. Aurélie Lemaire was funded BRAVO (ANR-14-CE19-0012-01). The preliminary data on sequencing were funded by the SILICOTIL project from INRA. The **Institut Jean-Pierre Bourgin** benefits from the support of the LabEx Saclay Plant Sciences-SPS (ANR-10-LABX-0040-SPS). Work done at the U.S. Department of Agriculture was supported by CRIS project 2030-21000-019.

References

Ahloowalia BS, Maluszynski M, Nichterlein K. Global impact of mutation-derived varieties. Euphytica. 2004;135(2):187–204.

Al-Qurainy F, Khan S. Mutagenic effects of sodium azide and its application in crop improvement. World Appl Sci J. 2009;6(12):1589–601.

Austin RS, Vidaurre D, Stamatiou G, Breit R, Provart NJ, Bonetta D, et al. Next-generation mapping of Arabidopsis genes. Plant J. 2011;67(4):715–25.

Belfield EJ, Gan X, Mithani A, Brown C, Jiang C, Franklin K, et al. Genome-wide analysis of mutations in mutant lineages selected following fast-neutron irradiation mutagenesis of *Arabidopsis thaliana*. Genome Res. 2012;22(7):1306–15.

Berthet S, Demont-Caulet N, Pollet B, Bidzinski P, Cézard L, Le Bris P, et al. Disruption of LACCASE4 and 17 results in tissue-specific alterations to lignification of *Arabidopsis thaliana* stems. Plant Cell. 2011;23(3):1124–37.

Bolon Y-T, Haun WJ, Xu WW, Grant D, Stacey MG, Nelson RT, et al. Phenotypic and genomic analyses of a fast neutron mutant population resource in soybean. Plant Physiol. 2011;156 (1):240–53.

Bouvier d'Yvoire M, Bouchabke-Coussa O, Voorend W, Antelme S, Cézard L, Légée F, et al. Disrupting the cinnamyl alcohol dehydrogenase 1 gene (*BdCAD1*) leads to altered lignification and improved saccharification in *Brachypodium distachyon*. Plant J. 2013;73 (3):496–508.

Bruce M, Hess A, Bai J, Mauleon R, Diaz MG, Sugiyama N, et al. Detection of genomic deletions in rice using oligonucleotide microarrays. BMC Genomics. 2009;10(1):129.

Bruggemann EP, Doan B, Handwerger K, Storz G. Characterization of an unstable allele of the Arabidopsis HY4 locus. Genetics. 1998;149(3):1575–85.

Brutnell TP, Bennetzen JL, Vogel JP. *Brachypodium distachyon* and *Setaria viridis*: model genetic systems for the grasses. Annu Rev Plant Biol. 2015;66(1):465–85.

Dalmais M, Antelme S, Ho-Yue-Kuang S, Wang Y, Darracq O, d'Yvoire MB, et al. A TILLING platform for functional genomics in *Brachypodium distachyon*. PLoS One. 2013;8(6):e65503.

Derbyshire P, Byrne ME. MORE SPIKELETS1 is required for spikelet fate in the inflorescence of *Brachypodium*. Plant Physiol. 2013;161(3):1291–302.

Engvild KC. Mutagenesis of the model grass *Brachypodium distachyon* with sodium azide. Roskilde, Denmark: Report; 2005.

Fekih R, Takagi H, Tamiru M, Abe A, Natsume S, Yaegashi H, et al. MutMap+: genetic mapping and mutant identification without crossing in rice. PLoS One. 2013;8(7):e68529.

Gady ALF, Hermans FWK, Van de Wal MHBJ, et al. Implementation of two high through-put techniques in a novel application: detecting point mutations in large EMS mutated plant populations. Plant Methods. 2009. doi:10.1186/1746-4811-5-13.

Gordon SP, Priest H, Marais Des DL, et al. Genome diversity in *Brachypodium distachyon*: deep sequencing of highly diverse inbred lines. Plant J. 2014;79:361–74. doi:10.1111/tpj.12569.

Greene EA, Codomo CA, Taylor NE, Henikoff JG, Till BJ, Reynolds SH, et al. Spectrum of chemically induced mutations from a large-scale reverse-genetic screen in Arabidopsis. Genetics. 2003;164(2):731–40.

Gruszka D, Szarejko I, Maluszynski M. Sodium azide as a mutagen. In: Shu QY, Forster BP, Nakagawa H, editors. Plant mutation breeding and biotechnology. Wallingford: CABI; 2012. p. 159–66.

Henikoff S, Till BJ, Comai L. TILLING. Traditional mutagenesis meets functional genomics. Plant Physiol. 2004;135:630–6. doi:10.1104/pp. 104.041061.

Henry IM, Nagalakshmi U, Lieberman MC, et al. Efficient genome-wide detection and cataloging of EMS-induced mutations using exome capture and next-generation sequencing. Plant Cell. 2014;26:1382–97. doi:10.1105/tpc.113.121590.

Hodgdon AL, Marcus AH, Arenaz P, Rosichan JL, Bogyo TP, Nilan RA. Ontogeny of the barley plant as related to mutation expression and detection of pollen mutations. Environ Health Perspect. 1981;37:5–7.

Hoffmann GR. Genetic effects of dimethyl sulfate, diethyl sulfate, and related compounds. Mutat Res. 1980;75(1):63–129.

Hou X, Li L, Peng Z, Wei B, Tang S, Ding M, et al. A platform of high-density INDEL/CAPS markers for map-based cloning in Arabidopsis. Plant J. 2010;63(5):880–8.

Kim Y, Schumaker KS, Zhu J-K. EMS mutagenesis of Arabidopsis. Methods Mol Biol. 2006;323:101–3.

Koornneef M, Dellaert LW, van der Veen JH. EMS- and radiation-induced mutation frequencies at individual loci in *Arabidopsis thaliana* (L.) Heynh. Mutat Res. 1982;93(1):109–23.

Kovalchuk I, Kovalchuk O, Hohn B. Genome-wide variation of the somatic mutation frequency in transgenic plants. EMBO J. 2000;19(17):4431–8.

Krieg DR. Ethyl methanesulfonate-induced reversion of bacteriophage T4rII mutants. Genetics. 1963;48(4):561–80.

Krzywinski MI, Schein JE, Birol I, Connors J, Gascoyne R, Horsman D, et al. Circos: an information aesthetic for comparative genomics. Genome Res. 2009;19(9):1639–45.

LaVelle JM, Mangold JB. Structure-activity relationships of the azide metabolite, azidoalanine, in *S. typhimurium*. Mutat Res. 1987;177(1):27–33.

Lee MB, Kim DY, Jeon WB, Hong MJ, Lee YJ, Bold O, et al. Effect of Gamma radiation on growth and lignin content in *Brachypodium distachyon*. J Crop Sci Biotechnol. 2013;16:105–10.

Li X, Lassner M, Zhang Y. Deleteagene: a fast neutron deletion mutagenesis-based gene knockout system for plants. Comp Funct Genomics. 2002;3(2):158–60.

Lochlainn SO, Amoah S, Graham NS, et al. High Resolution Melt (HRM) analysis is an efficient tool to genotype EMS mutants in complex crop genomes. Plant Methods. 2011. doi:10.1186/1746-4811-7-43.

Maple J, Møller SG. Mutagenesis in Arabidopsis. Methods Mol Biol. 2007;362:197–206.

McCallum CM, Comai L, Greene EA, Henikoff S. Targeting Induced Local Lesions IN Genomes (TILLING) for plant functional genomics. Plant Physiol. 2000a;123(2):439–42.

McCallum CM, Comai L, Greene EA, Henikoff S. Targeted screening for induced mutations. Nat Biotechnol. 2000b;18(4):455–7.

Muller HJ. Artifial transmutation of the gene. Science. 1927;66(1699):84–7.

Nilan RA. Increasing the effectiveness, efficiency, and specificity of mutation induction in flowering plants. Basic Life Sci. 1973;2:205–22.

O'Rourke JA, Iniguez LP, Bucciarelli B, Roessler J, Schmutz J, McClean PE, et al. A re-sequencing based assessment of genomic heterogeneity and fast neutron-induced deletions in a common bean cultivar. Front Plant Sci. 2013;4:210.

Olsen O, Wang X, von Wettstein D. Sodium azide mutagenesis: preferential generation of A.T→G.C transitions in the barley Ant18 gene. Proc Natl Acad Sci. 1993;90(17):8043–7.

Owais WM. The lack of L-azidoalanine interaction with DNA. In vitro studies. Mutat Res. 1993;302(3):147–51.

Owais WM, Kleinhofs A. Metabolic-activation of the mutagen azide in biological-systems. Mutat Res. 1988;197(2):313–23.

Owais WM, Ronald RC, Kleinhofs A, Nilan RA. Synthesis and mutagenicity of the 2 stereoisomers of an azide metabolite (azidoalanine). Mutat Res. 1986;175(3):121–6.

Page DR, Grossniklaus U. The art and design of genetic screens: *Arabidopsis thaliana*. Nat Rev Genet. 2002;3(2):124–36.

Pathirana R. Plant mutation breeding in agriculture. CAB Rev Perspect Agr Vet Sci Nutr Nat Res. 2011;6(032):1–20.

Perez-de-Castro AM, Vilanova S, Canizares J, et al. Application of genomic tools in plant breeding. Curr Genomics. 2012;13:179–95.

Qu L-J, Qin G. Generation and identification of Arabidopsis EMS mutants. Methods Mol Biol. 2014;1062:225–39.

Rines HW. Sodium-azide mutagenesis in diploid and hexaploid oats and comparison with ethyl methanesulfonate treatments. Environ Exp Bot. 1985;25(1):7–16.

Rogers C, Wen J, Chen R, Oldroyd G. Deletion-based reverse genetics in Medicago truncatula. Plant Physiol. 2009;151(3):1077–86.

Saballos A, Ejeta G, Sanchez E, Kang C, Vermerris W. A genome wide analysis of the cinnamyl alcohol dehydrogenase family in sorghum [*Sorghum bicolor* (L.) Moench] identifies SbCAD2 as the brown midrib6 gene. Genetics. 2009;181(2):783–95.

Sattler SE, Funnell-Harris DL, Pedersen JF. Efficacy of singular and stacked brown midrib 6 and 12 in the modification of lignocellulose and grain chemistry. J Agric Food Chem. 2010;58 (6):3611–6.

Schneeberger K, Ossowski S, Lanz C, Juul T, Petersen AH, Nielsen KL, et al. SHOREmap: simultaneous mapping and mutation identification by deep sequencing. Nat Methods. 2009;6 (8):550–1.

Stadler LJ. Mutations in barley induced by X-rays and radium. Science. 1928a;68(1756):186–7.

Stadler LJ. Genetic effects of x-rays in maize. Proc Natl Acad Sci. 1928b;14(1):69–75.

Sugihara N, Higashigawa T, Aramoto D, Kato A. Haploid plants carrying a sodium azide-induced mutation (fdr1) produce fertile pollen grains due to first division restitution (FDR) in maize (*Zea mays* L.). Theor Appl Genet. 2013;126(12):2931–41.

Till BJ, Colbert T, Tompa R, Enns LC, Codomo CA, Johnson JE, et al. High-throughput TILLING for functional genomics. Methods Mol Biol. 2003;236:205–20.

Till BJ, Cooper J, Tai TH, Colowit P, Greene EA, Henikoff S, et al. Discovery of chemically induced mutations in rice by TILLING. BMC Plant Biol. 2007;7(1):19.

Triques K, Sturbois B, Gallais S, Dalmais M, Chauvin S, Clepet C, et al. Characterization of Arabidopsis thaliana mismatch specific endonucleases: application to mutation discovery by TILLING in pea. Plant J. 2007;51(6):1116–25.

Vattem DA, Jang HD, Levin R, Shetty K. Synergism of cranberry phenolics with ellagic acid and rosmarinic acid for antimutagenic and DNA protection functions. J Food Biochem. 2006;30 (1):98–116.

Wang L, Hu W, Sun J, Liang X, Yang X, Wei S, et al. Genome-wide analysis of *SnRK* gene family in *Brachypodium distachyon* and functional characterization of *BdSnRK2.9*. Plant Sci. 2015;237:33–45.

Wang TL, Uauy C, Robson F, Till B. TILLING in extremis. Plant Biotechnol J. 2012;10:761–72. doi:10.1111/j.1467-7652.2012.00708.x.

Wu J-L, Wu C, Lei C, Baraoidan M, Bordeos A, Madamba MRS, et al. Chemical- and irradiation-induced mutants of indica rice IR64 for forward and reverse genetics. Plant Mol Biol. 2005;59 (1):85–97.

Yadav D, Anad G, Dubey AK, Gupta S, Yadav S. Patents in the era of genomics: an overview. Recent Pat DNA Gene Seq. 2012;6(2):127–44.

Zhang K, Qian Q, Huang Z, Wang Y, Li M, Hong L, et al. GOLD HULL AND INTERNODE2 encodes a primarily multifunctional cinnamyl-alcohol dehydrogenase in rice. Plant Physiol. 2006;140(3):972–83.

Library Resources: BACs, ESTs, Full-Length cDNAs, and Y2H

Keiichi Mochida and Kazuo Shinozaki

Abstract Library resources are essential tools for functional genomics studies in model species. In this chapter, we summarize the library resources available for *Brachypodium*, including libraries of bacterial artificial chromosomes (BACs), cDNAs, and yeast two-hybrid (Y2H) data. These resources enable the *Brachypodium* research community to explore useful genes and elucidate gene function involved in the biological properties of grasses.

Keywords BAC library • cDNA • Yeast two-hybrid • *Brachypodium*

BAC Libraries

Large-insert genomic DNA libraries are essential resources for genomic approaches in modern biology such as genome sequencing and positional gene cloning (Sasaki et al. 2005). The bacterial artificial chromosome (BAC) is a vector system for cloning and manipulating large DNA molecules. BACs have several advantages, such as the stability of the inserted DNA and ease of handling, compared to the alternative yeast artificial chromosome (YAC) system (O'Connor et al. 1989;

K. Mochida (✉)
Cellulose Production Research Team, Biomass Engineering Research Division, RIKEN Center for Sustainable Resource Science, 1-7-22 Suehiro-cho, Tsurumi-ku, Yokohama, Kanagawa 230-0045, Japan

Gene Discovery Research Group, RIKEN Center for Sustainable Resource Science, 1-7-22 Suehiro-cho, Tsurumi-ku, Yokohama, Kanagawa 230-0045, Japan

Kihara Institute for Biological Research, Yokohama City University, 641-12 Maioka-cho, Totsuka-ku, Yokohama, Kanagawa 244-0813, Japan
e-mail: keiichi.mochida@riken.jp

K. Shinozaki
Gene Discovery Research Group, RIKEN Center for Sustainable Resource Science, 1-7-22 Suehiro-cho, Tsurumi-ku, Yokohama, Kanagawa 230-0045, Japan

Biomass Research Platform Team, Biomass Engineering Research Division, RIKEN Center for Sustainable Resource Science, 3-1-1 Koyadai, Tsukuba, Ibaraki 305-0074, Japan

J. Vogel (ed.), *Genetics and Genomics of Brachypodium*,
Plant Genetics and Genomics: Crops Models, DOI 10.1007/7397_2015_4
Published online: 20 October 2015

Shizuya et al. 1992). Therefore, a number of BAC libraries have been constructed in various organisms and have become invaluable resources for promoting genome projects and genetic discoveries.

For genome sequencing, particularly when building chromosome-scale pseudomolecules that are usable as reference genome sequence data sets in higher eukaryotes, BAC libraries may serve as primary substrates. In a genome sequencing project, clones of BAC libraries are typically analyzed by restriction fragment "fingerprinting" to assemble overlapping genomic clones into contigs (Nelson and Soderlund 2009; Warren et al. 2006). BAC-end sequencing of the fingerprinted BAC clones is used to generate a physical map representing the target genome, which is usually integrated with a genetic map to provide sequence-ready tilling paths (Mascher and Stein 2014). Recently, the combined use of BAC-end sequencing and next-generation sequencer-based short-read sequencing has efficiently balanced the throughput of sequencing and quality of the resulting genome sequence (Ariyadasa and Stein 2012).

The physical mapping of a genomic region containing a genetically mapped gene is the primary process for positional gene cloning. Following the construction of a high-resolution genetic map containing a genetically allocated target gene, chromosome walking to establish a BAC contig that covers the genomic region is accomplished by screening BAC clones that include sequence-tagged markers genetically linked with the target gene (Tanksley et al. 1995). The entire DNA sequence of the BAC contig is essential for identifying the target gene and characterizing genomic features. A detailed comparison of the coding sequence and of flanking regions such as promoters between ecotypes is important to elucidate the molecular basis of the phenotypic change of interest.

Several BAC libraries have been constructed for *Brachypodium* species, as summarized in Table 1. Farrar and Donnison (2007) published a detailed protocol for the construction and screening of BAC libraries for *Brachypodium* species and related grasses. In 2006, two BAC libraries of the inbred diploid *Brachypodium distachyon* line Bd21 were constructed using the restriction enzymes *Hin*dIII and

Table 1 BAC libraries constructed in *Brachypodium* species

Accessions	Enzymes	Ave. insert size (kb)	Coverage	No. of clones	References
Bd21	*Hin*dIII	100	13.5[a]	36,864	Huo et al. (2006)
	*Bam*HI	105	14.2[a]	36,864	
Bd21	*Eco*RI	124	16.8[a]	36,864	Febrer et al. (2010)
Bd21	*Hin*dIII	128	17.3[a]	36,864	http://www.genome.arizona.edu/orders/
Bd3-1	*Hin*dIII	130	–	36,864	http://www.genome.arizona.edu/orders/
ABR1/ABR5	*Hin*dIII	88	2.9[a]	9100	Hasterok et al. (2006)
B. sylvaticum	*Hin*dIII	102	6.6[b]	30,228	Foote et al. (2004)

[a]Based on a genome size of 272 Mb
[b]Based on a genome size of 470 Mb

*Bam*HI (Huo et al. 2006). This library resource consists of 73,728 clones arrayed to one hundred ninety-two 384-well plates. For the *Hin*dIII and *Bam*HI libraries, the estimated average insert size was 100 kb and 105 kb, respectively, and the coverage of the haploid genome was 9.4- and 9.9-fold, respectively, based on the genome size reported at the time for diploid *Brachypodium* (Huo et al. 2006). A total of 64,696 BAC-end sequences (BES) derived from the Bd21 BAC libraries, covering 10.9 % of the *Brachypodium* genome, were analyzed. These results indicated that 11.0 % of the *Brachypodium* genome consists of known repetitive DNA and 21.2 % consists of coding sequences (Huo et al. 2008). Then, 67,151 clones of the Bd21 BAC libraries were subjected to fingerprinting analysis using the SNaPshot HICF BAC fingerprinting method (Gu et al. 2009). The constructed physical map consisted of 671 contigs and 2161 singleton clones spanning 414 Mb. A comparative analysis between *Brachypodium* and rice based on the BES, physical map, and the rice genome indicated high collinearities and the remnants of whole-genome duplication events in grass species (Gu et al. 2009). The information of the physical map has been made accessible as an online database (http://phymap.ucdavis.edu/brachypodium/). Another group constructed two BAC libraries of Bd21 by using *Eco*RI and *Hin*dIII. The BAC libraries were also used to carry out BES analysis and to construct a physical map (Febrer et al. 2010). The two BAC libraries of Bd21 and a library of Bd3-1 are available from the BAC/EST resource center of the Arizona Genomics Institute (http://www.genome.arizona.edu/orders/). The physical map and its associated database have become important genomic resources for completing *Brachypodium* genome sequencing. In the *Brachypodium* genome sequencing project, the Bd21 BAC libraries of Huo et al. (2006) and Febrer et al. (2010) were used to decipher the Bd21 genome. End sequencing of the BAC libraries, as well as of other types of DNA libraries with different-sized inserts, was used for whole-genome shotgun analysis. In total, 4,761,917 reads with 9.43-fold coverage of the estimated genome were assembled into 217 scaffolds covering 272.1 Mb of the *Brachypodium* genome (International Brachypodium Initiative 2010).

In another *Brachypodium* species, *B. sylvaticum*, a BAC library of 30,228 clones with an average insert size of 102 kb was constructed. This library achieved 6.6-fold coverage of the estimated genome and was arrayed onto two high-density colony filters (Foote et al. 2004). The *B. sylvaticum* BAC library was used to find marker sequences to screen BAC clones of hexaploid wheat based on a comparative mapping strategy. The Ph1 locus prevents the pairing and recombination of homoeologous chromosomes in hexaploid wheat, thereby generating its cytologically diploid-like behavior (Vega and Feldman 1998). Markers generated from the rice and *Brachypodium* comparative mapping were successfully used to form a physical map covering the wheat *Ph1* gene region, and 36 genes were identified in the sequenced region (Griffiths et al. 2006). Further analyses of this region revealed that the Ph1 locus consists of a cluster of Cdk-like genes, which are homologs of C2 in humans and mice, involved in preventing nonhomologous chromosome pairing at meiosis in hexaploid wheat (Al-Kaff et al. 2008).

Fluorescence in situ hybridization (FISH) using BACs as probes, known as "BAC landing" or "BAC-FISH," combines genomic sequences with chromosomal aspects. Hasterok et al. (2006) constructed a BAC library for the two *B. distachyon* genotypes, ABR1 and ABR5, and used the clones to BAC landing with FISH (Hasterok et al. 2006). A detailed FISH protocol using BAC for *Brachypodium* species has been described by Jenkins and Hasterok (2007). *Brachypodium* BAC clones were used in FISH analysis as probes to determine the positions of active and inactive centromeres in the five *B. distachyon* chromosomes (Qi et al. 2010). BAC-FISH was also applied to analyze the karyotype evolution of four *Brachypodium* species: diploid *B. distachyon* (2n = 10), diploid *B. sylvaticum* (2n = 18), allopolyploid (2n = 28) *B. pinnatum*, and *B. phoenicoides* (2n = 28) (Idziak et al. 2014). This study showed that dysploidy events commonly occurred in the diploid species and that two nested chromosome fusion events were specific to *B. distachyon* (Idziak et al. 2014).

cDNA Libraries and ESTs

Expressed sequence tags (ESTs) are generated by partial one-pass sequencing of randomly picked cDNA clones (Adams et al. 1993). The large-scale collection of cDNA clones and ESTs is an efficient approach for rapidly acquiring information on expressed genes. ESTs derived from various types of tissues at different developmental stages and under different stress conditions could provide significant information to facilitate a wider range of approaches in functional genomics (Mochida and Shinozaki 2010). For example, ESTs have been used for gene structural annotation in a whole-genome sequencing project, building a "gene atlas" to determine genome-scale gene expression patterns, designing expressed gene-oriented molecular markers and probes for microarrays, and conducting genome-scale intraspecific and interspecific comparative analysis of expressed genes (Mochida et al. 2003, 2006; Ogihara et al. 2003; Kawaura et al. 2005, 2006).

As of December 2014, 206,255 ESTs from *B. distachyon* are included in the National Center for Biotechnology Information (NCBI)'s dbEST, a public EST database (http://www.ncbi.nlm.nih.gov/dbEST/). In the first genome-scale collection of *Brachypodium* ESTs, 20,440 ESTs from five cDNA libraries–derived from leaves, stems plus leaf sheaths, roots, callus, and developing seeds–were generated using the Sanger sequencing method (Vogel et al. 2006). A subset of the *Brachypodium* ESTs was used for phylogenetic analysis with wheat and barley, demonstrating the close relationship of *Brachypodium* to the Triticeae crops.

The RNA-seq method, which uses the deep sequencing technologies of next-generation sequencers, allows for the precise measurement of transcript abundance and sequence structure of isoforms at a large scale with high throughput (Wang et al. 2009; Mochida and Shinozaki 2011). This approach rapidly establishes comprehensive EST resources. In the *Brachypodium* genome sequencing project, more than 20 cDNA libraries from various RNA sources, including both Sanger sequencing and next-generation sequencing techniques such as 454 pyrosequencing and Illumina sequencing, were subjected to EST analysis. In total, approximately 2.3 million 454 reads, 128,221 Sanger reads, and ~289 million Illumina reads were used to predict transcribed regions on the *Brachypodium* genome assembly (International

Brachypodium Initiative 2010). Brachypodium.org provides the EST data sets from Bd21 and several other accessions (ftp://brachypodium.org/brachypodium.org/ESTs/). The data set "454 sequencing of *Brachypodium distachyon* EST project" is available from the NCBI's SRA (SRX014502–SRX014507). The Illumina read data set is accessible in the SRA as SRA010177. Due to these EST resources, the structural annotation of expressed genes in the *Brachypodium* genome has greater accuracy than that of genes with only ab initio prediction.

Full-Length cDNA Library

Full-length cDNAs that include capped site sequences are suitable for investigating gene functions. The biotinylated cap trapper method for constructing full-length cDNA-enriched libraries, which uses trehalose-thermostabilized reverse transcriptase, was developed by the RIKEN Hayashizaki group about 15 years ago (Carninci et al. 1996; Carninci and Hayashizaki 1999). The large-scale collection and sequencing of full-length cDNA clones continue to be invaluable for life science projects studying various species (Hayashizaki 2003; Imanishi et al. 2004; Maeda et al. 2006).

The sequence resources of full-length cDNAs have been useful for precisely annotating gene structures such as transcription units, transcription start sites (TSSs), and isoforms in completed genome sequences (Seki et al. 2002; Iida et al. 2004; Rice Annotation Project et al. 2007; Yamamoto et al. 2009). In species with draft genomes or de novo genome assemblies, such as soybean, poplar, corn, and wheat, sequenced full-length cDNAs have contributed to the establishment of genomic tools for gene discovery studies and comparative analysis (Umezawa et al. 2008; Nanjo et al. 2007; Kawaura et al. 2009; Ralph et al. 2008; Soderlund et al. 2009). The clone resources of full-length cDNAs are useful for determining the three-dimensional (3D) structures of expressed proteins, as well as for investigating protein-ligand, protein-protein, and protein-nucleotide molecular interactions. Full-length cDNA libraries have also been used to analyze gene function *in planta* by creating overexpressors. Full-length cDNA overexpressor (FOX) gene hunting, which uses full-length cDNA transgenic plants as overexpressors, is an efficient method for the high-throughput discovery of functional genes with phenotypic effects (Higuchi-Takeuchi and Matsui 2014; Higuchi-Takeuchi et al. 2013; Kondou et al. 2009; Ichikawa et al. 2006).

In *Brachypodium*, a mixed full-length cDNA library was constructed using the biotinylated cap trapper method from 21 various tissues of the *B. distachyon* Bd21 accession by RIKEN (Table 2) (Mochida et al. 2013). These RNA source tissues included various normal developmental organ tissues, as well as root and leaf tissues under various abiotic stresses. The RIKEN *Brachypodium* full-length (RBFL) cDNAs clones were sequenced using the Sanger method. EST analysis using the full-length cDNA library identified 78,163 high-quality expressed sequence tags (ESTs) from both ends of approximately 40,000 clones, and 16,079 contigs were assembled as high-throughput cDNA (HTC) sequences. The ESTs and HTCs were used to update the gene structure annotations of *Brachypodium*.

Table 2 RNA samples for constructing a *Brachypodium* full-length cDNA library

	Tissue	Treatment	Growth condition
Normal tissues	Seed at germination		Hydroculture, growth chamber
	Shoot		Hydroculture, growth chamber
	Leaf at vegetative stage		Pot soil, greenhouse
	Leaf after flowering		Pot soil, greenhouse
	Root		Hydroculture, growth chamber
	Crown		Hydroculture, growth chamber
	Spikelet at flowering		Pot soil, greenhouse
	Spikelet (DAP1 ~ 5)		Pot soil, greenhouse
	Spikelet (DAP7 ~ 10)		Pot soil, greenhouse
	Spikelet (DAP20 ~ 30)		Pot soil, greenhouse
Callus			Culture on agarose gel, growth chamber
Stress tissues	Leaves at 2 weeks after germination	250 mM NaCl 5 h	Hydroculture, growth chamber
		250 mM NaCl 24 h	
		100 μM ABA 5 hr	
		100 μM ABA 24 h	
		Cold stress 4 °C 5 h	
		Cold stress 4 °C 24 h	
		Drought stress (on filter paper) 5 h	
		Drought stress (on filter paper) 24 h	
		Heat shock 42 °C 2 h	
		Wounding 1 h	

Approximately 10,000 nonredundant gene models were supported by the full-length cDNAs, and approximately 6000 models showed some modifications in their gene structures, particularly due to additions or extensions in the 5′ untranslated regions (UTRs). The full-length cDNA-based gene prediction indicated approximately 680 novel gene models, including 362 newly identified transcription units. The sequence information of the RBFL clones and their associated annotations have been released on the Web-accessible RIKEN *Brachypodium* FL cDNA database (RBFLDB) (http://brachy.bmep.riken.jp/ver.1/index.pl) (Mochida et al. 2013) (Fig. 1). The RBFL clones are also publicly available from the RIKEN Bio Resource Center (https://www.brc.riken.jp/lab/epd/catalog/b_distachyon.html).

The recent development of sequencing projects in the Triticeae plants, including wheat and barley, has furthered the creation of comprehensive sequence resources for full-length cDNAs and genomic sequence assemblies. In combination with the available sequence resources from the Pooideae, these resources increase the pace

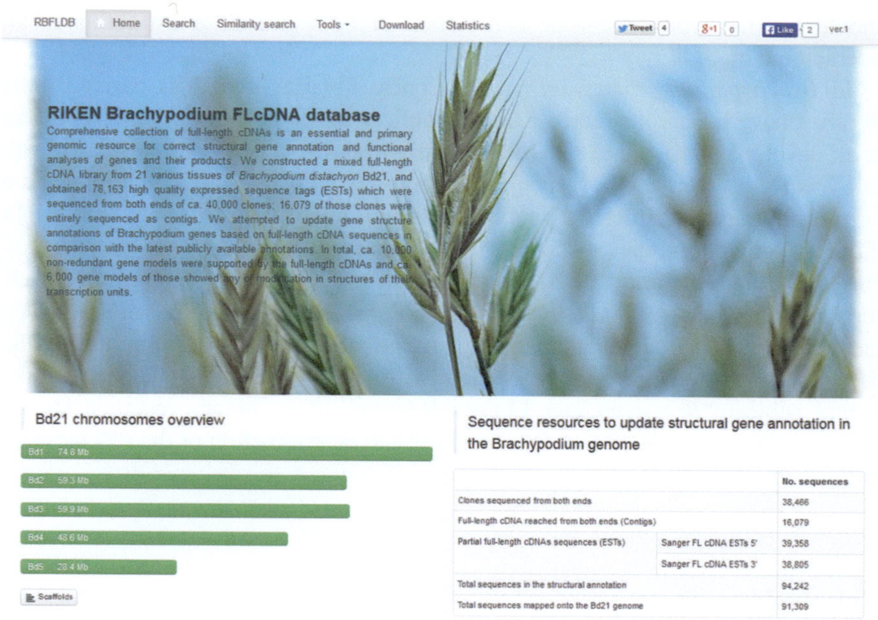

Fig. 1 Top page screenshot of the RIKEN *Brachypodium* full-length cDNA database (RBFLDB) website. Sequences and related information are accessible from the website at the following URL: http://brachy.bmep.riken.jp/ver.1/index.pl

of gene discovery and knowledge transfer between *Brachypodium* and Triticeae. In one such integration of Pooideae sequence resources, *Brachypodium*, wheat, and barley full-length cDNAs and genomic sequence assemblies were subjected to comparative analysis. This process allowed for the reciprocal integration of the RBFLDB and Triticeae full-length cDNA databases (TriFLDB) (Mochida et al. 2009). The resulting database represents a "one-stop" information resource for all Pooideae genomic information, allowing for easier functional analysis of genes in *Brachypodium* and the seamless application of this information to Triticeae crops (Mochida et al. 2013; Mochida and Shinozaki 2013).

Yeast Two-Hybrid Libraries

Protein-protein interactions (PPIs) mediate a variety of biological processes. The experimental investigation of protein interaction is essential for elucidating the molecular functions of genes. A genome-scale definition of PPIs that represents an "interactome map" is a useful resource for decoding cellular molecular networks (Zhang et al. 2010). The yeast two-hybrid (Y2H) assay is a widely used method for examining PPIs based on a reconstitution of a functional transcription factor by an interaction of two proteins (Rao et al. 2014). Therefore, Y2H-ready cDNA libraries should facilitate the large-scale exploration of PPIs and the construction of an interactome map.

In *Brachypodium*, Gateway™ cDNA entry libraries and their derived Y2H libraries have been constructed and are publicly available (Cao et al. 2011). Cao et al. developed two cDNA Gateway™ entry libraries: a photoperiod library with mRNAs from green shoots grown under long-day and short-day conditions and a hormone library with mRNAs from the roots and shoots of plants treated with eight different phytohormones. These entry libraries were then used to develop corresponding Y2H-ready cDNA libraries. These authors also demonstrated Y2H screening using *Brachypodium* NF-YC orthologs. According to Cao et al. (2011), these libraries are freely available for distribution to the research community (the user pays for the shipping) on request to Dr. Ben Holt (benholt@ou.edu) with the e-mail subject "*Brachypodium* Library Request." This library resource should facilitate the screening of protein interactors and determination of the PPI network in *Brachypodium*.

Conclusion

Library resources have been invaluable aids for genome sequencing, gene discoveries, and characterizations of gene function. Those resources should continue to facilitate various scientific projects that use *Brachypodium* as a model for grass species.

References

Adams MD, Kerlavage AR, Fields C, Venter JC. 3,400 new expressed sequence tags identify diversity of transcripts in human brain. Nat Genet. 1993;4(3):256–67. doi:10.1038/ng0793-256.

Al-Kaff N, Knight E, Bertin I, Foote T, Hart N, Griffiths S, et al. Detailed dissection of the chromosomal region containing the Ph1 locus in wheat Triticum aestivum: with deletion mutants and expression profiling. Ann Bot. 2008;101(6):863–72. doi:10.1093/aob/mcm252.

Ariyadasa R, Stein N. Advances in BAC-based physical mapping and map integration strategies in plants. J Biomed Biotechnol. 2012;2012:184854. doi:10.1155/2012/184854.

Cao S, Siriwardana CL, Kumimoto RW, Holt III BF. Construction of high quality Gateway entry libraries and their application to yeast two-hybrid for the monocot model plant Brachypodium distachyon. BMC Biotechnol. 2011;11:53. doi:10.1186/1472-6750-11-53.

Carninci P, Hayashizaki Y. High-efficiency full-length cDNA cloning. Methods Enzymol. 1999;303:19–44.

Carninci P, Kvam C, Kitamura A, Ohsumi T, Okazaki Y, Itoh M, et al. High-efficiency full-length cDNA cloning by biotinylated CAP trapper. Genomics. 1996;37(3):327–36. doi:10.1006/geno.1996.0567.

Febrer M, Goicoechea JL, Wright J, McKenzie N, Song X, Lin J, et al. An integrated physical, genetic and cytogenetic map of Brachypodium distachyon, a model system for grass research. PLoS One. 2010 Oct 18;5(10):e13461. doi:10.1371/journal.pone.0013461.

Farrar K, Donnison IS. Construction and screening of BAC libraries made from Brachypodium genomic DNA. Nat Protoc. 2007;2(7):1661–74.

Foote TN, Griffiths S, Allouis S, Moore G. Construction and analysis of a BAC library in the grass Brachypodium sylvaticum: its use as a tool to bridge the gap between rice and wheat in elucidating gene content. Funct Integr Genomics. 2004;4(1):26–33. doi:10.1007/s10142-003-0101-y.

Griffiths S, Sharp R, Foote TN, Bertin I, Wanous M, Reader S, et al. Molecular characterization of Ph1 as a major chromosome pairing locus in polyploid wheat. Nature. 2006;439(7077):749–52. doi:10.1038/nature04434.

Gu YQ, Ma Y, Huo N, Vogel JP, You FM, Lazo GR, et al. A BAC-based physical map of Brachypodium distachyon and its comparative analysis with rice and wheat. BMC genomics. 2009;10:496. doi:10.1186/1471-2164-10-496.

Hasterok R, Marasek A, Donnison IS, Armstead I, Thomas A, King IP, et al. Alignment of the genomes of Brachypodium distachyon and temperate cereals and grasses using bacterial artificial chromosome landing with fluorescence in situ hybridization. Genetics. 2006;173 (1):349–62. doi:10.1534/genetics.105.049726.

Hayashizaki Y. RIKEN mouse genome encyclopedia. Mech Ageing Dev. 2003;124(1):93–102.

Higuchi-Takeuchi M, Matsui M. Screening for gene function using the FOX (full-length cDNA overexpressor gene) hunting system. Methods Mol Biol. 2014;1056:201–10. doi:10.1007/978-1-62703-592-7_19.

Higuchi-Takeuchi M, Mori M, Matsui M. High-throughput analysis of rice genes by means of the heterologous full-length cDNA overexpressor (FOX)-hunting system. Int J Dev Biol. 2013;57 (6-8):517–23. doi:10.1387/ijdb.130176mm.

Huo N, Gu YQ, Lazo GR, Vogel JP, Coleman-Derr D, Luo MC, et al. Construction and characterization of two BAC libraries from Brachypodium distachyon, a new model for grass genomics. Genome. 2006;49(9):1099–108. doi:10.1139/g06-087.

Huo N, Lazo GR, Vogel JP, You FM, Ma Y, Hayden DM, et al. The nuclear genome of Brachypodium distachyon: analysis of BAC end sequences. Funct Integr Genomics. 2008;8 (2):135–47. doi:10.1007/s10142-007-0062-7.

Ichikawa T, Nakazawa M, Kawashima M, Iizumi H, Kuroda H, Kondou Y, et al. The FOX hunting system: an alternative gain-of-function gene hunting technique. Plant J. 2006;48(6):974–85. doi:10.1111/j.1365-313X.2006.02924.x.

Idziak D, Hazuka I, Poliwczak B, Wiszynska A, Wolny E, Hasterok R. Insight into the karyotype evolution of brachypodium species using comparative chromosome barcoding. PLoS One. 2014;9(3):e93503. doi:10.1371/journal.pone.0093503.

Iida K, Seki M, Sakurai T, Satou M, Akiyama K, Toyoda T, et al. Genome-wide analysis of alternative pre-mRNA splicing in Arabidopsis thaliana based on full-length cDNA sequences. Nucleic Acids Res. 2004;32(17):5096–103. doi:10.1093/nar/gkh845.

Imanishi T, Itoh T, Suzuki Y, O'Donovan C, Fukuchi S, Koyanagi KO, et al. Integrative annotation of 21,037 human genes validated by full-length cDNA clones. PLoS Biol. 2004;2(6), e162. doi:10.1371/journal.pbio.0020162.

International Brachypodium Initiative. Genome sequencing and analysis of the model grass Brachypodium distachyon. Nature. 2010;463(7282):763–8. doi:10.1038/nature08747.

Jenkins G, Hasterok R. BAC 'landing' on chromosomes of Brachypodium distachyon for comparative genome alignment. Nat Protoc. 2007;2(1):88–98.

Kawaura K, Mochida K, Enju A, Totoki Y, Toyoda A, Sakaki Y, et al. Assessment of adaptive evolution between wheat and rice as deduced from full-length common wheat cDNA sequence data and expression patterns. BMC Genomics. 2009;10:271. doi:10.1186/1471-2164-10-271.

Kawaura K, Mochida K, Ogihara Y. Expression profile of two storage-protein gene families in hexaploid wheat revealed by large-scale analysis of expressed sequence tags. Plant Physiol. 2005;139(4):1870–80. doi:10.1104/pp. 105.070722.

Kawaura K, Mochida K, Yamazaki Y, Ogihara Y. Transcriptome analysis of salinity stress responses in common wheat using a 22k oligo-DNA microarray. Funct Integr Genomics. 2006;6(2):132–42. doi:10.1007/s10142-005-0010-3.

Kondou Y, Higuchi M, Takahashi S, Sakurai T, Ichikawa T, Kuroda H, et al. Systematic approaches to using the FOX hunting system to identify useful rice genes. Plant J. 2009;57 (5):883–94. doi:10.1111/j.1365-313X.2008.03733.x.

Maeda N, Kasukawa T, Oyama R, Gough J, Frith M, Engstrom PG, et al. Transcript annotation in FANTOM3: mouse gene catalog based on physical cDNAs. PLoS Genet. 2006;2(4), e62. doi:10.1371/journal.pgen.0020062.

Mascher M, Stein N. Genetic anchoring of whole-genome shotgun assemblies. Front Genet. 2014;5:208. doi:10.3389/fgene.2014.00208.

Mochida K, Kawaura K, Shimosaka E, Kawakami N, Shin IT, Kohara Y, et al. Tissue expression map of a large number of expressed sequence tags and its application to in silico screening of stress response genes in common wheat. Mol Genet Genomics. 2006;276(3):304–12. doi:10. 1007/s00438-006-0120-1.

Mochida K, Shinozaki K. Genomics and bioinformatics resources for crop improvement. Plant Cell Physiol. 2010;51(4):497–523. doi:10.1093/pcp/pcq027.

Mochida K, Shinozaki K. Advances in omics and bioinformatics tools for systems analyses of plant functions. Plant Cell Physiol. 2011;52(12):2017–38. doi:10.1093/pcp/pcr153.

Mochida K, Shinozaki K. Unlocking Triticeae genomics to sustainably feed the future. Plant Cell Physiol. 2013;54(12):1931–50. doi:10.1093/pcp/pct163.

Mochida K, Uehara-Yamaguchi Y, Takahashi F, Yoshida T, Sakurai T, Shinozaki K. Large-scale collection and analysis of full-length cDNAs from Brachypodium distachyon and integration with Pooideae sequence resources. PLoS One. 2013;8(10), e75265. doi:10.1371/journal.pone. 0075265.

Mochida K, Yamazaki Y, Ogihara Y. Discrimination of homoeologous gene expression in hexaploid wheat by SNP analysis of contigs grouped from a large number of expressed sequence tags. Mol Genet Genomics. 2003;270(5):371–7. doi:10.1007/s00438-003-0939-7.

Mochida K, Yoshida T, Sakurai T, Ogihara Y, Shinozaki K. TriFLDB: a database of clustered full-length coding sequences from Triticeae with applications to comparative grass genomics. Plant Physiol. 2009;150(3):1135–46. doi:10.1104/pp. 109.138214.

Nanjo T, Sakurai T, Totoki Y, Toyoda A, Nishiguchi M, Kado T, et al. Functional annotation of 19,841 Populus nigra full-length enriched cDNA clones. BMC Genomics. 2007;8:448. doi:10. 1186/1471-2164-8-448.

Nelson W, Soderlund C. Integrating sequence with FPC fingerprint maps. Nucleic Acids Res. 2009;37(5):e36. doi:10.1093/nar/gkp034.

O'Connor M, Peifer M, Bender W. Construction of large DNA segments in Escherichia coli. Science. 1989;244(4910):1307–12.

Ogihara Y, Mochida K, Nemoto Y, Murai K, Yamazaki Y, Shin IT, et al. Correlated clustering and virtual display of gene expression patterns in the wheat life cycle by large-scale statistical analyses of expressed sequence tags. Plant J. 2003;33(6):1001–11.

Qi L, Friebe B, Wu J, Gu Y, Qian C, Gill BS. The compact Brachypodium genome conserves centromeric regions of a common ancestor with wheat and rice. Funct Integr Genomics. 2010;10(4):477–92. doi:10.1007/s10142-010-0190-3.

Ralph SG, Chun HJ, Cooper D, Kirkpatrick R, Kolosova N, Gunter L, et al. Analysis of 4,664 high-quality sequence-finished poplar full-length cDNA clones and their utility for the discovery of genes responding to insect feeding. BMC Genomics. 2008;9:57. doi:10.1186/1471-2164-9-57.

Rao VS, Srinivas K, Sujini GN, Kumar GN. Protein-protein interaction detection: methods and analysis. Int J Proteomics. 2014;2014:147648. doi:10.1155/2014/147648.

Rice Annotation Project, Itoh T, Tanaka T, Barrero RA, Yamasaki C, Fujii Y, et al. Curated genome annotation of Oryza sativa ssp. japonica and comparative genome analysis with Arabidopsis thaliana. Genome Res. 2007;17(2):175–83. doi:10.1101/gr.5509507.

Sasaki T, Matsumoto T, Antonio BA, Nagamura Y. From mapping to sequencing, post-sequencing and beyond. Plant Cell Physiol. 2005;46(1):3–13. doi:10.1093/pcp/pci503.

Seki M, Narusaka M, Kamiya A, Ishida J, Satou M, Sakurai T, et al. Functional annotation of a full-length Arabidopsis cDNA collection. Science. 2002;296(5565):141–5. doi:10.1126/science.1071006.

Shizuya H, Birren B, Kim UJ, Mancino V, Slepak T, Tachiiri Y, et al. Cloning and stable maintenance of 300-kilobase-pair fragments of human DNA in Escherichia coli using an F-factor-based vector. Proc Natl Acad Sci U S A. 1992;89(18):8794–7.

Soderlund C, Descour A, Kudrna D, Bomhoff M, Boyd L, Currie J, et al. Sequencing, mapping, and analysis of 27,455 maize full-length cDNAs. PLoS Genet. 2009;5(11), e1000740. doi:10.1371/journal.pgen.1000740.

Tanksley SD, Ganal MW, Martin GB. Chromosome landing: a paradigm for map-based gene cloning in plants with large genomes. Trends Genet. 1995;11(2):63–8.

Umezawa T, Sakurai T, Totoki Y, Toyoda A, Seki M, Ishiwata A, et al. Sequencing and analysis of approximately 40,000 soybean cDNA clones from a full-length-enriched cDNA library. DNA Res. 2008;15(6):333–46. doi:10.1093/dnares/dsn024.

Vega JM, Feldman M. Effect of the pairing gene Ph1 on centromere misdivision in common wheat. Genetics. 1998;148(3):1285–94.

Vogel JP, Gu YQ, Twigg P, Lazo GR, Laudencia-Chingcuanco D, Hayden DM, et al. EST sequencing and phylogenetic analysis of the model grass Brachypodium distachyon. Theor Appl Genet. 2006;113(2):186–95. doi:10.1007/s00122-006-0285-3.

Wang Z, Gerstein M, Snyder M. RNA-Seq: a revolutionary tool for transcriptomics. Nat Rev Genet. 2009;10(1):57–63. doi:10.1038/nrg2484.

Warren RL, Varabei D, Platt D, Huang X, Messina D, Yang SP, et al. Physical map-assisted whole-genome shotgun sequence assemblies. Genome Res. 2006;16(6):768–75. doi:10.1101/gr.5090606.

Yamamoto YY, Yoshitsugu T, Sakurai T, Seki M, Shinozaki K, Obokata J. Heterogeneity of Arabidopsis core promoters revealed by high-density TSS analysis. Plant J. 2009;60(2):350–62. doi:10.1111/j.1365-313X.2009.03958.x.

Zhang Y, Gao P, Yuan JS. Plant protein-protein interaction network and interactome. Curr Genomics. 2010;11(1):40–6. doi:10.2174/138920210790218016.

Brachypodium distachyon Genetic Resources

David F. Garvin

Abstract *Brachypodium distachyon* is a well-established model species for the grass family Poaceae. It possesses an array of features that make it suited for this purpose, including a small sequenced genome, simple transformation methods, and additional functional genomics tools. However, the most critical tool for any model species is an accumulation of genetic resources that can be exploited to explore the genetic and molecular basis of traits of interest that exhibit natural variation. The recent emergence of *B. distachyon* as a model species has led to an inversion in the normal progression of resource development, with genome resource development emerging faster than genetic resource development for *B. distachyon*. During the last several years, the scientific community has been collecting new *B. distachyon* germplasm and releasing freely available inbred lines from many regions of Europe. These resources will be critical for *B. distachyon* research. Despite having small flowers, crossing methods have been developed for *B. distachyon*, and this has led to the development of segregating populations that have been, and will continue to be, important resources for mapping and isolating genes of interest. This chapter provides a history of *B. distachyon* genetic resource development.

Keywords Grass crops • Genomics • Germplasm collections • Genetic diversity • Phenotypic variation

Abbreviations

AFLP	Amplified fragment length polymorphism
BAC	Bacterial artificial chromosome
BSMV	Barley stripe mosaic virus
cM	Centimorgan
EST	Expressed sequence tag
kb	Kilobase
RIL	Recombinant inbred line

D.F. Garvin, Ph.D. (✉)
USDA-ARS Plant Science Research Unit, University of Minnesota, 411 Borlaug Hall, 1991 Upper Buford Circle, St. Paul, MN 55108, USA
e-mail: David.Garvin@ars.usda.gov

J. Vogel (ed.), *Genetics and Genomics of Brachypodium*,
Plant Genetics and Genomics: Crops Models, DOI 10.1007/7397_2015_5
Published online: 02 October 2015

SNP	Single nucleotide polymorphism
SSR	Simple sequence repeat
USDA-NPGS	United States Department of Agriculture National Plant Germplasm System

Introduction

Biological research benefits greatly from the availability of shared community genetic stocks. The ability to validate results, and to unify and integrate findings both across labs and over time provides a strong biological impetus for the development, maintenance, and distribution of reference genotypes. Throughout the history of biology, reference genotypes have played a prominent research role, spanning microbes to more complex eukaryotes throughout the tree of life. For species that are considered model organisms, typically there is extensive background knowledge on all biological aspects of the organism itself. Then, as new biological disciplines emerge, reference genotypes can be employed for new research in a seamless manner, thus building upon a strong and deep foundation of knowledge and resources.

Unlike most other species considered model organisms, *Brachypodium distachyon* has followed a somewhat unusual historical pattern of development that deviates from organisms which, in some cases, have more than a century of accumulated research results behind them. Indeed, before the publication of the seminal paper by Draper, Mur and colleagues (2001), it is fair to state that the existence of *B. distachyon* was unknown to essentially all of the plant research community. For that reason, there was almost no biological knowledge on this species to aid scientists interested in adopting *B. distachyon* for scientific research, nor were there typical resources that, in other species, had built up from previous research. In an era of rapid technological advances in techniques such as DNA and RNA sequencing, computational biology, and a plethora of 'omics methods, the time it takes to develop significant genome resources is short and increasingly inexpensive. In contrast, the basic biology of an organism still governs the development of some other key resources for a model organism, most importantly genetic resources. Thus, with the Draper et al. (2001) publication proposing *B. distachyon* as a model plant, both genome and genetic resource development began simultaneously, which contrasts greatly from the most widely recognized model plant, *Arabidopsis thaliana*, which has been the subject of research dating back more than a century (Somerville and Koornneef 2002). This chapter aims to give an overview of *B. distachyon* natural genetic resource development.

Early Genetic Resources

In 2001, 52 *B. distachyon* ecotypes, designated "ABR", and of varying ploidy, were maintained at the University of Aberystwyth in Wales, and of these 12 were considered diploid, with a base chromosome number of 5; the others were considered either tetraploid or hexaploid (Draper et al. 2001). With the recent revision of the systematics of this polyploidy series into three distinct species, *B. hybridum* (2n = 30), *B. staceii* (2n = 20), and *B. distachyon* (2n = 10) (Catalan et al. 2012), the original 12 "diploid" ABR ecotypes represented the entire germplasm base for what is now considered *B. distachyon*, upon which to develop this new model species. Obtaining seeds from the Aberystwyth collection was possible under a formal agreement with the university.

In an effort to expand the *B. distachyon* germplasm base, in 2002 a search of the United States Department of Agriculture's National Plant Germplasm System (USDA-NPGS) was conducted for *B. distachyon* germplasm that may have been deposited, to identify additional *B. distachyon* collections that would complement the germplasm resources to the Aberystwyth collection; 28 accessions listed as *B. distachyon* were obtained, although one was misclassified (this has since been corrected). While compiling details of the provenance of these USDA-NPGS accessions, similarity was observed between their collection locations and those listed for the Aberystwyth ecotype collection (Jenkins et al. 2003), which suggested some overlap. After various lines of investigation and communication, it was possible to determine that the majority of the Aberystwyth ecotypes derive from the USDA-NPGS *B. distachyon* accessions (Table 1). In fact each USDA-NPGS *B. distachyon* accession was represented by an ABR ecotype in the Aberystwyth collection. The route by which the USDA-NPGS *B. distachyon* accessions—which are freely available as is the case for all USDA-NPGS germplasm—found their way to Europe was via the plant taxonomist Clive Stace (Jenkins et al. 2003) who shared this material with others. While the details of the origin of the Aberystwyth ecotypes may seem insignificant, it was in fact important to discover their provenance because a firm understanding of the origin of germplasm is essential, particularly at the beginning stages of the development of a new model species with so little available germplasm.

Table 1 Relationships between early *Brachypodium distachyon* germplasm deposited in the United States Department of Agriculture National Plant Germplasm System (USDA-NPGS) and ecotypes and inbred lines derived from them

Original USDA-NPGS accession	Origin[a]	Garvin inbred line[b]	Inbred line USDA-NPGS designation	ABR ecotype[c]	Additional hypothesized ABR ecotype[d]
PI 170218	Turkey	Bd1-1	W6 46201	ABR11	
PI 185133	Iraq	Bd2-3	W6 46202	ABR12	
PI 185134	Iraq	Bd3-1	W6 46203	ABR13	
PI 245730	Turkey	Bd18-1	W6 46204	ABR14	ABR1[f]
PI 254867	Iraq	Bd21[e]	W6 36678	ABR15	

[a]Based on original USDA-NPGS Plant Inventory records
[b]Vogel et al. (2006)
[c]Based on descriptions in Jenkins et al. (2003)
[d]Aberystwyth B. distachyon collection designation
[e]PI 254867 is also the source of inbred line Bd21-3 (Vogel and Hill 2008). It has the USDA-NPGS accession number W6 39233
[f]Based on collection location similarity between USDA Plant Inventory records ("Kaman Kirschir Province, Turkey") and description in Jenkins et al. (2003) ("Kaman, Kiresehir, Turkey")

It was fortuitous that chromosome numbers were available for the Aberystwyth *B. distachyon* collection, because cross referencing the ecotypes in this collection with the USDA-NPGS accessions allowed a prediction of the ploidy of the latter germplasm. From this, it was predicted that just five of the USDA-NPGS accessions are what is now, after the 2012 taxonomic revision, considered *B. distachyon*. Flow cytometry coupled with select chromosome counts confirmed this prediction (Vogel et al. 2006). These five USDA-NPGS accessions are the result of collecting trips conducted by scientists who sought to gather many plant species from diverse locations, with the collected germplasm then deposited in the USDA-NPGS and given an official accession designations.

The first *B. distachyon* germplasm deposited in the USDA-NPGS was, perhaps appropriately, collected by the distinguished scientist Jack Harlan, a prolific seed and plant collector who conducted many collecting expeditions during his career, and contributed nearly 10,000 accessions into the USDA-NPGS. In 1948, as an Agricultural Explorer for the U.S. government, he collected *B. distachyon* in the vicinity of the city of Soma located in the western province of Manisa, Turkey, and this was deposited into the USDA-NPGS and given the accession number PI 170218. The published record of this accession provides no additional information. Two other accessions (deposited in the USDA-NPGS as PI 185133 and PI 185134) simply represent seeds provided by an anonymous donor to the USDA via the American Embassy in Baghdad, Iraq, in 1949, with no additional descriptive information. The last of the five USDA-NPGS *B. distachyon* accessions were the result of collections in 1958 and 1959, respectively. One (PI 245730) was originally collected in Turkey and another, PI 254867, in Iraq. Again, there are few additional details accompanying the description of these seeds, although the latter accession has a well-documented collection location listed as 4 kilometers from Salakudin (presumably a phonetic misspelling of the city of Salah ad Din) northeast of Mosul, and the collector, UC-Davis scientist Paulden Knowles, noted "Plants to 12 cm high, scarce" (USDA 1966). These *B. distachyon* accessions are the source of five, or possibly six of the Aberystwyth *B. distachyon* ecotypes (Table 1). The fact that *B. distachyon* was collected at all by the early seed collectors is a testament to their foresight in recognizing that the value of a species cannot be known in advance, and thus collection and *ex situ* preservation of as many species as possible is a valuable goal.

Development of Single Seed Descent Inbred *Brachypodium distachyon* Genetic Stocks

Without knowing how the seeds comprising each USDA-NPGS *B. distachyon* accession were collected, it is prudent to assume that each is a composite of many plants from their site of collection, since the goal of the collecting trips was to capture genetic diversity. Thus it was determined that development of single seed-derived homozygous inbred lines from these accessions was important to ensure that potential genetic heterogeneity residing in the accessions would not be a confounding factor in future research. While it may be desirable to develop a series of single seed-derived inbred lines from each *B. distachyon* accession to capture a proportion of the intrinsic molecular diversity present in each, at the time that the effort to develop inbred lines was initiated, very little was known about efficient growth methods for *B. distachyon*, and so the focus fell solely on successfully obtaining at least one such inbred line per accession. While previous research (Khan and Stace 1998) demonstrated that *B. distachyon* exhibited a high degree of self-fertility, this result—obtained by bagging spikes of two different ecotypes—did not preclude the possibility that the species can outcross as well. Further, the research was actually completed with what is now considered *B. hybridum*. Thus, with knowledge neither of the degree of outcrossing rates in wild populations of *B. distachyon* nor the level of heterozygosity in *B. distachyon*, it was deemed imperative to complete more than one generation of single seed descent to develop inbred lines that would be highly homozygous. During the course of this effort, a fortuitous discovery that long daylengths eliminated the need for vernalization of lines being developed from three of the five *B. distachyon* accessions provided a valuable mechanism to accelerate development of some of the inbred lines (Vogel et al. 2006).

Ultimately, an initial series of single seed descent-derived inbred lines was developed from the five USDA-NPGS *B. distachyon* accessions (Vogel et al. 2006). These lines were given the designations Bd1-1, Bd2-3, Bd3-1, Bd18-1, and Bd21. While Bd2-3, Bd3-1, and Bd21 will, under long daylengths, exhibit a spring habit, Bd1-1 and Bd18-1 do not exhibit this response and so require a vernalization period to flower in a timely manner. An independent single seed-derived inbred line from accession PI 254867, designated Bd21-3 to differentiate it from Bd21, was also developed during the same approximate time frame (Vogel and Hill 2008). In addition, two more inbred lines, one from USDA-NPGS accession PI 639818 collected in Ukraine in 1999, and one from *B. distachyon* germplasm collected in southern Spain by Antonio Manzaneda at the University of Jaen, were developed shortly thereafter. These lines are designated Bd29-1 and Bd30-1, respectively. These eight lines, often referred to as the "core inbred line set", have each been deposited in the USDA NPGS and given their own individual accession numbers (Table 1), and can be freely requested through the USDA-NPGS website (www.npgs.usda.gov). It is worth noting that the original accession from which Bd29-1 derives, as well as Bd29-1 itself, falls under the conditions of a Memorandum of Understanding with Ukraine. While it can be used for educational and research purposes, no intellectual property claims can be obtained on discoveries associated with this germplasm, although this may be possible after negotiation with Ukraine. There are no such restrictions for the other inbred lines.

The intent of developing these single seed descent inbred lines was to ensure that potential heterozygosity was eliminated in the lines, and to ensure that researchers have access to these homozygous genetic stocks rather than working with the original USDA-NPGS accessions from which they derive since, as noted earlier, the original accessions most certainly represent admixtures of seed from multiple plants in a population, and thus are very likely to harbor molecular heterogeneity. This has since been confirmed in several studies. For instance, 157 SSR markers were developed and used to compare marker haplotypes of three of the inbred lines (Bd2-3, Bd3-1, and Bd18-1) to the USDA-NPGS accessions from which they derive, and the polymorphism rate ranged between 4.2 and 4.8 % (Azhaguvel et al. 2008). This indicates that while the genomes of the plants collected in the original accessions are similar, molecular variation is present within the accessions. Further, Bd21 and Bd21-3 derive from the same USDA-NPGS accession, but they exhibit differences in *Agrobacterium*-mediated transformation competence (Vogel and Hill, 2008), and phylogenetic analysis with SSR markers revealed that while these lines cluster together, they are not the same (Vogel et al. 2009). These two lines also exhibit, among other things, variation for responses to *Puccinia graminis* (Figueroa et al. 2013), variation in seed dormancy (Barrero et al. 2012) and other morphological traits (Tyler et al. 2014). Recent genome resequencing of Bd21-3 revealed a surprising level of molecular polymorphism when compared to the Bd21 reference genome sequence (TIBI 2010), with approximately two single nucleotide polymorphisms (SNPs) on average present per kilobase (kb) of nuclear sequence (Gordon et al. 2014). Thus, any research conducted with the original accessions may be confounded by this molecular heterogeneity, while the available inbred lines eliminate this problem. As a future goal it may be interesting to develop more inbred lines from the original accessions to capture more of this molecular variation. Until then, should a researcher be interested in examining genetic stocks from the same USDA-NPGS accession, consideration may be given to comparing the core inbred lines to their counterpart Aberystwyth ecotypes (Table 1), since they most likely will be different genotypically.

New *Brachypodium distachyon* Collections and Inbred Lines

As interest in *B. distachyon* burgeoned internationally, it became clear that more collecting was needed to increase the genetic diversity of *B. distachyon* available to researchers. Thus, a major step toward increasing the genetic diversity present in *B. distachyon* collections came from new collection efforts by Metin Tuna and Hikmet Budak, scientists at Sebanci University and Namik Kemal University in Turkey, respectively, with an early interest in the potential utility of *B. distachyon* as a model grass. Collection of *B. distachyon* was undertaken across the majority of Turkey, ranging from the extreme southeast to the extreme northwest regions of the country, at 53 locations in total between the two scientists (Filiz et al. 2009; Vogel et al. 2009). The locations are well-documented both by latitude and longitude coordinates and altitude (Fig. 1). Single seed descent inbreeding was used to develop 200 inbred lines representing all 53 locations. Based on flowering behavior, the lines may not possess an obligate vernalization requirement but rather may possess a facultative winter habit with significant variation for this trait (Filiz et al. 2009; Vogel et al. 2009). These collections represent the first effort to develop new genetic resources for *B. distachyon* specifically because of its emergence as a model species.

Fig. 1 New collection locations of *Brachypodium distachyon* germplasm from Turkey. *Red* locations were collected by Hikmet Budak (Filiz et al. 2009), and *blue* locations were collected by Mehtin Tuna (Vogel et al. 2009). Inbred lines from these collections are deposited in the United States Department of Agriculture's National Plant Germplasm System. Figure credit: Vogel et al. (2009)

A molecular analysis of 116 of the inbred lines using amplified fragment length polymorphism (AFLP) markers revealed extensive diversity, with genetic similarity not necessarily associated with geographic proximity (Filiz et al. 2009). Phenotypic variation for various morphological traits was also evaluated in this study, and the grouping of lines based on these traits was concordant with the results of the molecular diversity analysis. Another analysis of molecular diversity that included most of the new inbred lines developed from both collections using 43 newly developed simple sequence repeat (SSR) markers revealed how much the molecular diversity of the publically available *B. distachyon* was expanded by this collection (Vogel et al. 2009) (Fig. 2). Interestingly, this analysis also included several of the core inbred lines (Bd1-1, Bd2-3, Bd3-1, Bd18-1, Bd21, Bd21-3), and revealed that these few inbred lines capture significant molecular diversity. Further, this study was the first to examine whether *B. distachyon* exhibits a significant degree of outcrossing in the wild. Using SSR markers, 62 founder plants of the inbred lines that had not yet been subjected to any single seed descent inbreeding were found to exhibit an extremely high degree of homozygosity, with just one heterozygous locus detected in one of the plants examined (Vogel et al. 2009). This has important ramifications for developing inbred lines for community use, because it suggests that fewer generations of single seed descent inbreeding will likely be adequate to obtain highly homozygous lines. This will be particularly valuable for newly collected *B. distachyon* germplasm with long vernalization requirements. The inbred lines developed from these Turkish collecting efforts have been deposited in the USDA NPGS, and represent a highly valuable expansion of freely available inbred *B. distachyon* lines.

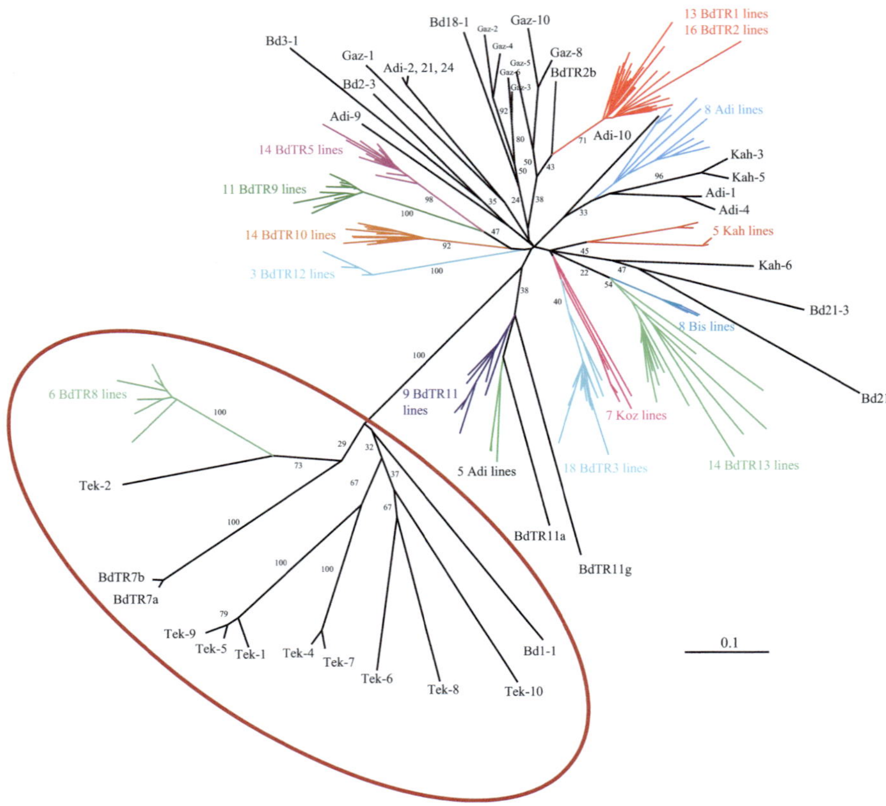

Fig. 2 Genetic diversity in Iraqi and Turkish *Brachypodium distachyon* inbred lines. Diversity was assessed using simple sequence repeat marker genotypes at 43 loci. The *red circle* delineates a set of Turkish lines that share a number of characteristics such as a long vernalization requirement. *Multiple lines* from the early *B. distachyon* "core inbred line set", including Bd1-1, Bd2-3, Bd3-1, Bd18-1, Bd21, Bd21-3, are included in the analysis, in addition to the lines developed from new Turkish collections described in Filiz et al. (2009) and Vogel et al. (2009). Figure credit: Vogel et al. (2009)

With the split of *B. distachyon* into three species, one unanswered question is the extent of the geographic distribution of the single species known as *B. distachyon*. While *B. distachyon* had been collected from Spain, France, Turkey, Iraq, Ukraine, Italy, and Croatia (Jenkins et al. 2003), aside from the extensive collection completed in Turkey, there was a dearth of germplasm collected from other potential *B. distachyon* habitats. Fortunately, additional collecting has now been completed in other regions of Europe by a number of groups, and has helped provide more clarity on this issue, as well as increase *B. distachyon* genetic stocks for community use. Like Turkey, *B. distachyon* appears to be broadly distributed in Spain, and Luis Mur of Aberystwyth University in Wales collected *B. distachyon* from 46 locations across a large transect running from Barcelona and northwest towards Bilbao over highly diverse environments, as well as the island of Ibiza (Mur et al. 2011). From this collection, he developed 80 inbred lines with detailed notes on the collection site (latitude/longitude, elevation, ecological niche). A phylogenetic analysis on a proportion of these lines using a subset of SSR markers previously used to characterize the large collection of Turkish inbred lines revealed extensive molecular variation in the Spanish inbred lines that is distinct from the Turkish and Iraqi lines included in the study (Mur et al. 2011). Thus, as for the Turkish inbred lines, these Spanish lines are a significant expansion of experimental germplasm representing a distinctly different region of the species' distribution. They will, in time, be deposited into the USDA-NPGS, to provide more freely available genetic resources for exploring natural variation in *B. distachyon*.

Additional collecting efforts from regions of Europe in which *B. distachyon* had not been collected before, or for which perhaps just a single ecotype had been collected, have further refined our knowledge of the geographic distribution of *B. distachyon*. Researchers at the University of Massachusetts-Amherst and colleagues have collected new *B. distachyon* germplasm from Albania, Armenia, Georgia, and Sicily, as well as additional germplasm from both northern and southern Spain. Inbred lines are being developed from this material for subsequent deposition in the USDA-NPGS to be further increase germplasm available to the scientific community (Ludmila Tyler personal communication). Based on the sum of the information to date on collection locations, it is likely that *B. distachyon* is native to all of the countries on the northern and eastern borders of the Mediterranean Sea, as well as the countries that border the Black Sea. Thus, prospecting for new germplasm in countries for which *B. distachyon* has not yet been collected but border on one of these bodies of water, such as Romania, Greece, and Bulgaria, may be expected to yield novel *B. distachyon* genotypes to add to the currently available inventory.

It is clear from the collection efforts completed to date that there is broad recognition of the need for a large and diverse collection of well-defined inbred *B. distachyon* lines that are freely available for research. It is commendable that so many scientists have made careful collections of *B. distachyon* and have undertaken the sometimes arduous chore of developing inbred lines for community use. The inbred lines developed to date have been, and will continue to be, a central resource for exploring various aspects of grass biology using natural variation in *B. distachyon* as a surrogate for other species.

Crossing *Brachypodium distachyon* and Development of Segregating Populations

While natural variation in inbred lines can directly provide a foundation for the genetic dissection of traits via such analytical methods as genome-wide association studies (Ozaki et al. 2002), segregating populations developed by controlled crosses between selected genotypes remain a research mainstay. For several years following the 2001 publication by Draper and colleagues, the only mention of a segregating population of *B. distachyon* was by Routledge and colleagues (2004), who examined segregation for resistance responses to *Magnaporthe grisea*, the fungus which causes rice blast, in a number of small F_2 populations from reciprocal crosses between ecotypes ABR1 and ABR5. Thus segregating populations derived from more diverse germplasm was a needed resource.

The small size of *B. distachyon* floral structures poses an obstacle to manipulating the plant for cross-pollination. One method for crossing was developed by adapting a common method used in wheat hybridization. This entails (a) removing all but the basal floret of an inflorescence of the female plant, (b) trimming the palea and lemma to a level above the stigma of the floret, (c) using a small needle to emasculate the floret, (d) identifying a floret from the male parent with plump anthers close to dehiscence, (e) removing the anther from the floret and placing it into the emasculated floret above the stigma to allow shed pollen to drop onto the stigma. A number of F_1 hybrids were successfully established between several of the core inbred lines soon after they were developed. A detailed guide of this method is available online http://jgi.doe.gov/our-science/science-programs/plant-genomics/brachypodium/. A second crossing method has also been developed that follows a similar strategy, but employs somewhat different manipulations (John Vogel personal communication). In this method, (a) florets of prospective female plants are emasculated a day in advance by gently folding back the lemma and teasing out the immature anthers, (b) the following day, approximately 5 h after lights come on in the growth chamber, anthers near dehiscence are identified on male parents and removed, (c) after waiting for anthers to dehisce, the lemma of the female floret is folded back and the stigma is gently brushed with a dehiscing anther to deposit pollen and then folded back. A detailed guide for this method can also be found at: http://jgi.doe.gov/our-science/science-programs/plant-genomics/brachypodium/. Obtaining pollen is the rate limiting step in crossing *B. distachyon*, and one important finding that emerged during the development of both methods is that anthers, under growth chamber conditions, tend to reach an optimal developmental stage for use in crosses approximately 5–6 h after lights come on in the chamber, and that pollen viability is short. It is more difficult to identify useable anthers for crosses outside this general window of time. Both methods have been used with success, and so a novice may choose to read both guides to gain an understanding of *B. distachyon* floral development and then determine which is more comfortable, or develop a hybrid from both methods that suits the experimentalist's needs. The second method may be more useful if one is interested in making many crosses between a pair of genotypes, because the pollen is used more parsimoniously and so an anther can be used to pollinate more florets. This method may also be more successful if attempting to cross between genotypes with small florets.

Obtaining F_1 hybrid seeds from crosses between many of the core inbred lines allowed the development of many segregating F_2 populations. Good fertility was observed for nearly all crosses. One exception is crosses with Bd18-1, where varying degrees of F_1 hybrid necrosis were observed depending on the alternate parent, and sterility was present to varying degrees but was present in all of the F_1 hybrids. The effect is associated with the nuclear genome because F_2 plants from crosses were observed to segregate for their degree of sterility (David Garvin unpublished data). The molecular basis for the sterility associated with crosses with Bd18-1 may be revealed from genome resequencing data for Bd18-1, which is being completed along with many other inbred lines (John Vogel personal communication).

The first use of one of the F_2 populations obtained from crosses between the core inbred lines was for development of genetic linkage maps of *B. distachyon*. One hundred and eighty-three F_2 individuals, all from a single F_1 plant of a cross between Bd3-1 as the female and Bd21, were used to develop the first molecular map of *B. distachyon*, a low density molecular map comprised of 139 markers from different molecular methods that produced a map of 20 linkage groups (Garvin et al. 2010). This population was subsequently pooled with F_2 individuals from a separate F_1 plant from a cross between the same parents to develop a new molecular map from nearly 500 individuals, using SNP markers (Huo et al. 2011). This map, containing more than 500 markers, resolved five large linkage groups, and played an important role in the *B. distachyon* genome sequencing project, because it allowed the identification of several false joins in the initial genome assembly and thus guided the correct final assembly (TIBI 2010). This map also revealed an important aspect of *B. distachyon* chromosome biology, namely that the species appears to exhibit a high recombination rate, something suggested previously based on microscopic observations of metaphase chromosomes of pollen mother cells (Draper et al. 2001).

While F_2 populations are useful, recombinant inbred line populations (RILs) bring a new dimension to genetic analysis due to their fixed recombinant genotypes, and thus they have long been a staple of plant genetic research. In *B. distachyon*, several RILs have been developed from crosses between different core inbred lines. The first of these came from single seed descent advancement of the Bd3-1 × Bd21 F_2 population used to develop the first molecular map of *B. distachyon* mentioned above. This population was advanced to the F_6 generation by single seed descent, and then these plants were individually selfed to generate $F_{6:7}$ RILs. The expected level of heterozygosity for any given locus in this population is expected to be approximately 3 %. The reason that more inbreeding was not conducted was that having residual heterozygosity in individual RILs permits the identification of heterogeneous inbred families, in which a locus or chromosome interval of interest segregates, but otherwise the rest of the genome will be largely homozygous. It is therefore possible to identify near-isogenic lines for a locus or interval of interest from such families (Tuinstra et al. 1997), which has great value for validation of the effect of a genome interval on a trait of interest.

The first use of a *B. distachyon* RIL population was to map a gene controlling natural variation for resistance to barley stripe mosaic virus (BSMV). In this study (Cui et al. 2012), the Bd3-1 × Bd21 RIL population was employed to map a BSMV resistance gene with SNP markers previously employed to develop an SNP map in the F_2 generation of this same population (Huo et al. 2011). The gene was mapped to the terminal region of the short arm of chromosome 3, distal to the most terminal SNP marker. Additional markers were subsequently added to this region, and the gene was found to fall within a 705 kb interval corresponding to a 7.7 centimorgan (cM) interval. Remarkably, using this population of just 165 RILs and additional markers, ultimately it was possible to fine map this gene to an interval of approximately 25 kb, containing just five gene models. It is not clear whether it is purely chance that such fine mapping was achieved with so few RILs, or whether it reflects the high recombination observed in *B. distachyon* (Huo et al. 2011) amplifying the level of recombination that accrued during RIL development. However, it is worth noting that a comparably sized population from the cross Bd2-3 × Bd21 has been used to map a gene conferring resistance to *Puccinia graminis*, the fungal pathogen causing stem rust, to a 120 kb interval (Garvin 2013). Higher genetic resolution was only limited due to the fact that half of this interval exhibits the same haplotype in Bd2-3 and Bd21 and so more markers cannot be developed within it (Sean Gordon and John Vogel personal communication). Additional RILs have been developed for the crosses Bd21 × Bd1-1, Bd3-1 × Bd2-3, and Bd21 × Bd30-1 (David Garvin unpublished data), and more are in development by other researchers. While the RIL populations developed from crosses between several of the core inbred lines are not particularly large, averaging approximately 150 lines per population, the high recombination rate of *B. distachyon* may still make them quite useful for high resolution mapping. Many of these RILs are currently being used by scientists to dissect the genetics of diverse aspects of natural variation in *B. distachyon*, and so in the near future additional results obtained that provide details on how well loci can be genetically resolved in these populations may provide a guide to the research community about the population size of RILs that will be adequate for the molecular genetic dissection of trait variation.

The rapid embrace of *B. distachyon* as a model grass is notable for many reasons. Even without many basic resources such as community genetic stocks, segregating populations, genetic linkage maps, molecular markers, and other genome resources such as expressed sequence tags (ESTs) and bacterial artificial chromosome (BAC) libraries, which cumulatively are typically a feature of a model species, many scientists rapidly contributed to their development because of the potential value of what is likely to be the closest that we will come to a grass version of *Arabidopsis thaliana*. This has propelled *B. distachyon* research forward rapidly in the last decade to *bona fide* model plant status. Inbred lines and RILs are one important component of the resource toolkit that has been developed for *B. distachyon*.

References

Azhaguvel P, Li W, Rudd JC, Gill B, Michels Jr GJ, Weng Y. Aphid feeding response and microsatellite-based genetic diversity among diploid *Brachypodium distachyon* (L.) Beauv accessions. Plant Genet Resour. 2008;7:72–9.

Barrero JM, Jacobsen JV, Talbot MJ, White RG, Swain SM, Garvin DF, et al. Grain dormancy and light quality effects on germination in the model grass *Brachypodium distachyon*. New Phytol. 2012;193:376–86.

Catalan P, Muller J, Hasterok R, Jenkins G, Mur LAJ, Langdon T, et al. Evolution and taxonomic split of the model grass *Brachypodium distachyon*. Ann Bot. 2012;109:385–405.

Cui Y, Lee MY, Huo N, Bragg J, Yan L, Yuan C, et al. Fine mapping of the *Bsr1* barley stripe mosaic virus resistance gene in the model grass *Brachypodium distachyon*. PLoS One. 2012;7:e38333. doi:10.1371/journal.pone.0038333.

Draper J, Mur LAJ, Jenkins G, Ghosh-Biswas GC, Bablak P, Hasterok R, et al. *Brachypodium distachyon*. A new model system for functional genomics in grasses. Plant Physiol. 2001;127:1539–55.

Figueroa M, Alderman S, Garvin DF, Pfender WF. Infection of *Brachypodium distachyon* by *formae speciales* of *Puccinia graminis*: early infection events and host-pathogen incompatibility. 2013. PLoS One. 2013;8:e56857. doi:10.1371/journal.pone.0056857.

Filiz E, Ozdemir BS, Budak F, Vogel JP, Tuna M, Budak H. Molecular, morphological, and cytological analysis of diverse *Brachypodium distachyon* inbred lines. Genome. 2009;52:876–90.

Garvin DF. Genome analysis of nonhost resistance to stem rust in Brachypodium. Proceedings of the 12th International Wheat Genetics Symposium, 8–14 Sep 2013; Yokohama, Japan. Yokohama: Yokohama City University; 2013.

Garvin DF, McKenzie N, Vogel JP, Mockler TC, Blankenheim ZJ, Wright J, et al. An SSR-based genetic linkage map of the model grass *Brachypodium distachyon*. Genome. 2010;53:1–13.

Gordon SP, Priest H, Des Marais DL, Schackwitz W, Figueroa M, Martin J, et al. Genome diversity in *Brachypodium distachyon:* deep sequencing of highly diverse inbred lines. Plant J. 2014;79:361–74.

Griffith E, Hyland HL, editors. United States Department of Agriculture, National Plant Germplasm System, Plant Inventory No. 167, Plant material introduced January 1 to December 31, 1959. Beltsville, MD: USDA; 1966.

Huo N, Garvin DF, You FM, McMahon S, Luo MC, Gu YQ, et al. Comparison of a high-density genetic linkage map to genome features in the model grass *Brachypodium distachyon*. Theor Appl Genet. 2011;123:455–64.

Jenkins G, Mikhailova E, Phillips D, Hasterok R, Draper J, Jones N. Models and meiosis in the 'omics era. In: Zwierzykowski Z, Kosmala A, editors. Recent advances in genetics and breeding of grasses. Poznan: Institute of Plant Genetics; 2003. p. 97–104.

Khan MA, Stace CA. Breeding relationships in the genus *Brachypodium* (Poaceae: Pooideae). Nordic J Bot. 1998;19:257–69.

Mur LAJ, Allainguillaume J, Catalan P, Hasterok R, Jenkins G, Lesniewska K, et al. Exploiting the Brachypodium tool box in cereal and grass research. New Phytol. 2011;191:334–47.

Ozaki K, Ohnishi Y, Iida A, Sekine A, Yamada R, Tsunoda T, et al. Functional SNPs in the lymphotoxin-alpha gene that are associated with susceptibility to myocardial infarction. Nat Genet. 2002;32:650–4.

Routledge APM, Shelley G, Smith JV, Talbot NJ, Draper J, Mur LAJ. *Magnaporthe grisea* interactions with the model grass *Brachypodium distachyon* closely resemble those with rice (*Oryza sativa*). Mol Plant Pathol. 2004;5:253–65.

Somerville C, Koornneef M. A fortunate choice: the history of *Arabidopsis* as a model plant. Nat Genet. 2002;3:883–9.

The International Brachypodium Initiative (TIBI). Genome sequencing and analysis of the model grass *Brachypodium distachyon*. Nature. 2010;463:763–8.

Tuinstra MR, Ejeta G, Goldsborough PB. Heterogeneous inbred family (HIF) analysis: a method for developing near-isogenic lines that differ at quantitative trait loci. Theor Appl Genet. 1997;95:1005–11.

Tyler L, Fangel JU, Fagerstrom AD, Steinwand MA, Raab TK, Willats WGT, et al. Selection and phenotypic characterization of a core collection of *Brachypodium distachyon* inbred lines. BMC Plant Biol. 2014a;14:25. doi:10.1186/1471-2229-14-25.

Vogel JP, Garvin DF, Leong OM, Hayden DM. *Agrobacterium*-mediated transformation and inbred line development in the model grass *Brachypodium distachyon*. Plant Cell Tiss Organ Cult. 2006;84:199–211.

Vogel J, Hill T. High-efficiency *Agrobacterium*-mediated transformation of *Brachypodium distachyon* inbred line Bd21-3. Plant Cell Rep. 2008;27:471–8.

Vogel JP, Tuna M, Budak H, Huo N, Gu YQ, Steinwand MA. Development of SSR markers and analysis of diversity in Turkish populations of *Brachypodium distachyon*. BMC Plant Biol. 2009;9:88. doi:10.1186/1471-2229-9-88.

Brachypodium distachyon as a Model Species to Understand Grass Cell Walls

Joshua H. Coomey and Samuel P. Hazen

Abstract The utilization of *Brachypodium distachyon* as a model system has allowed for a synthesis of known aspects of grass cell wall biosynthesis and provided a platform to investigate new areas of the field. Compositional analysis of *B. distachyon* cell walls shows many similarities with the walls of major food and energy crop species. This chapter presents a summary of these comparisons, as well as a review of work done in *B. distachyon* characterizing genes involved with cell wall biogenesis. Aspects of lignin biosynthesis and polymerization, cellulose and hemicellulose synthesis, and transcriptional regulation of secondary walls have all been characterized in *B. distachyon*, with genetic, biochemical, and phenotypic data outlined herein. Finally, the use of *B. distachyon* in identifying saccharification and digestibility traits relatable to biofuel feedstock quality in grasses are discussed. Taken together, the reviewed material demonstrates the utility of *B. distachyon* as a model for grass cell wall research, highlighting known and novel facets of cell wall biosynthesis.

Keywords *Brachypodium distachyon* • Lignin • Cellulose • Mixed linkage glucan • Cell wall • Grasses • Biofuel feedstock • Digestibility • Saccharification

Introduction

The importance of grasses as food and forage crops, as well as their more recent appropriation as bioenergy feedstocks, is evident the world over. Thus, understanding the dynamics of growth in these plants is key to maximizing their use. One major aspect of these dynamics is the synthesis and deposition of cell walls. As in all plants, a primary wall surrounds grass cells, with certain cell types also developing a secondary wall. The composition and regulation of cell walls has been investigated in many plant species, most rigorously in the model eudicot *Arabidopsis thaliana*. However, because grasses belong to an evolutionarily distinct group, specific knowledge is needed to appropriately understand their cell walls. The prominent

J.H. Coomey • S.P. Hazen (✉)
Plant Biology Graduate Program, Biology Department,
University of Massachusetts, Amherst, MA, USA
e-mail: hazen@bio.umass.edu

J. Vogel (ed.), *Genetics and Genomics of Brachypodium*,
Plant Genetics and Genomics: Crops Models, DOI 10.1007/7397_2015_11
Published online: 06 October 2015

rise of *Brachypodium distachyon* as a model for the grasses has allowed for a unified system in which to discuss the known aspects of grass cell wall biology, and moreover has created a platform to expand this knowledge base. In this chapter we will review the use of *B. distachyon* as a model for grass cell walls. Compositional studies have characterized the contents of *B. distachyon* cell walls and compared them with those of other grass species. Significant work has been done in characterizing the enzymes involved in monolignol biosynthesis and incorporation, uncovering new aspects of secondary wall chemistry. Cellulose synthesis has also been investigated, as well as the formation of the distinctive wall polysaccharide, mixed-linkage glucans (MLG). Transcriptional regulation of secondary wall deposition is well defined in *A. thaliana*, and a similar network has begun to be understood in *B. distachyon*. Finally, screens for altered biomass digestibility and saccharification efficiency have been used to identify *B. distachyon* genetic variation for traits that may be applicable to biofuel feedstock engineering.

Brachypodium Cell Wall Composition

Grasses, as a subset of the commelinoid monocots, have salient differences in their cell wall composition from other angiosperms. Grass secondary cell walls are composed primarily of cellulose, hemicellulose, and lignin. Cellulose comprises the majority of grass secondary walls and is the major target of saccharification and fermentation efforts for biofuel production. The hemicelluloses in these species consist of glucuronoarabinoxylans, accompanied by high levels of mixed linkage glucans, as opposed to the xyloglucans common to non-commelinoid monocots, eudicots, and gymnosperms (Vogel 2008). Grasses also have much higher levels of hydroxycinnamates and significantly lower levels of pectins and structural proteins (Vogel 2008).

Grass cell wall composition varies across tissues and throughout development. For *B. distachyon*, these changes have been characterized in leaf, sheath, and stem tissue in plants at the seedling, elongating stem, and reproductive maturity stages. Lignin levels varied among the tissues examined, with the lowest levels in the leaf and sheath and significantly more in the stem. These levels increased with development (Rancour et al. 2012; Matos et al. 2013). Stem cross section analysis showed that lignification occurred early in the development of vascular bundles, concurrent with the presence of stem tissue, while the interfasicular fibers lignified later as the plants reached maturity (Matos et al. 2013). The levels of non-lignified hydroxycinnamates, comprising monomeric and dimeric ferulic acid as well as *p*-coumaric acid, also varied with tissue and development (Rancour et al. 2012). Dimeric ferulates accumulated at higher levels in the sheath and stem compared to leaves, where levels actually decreased from seedling to elongating stages. A similar pattern was observed for monomeric ferulic acid (Rancour et al. 2012). Ferulates are proposed nucleators of lignin polymerization, and as such the commensurate increase of ferulate levels with lignification in the maturing sheath and stem is understandable, while the observed decrease in leaves remains to be explained. Much like ferulic acid, *p*-coumaric acid levels increased sequentially from leaf to sheath to stem, and likewise from the seedling to the mature plant (Rancour et al. 2012). Ferulic acid and *p*-coumaric acid form linkages between lignin and hemicelluloses through ester bonds. The accumulation of these hydroxycinnamates indicates increased cross linking between the lignin and hemicellulose polymers, a sign of greater secondary wall deposition and maturation.

The major hemicelluloses in grasses are arabinoxylan and MLGs (Vogel 2008). Xyloglucans may be present in low abundance, but their roles as seen in other plant groups seem to be performed by arabinoxylan and MLG in grasses (Vogel 2008). Mixed-linkage glucans are unsubstituted D-glucopyranosyl chains in which segments of 1,4-β-linkages are occasionally interrupted with a 1,3-β linkage. These polymers are deposited primarily during cell elongation, with a sharp decrease at the onset of secondary wall thickening, indicating that they may play a key structural role prior to cell maturation (Christensen et al. 2010; Fincher 2009).

Arabinoxylans consist of xylan chains highly substituted with α-L-arabinofuranosyl units at the O-2 and O-3 positions (Carpita 1996; Urahara et al. 2004). The xylan backbone can also be amended with α-D-glucuronopyranosyl acid and α-4-O-methyl-D-glucuronopyranosyl acid, yielding glucuronoarabinoxylans. Typically, glucuronoarabinoxylans are found in vegetative tissues, while arabinoxylans without glucuronic acids are typical of grain endosperm cell walls (Carpita 1996; Urahara et al. 2004). Higher degrees of substitution on the xylan backbone reduce bonding between arabinoxylans and other polymers, making the walls more soluble and fluid. As the cell matures, the degree of substitution decreases, allowing unsubstituted regions of the xylan backbone to crosslink with other polymers and stiffen the wall. Maturation of the cell wall is also tied to the esterification of O-5 arabinoxylan residues with the hydroxycinnamates ferulic acid and *p*-coumaric acid (Christensen et al. 2010; Grabber et al. 2004). The degree of this cross linking can be estimated over development by the ratios of arabinose: xylose compared to ferulic acid:xylose. Arabinose:xylose is higher in leaves and low in stems, while ferulic acid:xylose is high in stems and increases with maturity, commensurate with secondary wall deposition.

Chemical analysis of cell wall polysaccharides is often achieved by measurement of their constituent monomers. The saccharide components of these polymers are either uronic acids or neutral sugars. The uronic acids represent the oxidized forms of glucose and galactose as glucuronic and galacturonic acid, respectively. These compounds are present in the hemicellulose glucuronoarabinoxylan, as well as in pectin. Total uronic acid levels do not vary between tissue types or developmental stages.

Neutral sugars represent a wider range of compounds, including glucose, galactose, xylose, and arabinose, as well as the minor fractions of rhamnose, fucose, and mannose. The most abundant of these in both tissue types and developmental stages is glucose, derived from cellulose. Glucose levels in shoots at the elongating and mature stages were lowest in leaves and highest in stems; however, as a molar percentage of total neutral sugar content in these tissues, leaves contained 62 %, sheaths 51 %, and stems 54 % (Rancour et al. 2012). Xylose, arabinose, and galactose rank as the second, third, and fourth most abundant monosaccharides, respectively, and are derived from arabinoxylan and xyloglucan hemicelluloses and arabinogalactan pectins. Xylose levels do not vary with developmental stage, but rather between tissue types in a manner inverse with glucose, with the highest levels of xylose in the leaves (Rancour et al. 2012). Arabinose in the elongating and mature plant reached its highest levels in sheaths and its lowest levels in leaves, but as a molar percentage of neutral sugars, arabinose does not vary between the organs. Galactose levels were highest in young seedlings, indicative of mostly primary walls where this sugar is derived from galactans and arabinogalactans. There was no difference in galactose levels between tissues. Finally, the minor neutral sugars rhamose, fucose, and mannose existed only at very low levels, less than 2 % of the molar percentage, and did not vary in their abundance (Rancour et al. 2012).

Comparison of *B. distachyon* with Other Grass Species

Aspects of *B. distachyon* stem anatomy are highly comparable to other grasses. Vascular bundles are arranged in two rings with the inner set being larger than the outer. A region of sclerenchyma cells known as the interfasicular fibers surrounds the vascular bundles (Matos et al. 2013). Interfasicular fibers and the vascular bundles make up the cortex, the region between the epidermis and inner pith parenchyma. As the plant matures, the stem thickens considerably. This gain in size can be attributed to thickening of the sclerenchyma secondary walls, which occurs from elongation to senescence, whereas vascular bundles are relatively unchanged following internode elongation (Matos et al. 2013). The xylem tends to have several large vessels separated by tracheids, and in some cases there is evidence of a lacuna, the remains of the protoxylem. Anatomically, these features of the *B. distachyon* stem are very similar to those of other related grass species. The vascular patterns of rice (*Oryza sativa*), wheat (*Triticum aestivum*), perennial rye-grass (*Lolium perenne*), and tall fescue (*Festuca arundinacea*) all exhibit two rings of bundles, although the specific arrangement of the bundles in these rings varies slightly between species (Chen et al. 2002; Tufarelli et al. 2003; Patrick 1972; Tu et al. 2010). Larger grasses such as maize (*Zea mays*) and sorghum (*Sorghum bicolor*) have significantly more vascular bundles distributed throughout their stems; however, the anatomy of the bundles themselves, as well as the sclerenchyma tissue surrounding them, is highly comparable to *B. distachyon* (Kiesselbach 1949; Wilson et al. 1993).

Cell wall composition has been compared between *B. distachyon* and several agricultural grasses to determine its suitability as a model for growth dynamics in related grass crops. In comparison with the cereals barley (*Hordeum vulgare*) and wheat at the seedling stage, *B. distachyon* showed similar levels of arabinose and xylose, indicators of arabinoxylan content. The degree of arabinoxylan substitution decreased over the development of all three species, with the only difference being a generally lower arabinose:xylose ratio in *B. distachyon* (Christensen et al. 2010). Mixed-linkage glucan levels were also similar across the three species (Christensen et al. 2010). There was an increase in esterified hydroxycinnamates that correlated with the decreasing degree of substitutions on the arabinoxylans. While this trend was observed in all three species, *B. distachyon* had overall higher levels of hydroxycinnamates compared to barley and wheat. Interestingly, the percentage of ferulic acid remained constant in each species, with *B. distachyon* having the highest at 34 %, compared to about 27 % for the other two species (Christensen et al. 2010). The overall higher hydroxycinnamates levels in *B. distachyon*, which can be primarily attributed to its strikingly higher *p*-coumaric acid level, are correlated with the relatively lower degree of arabinoxylan substitution in the model grass.

B. distachyon was also compared to switchgrass (*Panicum virgatum*), a proposed energy crop. *B. distachyon* has an overall higher level of hydroxycinnamates, but lower relative lignin abundance in leaves, sheaths, and stems (Rancour et al. 2012). Both have similar levels of *p*-coumaric acid, but *B. distachyon* has significantly more ferulic acid, which may account for the overall difference in hydroxycinnamates.

Compared to several C3 forage grasses (tall fescue, bromegrass, orchardgrass, and reed canarygrass), *B. distachyon* has comparable composition, although there are some differences in the relative levels of neutral sugars as well as overall lignin content (Rancour et al. 2012; Hatfield et al. 2009). *B. distachyon* has more non-glucose sugars, as well as an overall greater glucose content and lower lignin levels. These differences in sugar and lignin content may be due to the small stature of *B. distachyon* compared to the aforementioned forage grasses, which may require greater lignin to support their size.

Comparison of senesced leaf and stem tissue from *B. distachyon*, wheat, *Miscanthus x giganteus*, and maize showed largely similar composition and saccharification (Gomez et al. 2008; Meineke et al. 2014). However, some differences were also observed that suggested phylogeny alone is not a sufficient indicator of relatable biomass productivity. Lower lignin was found in both leaves and stems of *B. distachyon* and wheat compared to *M. x giganteus* and maize. *B. distachyon* had lower cellulose and higher hemicellulose content compared to wheat (Meineke et al. 2014). After acid and enzyme pretreatment, saccharification and ethanol yield were highest in wheat and lowest in *B. distachyon*, with *M. x giganteus* and maize performing similarly to each other (Meineke et al. 2014). *B. distachyon* and wheat are more closely related, both being C3 grasses, while *M. x giganteus* and maize are C4. The differences in content and yield between the C3 grasses in this study suggest that phylogeny alone does not guarantee similar biofuel-related traits. Despite these differences, *B. distachyon* is far more relatable to the grasses than any eudicot models, making it useful model for grass cell wall traits.

Lignin

The term lignin refers broadly to 4-hydroxyphenylpropanoid polymers deposited in secondary plant cell walls. These aromatic polymers confer strength and hydrophobicity to structural and vascular tissues and are highly recalcitrant to mechanical and enzymatic degradation. Lignin polymers form from the combinatorial oxidative coupling of hydroxycinnamate subunits. Three hydroxycinnamyl alcohols, *p*-coumaryl, coniferyl, and sinapyl alcohols, give rise to *p*-hydroxyphenyl (H), guaiacyl (G), and syringyl (S) monolignol subunits respectively (Ralph 2010; Vanholme et al. 2010). The hydroxycinnamates *p*-coumaric acid and ferulic acid are also incorporated into secondary walls in linkages between lignin and hemicellulose polymers. The biochemistry and genetics of grass lignin are extensively reviewed in Harrington et al. (Harrington et al. 2012). Lignin biosynthesis has been defined largely in terms of monolignol biosynthesis, with extensive characterization of the enzymes and transcriptional regulators governing this branch of the phenylpropanoid pathway (Vanholme et al. 2010). Incorporation of these monomers has only recently been investigated, and the exact mechanisms controlling the varied linkages found within the lignin polymer remain to be uncovered. Many of the first lignin biosynthesis mutants were identified in grasses, and while the genetics and biochemistry of lignification have been extensively explored in *A. thaliana*, the understanding of the grasses has kept pace to some extent. *B. distachyon* provides an excellent system in which to further these studies.

The Lignin Methyltransferases, COMT and CCoAOMT

Two of the key steps in lignin biosynthesis involve methyl transferases: caffeoyl CoA 3-*O*-methyltransferase (CCoAOMT) and caffeic acid *O*-methyltransferase (COMT). CCoAOMT acts in the conversion of caffeoyl-CoA to feruloyl-CoA (Vanholme et al. 2010). COMT acts further downstream in the pathway, methylating 5-hydroxy-coniferaldehyde to sinapylaldehyde as part of the shift from coniferyl alcohol to sinapyl alcohol (Vanholme et al. 2010).

The first characterized *COMT* gene was identified in maize, and subsequent discovery in several other species solidified COMT as a key player in lignin biosynthesis (Collazo et al. 1992). In *B. distachyon*, eight genes were identified as putative COMTs numbered *BdCOMT1-8* (Dalmais et al. 2013). These genes were assigned numbers as they appear in the genome, *BdCOMT1-8*, respectively, by Dalmais et al. (Dalmais et al. 2013); however, other publications name *BdCOMT6* (Bradi3g16530) as *BdCOMT4* (Trabucco et al. 2013; Wu et al. 2013) or *BdCOMT1* (Lee et al. 2014). This difference in naming is due to the fact that Bradi3g16530 clusters with four other BdCOMTs as phylogenetically similar to characterized COMT proteins from *A. thaliana*, poplar, rice, maize, ryegrass, switchgrass, sorghum, and fescue (Dalmais et al. 2013; Trabucco et al. 2013; Wu et al. 2013). BdCOMT1-3 (Bradi1g14870, Bradi2g02380, and Bradi2g02390) also fall into this clade, although they do not group as closely with characterized grass COMTs as Bradi3g16530 does (Dalmais et al. 2013; Trabucco et al. 2013; Wu et al. 2013). Ostensibly, these four genes form a lignin-biosynthesis-related subfamily within the *B. distachyon* COMTs, and analysis of their function by other researchers has resulted in naming these genes BdCOMT1-4. The remaining four *BdCOMT*s (Bradi2g19830, Bradi2g19850, Bradi3g55890, and Bradi4g20020) fall amongst clades containing *O*-methyltransferases in other species that are not known to be involved with lignin biosynthesis. The nomenclature used in this publication will follow Dalmais et al., with all genes and aliases listed in Table 1.

Analysis of the genomic regions surrounding *BdCOMT6* shows synteny to the genomic regions of other characterized grass COMTs (Wu et al. 2013). Beyond genomic collinearity with characterized *COMT* genes, *BdCOMT6* was identified as a COMT functioning in the *B. distachyon* lignin pathway by phylogeny with characterized COMTs and substantiated with enzymatic assays and genetics (Dalmais et al. 2013; Trabucco et al. 2013; Wu et al. 2013; Lee et al. 2014). From the crystal structure of alfalfa (*Medicago sativa*) COMT, catalytic and substrate binding sites have been identified (Zubieta et al. 2002), all of which are preserved in BdCOMT6 (Trabucco et al. 2013; Wu et al. 2013). Other members of the BdCOMT family contain the S adenosyl methyltransferase domain characteristic of OMTs, but vary in their conserved substrate binding residues.

In vitro enzymatic activity assays have shown that recombinant BdCOMT6 is highly active in the methylation of caffeic acid, as is expected in the canonical role of COMTs in lignin biosynthesis (Wu et al. 2013). However, some activity was also observed when BdCOMT6 was challenged with caffeoyl aldehyde and the flavonoid luteolin, indicating that COMT activity may not be strictly dedicated to one metabolic pathway. It has been suggested that single amino acid changes in the COMT substrate binding domains may dictate substrate specificity. For example, both BdCOMT6 and AtCOMT have a valine residue at comparable binding locations, and both show a preference for caffeic acid. Conversely, the alfalfa and *Vanilla planifolia* COMTs have a leucine and isoleucine, respectively, at this location and prefer caffeoyl aldehyde (Wu et al. 2013; Li et al. 2006). The influence of single amino acid changes on substrate specificity remains a topic for further investigation.

Genetic manipulation of the *BdCOMT6* allele revealed that mutations or down regulation resulted in reduced lignin content and modified lignin composition (Dalmais et al. 2013; Trabucco et al. 2013). *Bdcomt6* mutants had 24–31 % less total soluble lignin, with a compositional 40 % loss of S lignin (Dalmais et al. 2013). Also, plants with disrupted *BdCOMT6* activity were more digestible by the bacterium *Clostridium phytofermentans*, yielding more ethanol than control plants (Trabucco et al. 2013). These results solidify the importance of *BdCOMT6* in modulating S lignin levels in *B. distachyon*, and consequently influencing the availability of cell wall polysaccharides for conversion to ethanol.

Table 1 *Brachypodium distachyon* genes described as encoding cell wall related enzymes

Bradi ID#	Alias	Alternative aliases
Bradi1g14870	COMT1 (Dalmais et al. 2013)	
Bradi2g02380	COMT2 (Dalmais et al. 2013)	
Bradi2g02390	COMT3 (Dalmais et al. 2013)	
Bradi2g19830	COMT4 (Dalmais et al. 2013)	
Bradi2g19850	COMT5 (Dalmais et al. 2013)	
Bradi3g16530	**COMT6** (Dalmais et al. 2013)	COMT4 (Trabucco et al. 2013; Wu et al. 2013; Lee et al. 2014), COMT1 (Lee et al. 2014)
Bradi3g55890	COMT7 (Dalmais et al. 2013)	
Bradi4g20020	COMT8 (Dalmais et al. 2013)	
Bradi3g06480	**CAD1** (Trabucco et al. 2013; Bouvier et al. 2013)	CAD5 (Bukh et al. 2012)
Bradi3g17920	CAD2 (Trabucco et al. 2013; Bouvier et al. 2013)	CAD7 (Bukh et al. 2012)
Bradi3g22980	CAD3 (Trabucco et al. 2013; Bouvier et al. 2013)	CAD8 (Bukh et al. 2012)
Bradi4g29770	CAD4 (Trabucco et al. 2013; Bouvier et al. 2013)	CAD3 (Bukh et al. 2012)
Bradi4g29780	CAD5 (Trabucco et al. 2013; Bouvier et al. 2013)	CAD4 (Bukh et al. 2012)
Bradi5g04130	CAD6 (Trabucco et al. 2013; Bouvier et al. 2013)	CAD1 (Bukh et al. 2012)
Bradi5g21550	CAD7 (Trabucco et al. 2013; Bouvier et al. 2013)	CAD2 (Bukh et al. 2012)
Bradi5g15380	PAL1 (Lee et al. 2014)	
Bradi2g53470	C4H1 (Lee et al. 2014)	
Bradi4g08650	CCR3 (Lee et al. 2014)	
Bradi3g39420	CCoACOMT1 (Lee et al. 2014)	
Bradi3g39390	CCoACOMT2 (Lee et al. 2014)	
Bradi1g48370	CCoACOMT3 (Lee et al. 2014)	
Bradi2g36910	**PMT** (Petrik et al. 2014)	
Bradi1g66720	**LAC5** (Wang et al. 2015)	
Bradi1g74320	**LAC6** (Wang et al. 2015)	
Bradi2g23370	LAC8 (Wang et al. 2015)	
Bradi2g54740	LAC12 (Wang et al. 2015)	
Bradi2g34240	CESA1 (Handakumbura et al. 2013; Burton et al. 2008)	
Bradi1g04597	CESA2 (Handakumbura et al. 2013)	
Bradi1g54250	CESA3 (Handakumbura et al. 2013)	
Bradi3g28350	**CESA4** (Handakumbura et al. 2013)	
Bradi1g29060	CESA5 (Handakumbura et al. 2013)	
Bradi1g53207	CESA6 (Handakumbura et al. 2013)	
Bradi4g30540	**CESA7** (Handakumbura et al. 2013)	

(continued)

Table 1 (continued)

Bradi ID#	Alias	Alternative aliases
Bradi2g49912	CESA8 (Handakumbura et al. 2013)	CESA4 (Valdivia et al. 2013)
Bradi1g02510	CESA9 (Handakumbura et al. 2013)	
Bradi1g36740	CESA10 (Handakumbura et al. 2013)	
Bradi3g16310	**CSLF6** (Kim et al. 2015)	
Bradi1g76730	**SWN1** (Valdivia et al. 2013)	
Bradi5g16917	**SWN2** (Valdivia et al. 2013)	
Bradi3g50067	**SWN3** (Valdivia et al. 2013)	
Bradi1g52187	**SWN4** (Valdivia et al. 2013)	
Bradi5g27467	**SWN5** (Valdivia et al. 2013)	
Bradi3g13117	**SWN6** (Valdivia et al. 2013)	
Bradi1g50057	**SWN7** (Valdivia et al. 2013)	
Bradi3g13727	**SWN8** (Valdivia et al. 2013)	
Bradi2g39320	**BdXCP1** (Valdivia et al. 2013)	
Bradi4g06317	**BdMYB1** (Valdivia et al. 2013)	

Gene numbers and aliases are listed with references. Alternative aliases are also listed along with their respective references. Bolded genes have been functionally characterized

BdCAD Family Characterizations

The final step in monolignol biosynthesis is mediated by the cinnamyl alcohol dehydrogenase (CAD) enzyme. Some of the first mutants with lignin-associated defects were identified as having causative mutant *CAD* alleles, such as in the *brown midrib* mutants found in sorghum and maize (Saballos et al. 2009; Sattler et al. 2009; Fornalé et al. 2011; Vignols et al. 1995). In *B. distachyon*, the CAD gene family has seven members as determined by amino acid sequence alignment with known CAD enzymes from sorghum, rice, wheat, and tall fescue (Sattler et al. 2009; Chen et al. 2003; Hirano et al. 2012; Ma 2010). This family was numbered BdCAD1-7 by Trabucco et al. 2013 and Bouvier et al. 2013 (Trabucco et al. 2013; Bouvier et al. 2013); however, alternative numbering appears in publications by Guo et al. and Buhk et al. (Bukh et al. 2012; Guo et al. 2010). Refer to Table 1 for gene numbers and aliases used in this and other publications.

Amino acid sequence comparison between the BdCADs and characterized CAD proteins shows high similarity with the conserved binding and catalytic domains (Trabucco et al. 2013; Bouvier et al. 2013; Bukh et al. 2012). BdCAD1 was found to have the highest amino acid sequence similarity with characterized CAD proteins from other species, containing 10 of the 12 conserved amino acid residues known to be involved in substrate binding, including those that convey specificity for aromatic alcohol substrates, as well as NADP(H) binding (Trabucco et al. 2013; Bukh et al. 2012). *BdCAD1* is expressed primarily in stem tissue, with tenfold higher expression than other *BdCAD*s. Recombinant BdCAD1 and BdCAD4 enzyme were capable of reducing coniferyl aldehyde and sinapyl aldehyde to coniferyl and sinapyl alcohol, respectively, with significantly higher activity seen in BdCAD1 (Bukh et al. 2012). Further genetic characterization of *BdCAD1* has reinforced the idea that this gene plays a prominent role in lignin biosynthesis. Similar to the analysis of *BdCOMT4*, knockdown lines expressing amiRNAs as well as TILLING mutant lines have been generated for the *BdCAD1* locus (Bouvier et al. 2013). One of the *Bdcad1* TILLING mutants has a change at amino acid 192 of glycine to aspartic acid; a substitution of glycine to serine at the identical site in sorghum was shown to be the causative mutation of the *bmr6* mutant (Saballos et al. 2009). This alteration may disrupt the ability of the CAD protein to bind NADPH as well as the substrate and may also influence the zinc-binding catalytic site. Two other lines with altered BdCAD1 protein structure also showed a mutant phenotype (Bouvier et al. 2013).

In mutant *Bdcad1* lines, the identifying visible phenotypes showed a range of brown to reddish discoloration of the spikelet, flowers, rachilla, nodes and lemma (Bouvier et al. 2013), as well as a classic brown leaf midrib (Trabucco et al. 2013). This type of discoloration is typical of lignin mutants with defects in CAD function, particularly the browning of the leaf midrib seen in C4 grass lignin mutants (Sattler et al. 2009; Vignols et al. 1995). In addition to the visible phenotypes, transverse cross-sections of the stem internode revealed a decrease in S lignin in the sclerenchyma, epidermis, and vascular bundle sheath (Trabucco et al. 2013; Bouvier et al. 2013). This was reflected in compositional analysis of the *Bdcad1* mutants, which had 20–26 % less total lignin and proportionally a 60 % reduction in S units (Trabucco et al. 2013; Bouvier et al. 2013). Consistent with the reduction of S lignin, there was also a reduction in the *p*-coumaric acid levels in *Bdcad1* mutants (Bouvier et al. 2013). *P*-coumaric acids are involved in the cross linking between hemicellulose arabinoxylans and lignin polymers, particularly through ester linkage to S lignin monomers (Ralph 2010; Vanholme et al. 2010; Harrington et al. 2012; Bouvier et al. 2013). Ferulic acid levels, which are ester-linked to hemicelluloses much like *p*-coumaric acid is, were not affected (Bouvier et al. 2013). The inter-unit bonds in the *Bdcad1* mutants were also affected, with higher levels of biphenyl ether (4-*O*-5) and biphenyl G-G dimer (5-5) linkages. Higher frequencies of these inter-unit bonds indicate a greater degree of branching within the polymer. Consistent with this observation, a greater frequency of terminal phenolic groups was observed through thioacidolysis in the *Bdcad1* mutants (Bouvier et al. 2013). The relative levels of phenolic terminal units on the lignin polymer are positively associated with the degree of reactivity to alkaline and oxidative degradation.

Interestingly, *Bdcad1* mutants also showed increased levels of β-*O*-4-coupled sinapyl aldehyde in the lignin polymer, while coniferyl aldehyde levels remained unchanged (Bouvier et al. 2013). Altered incorporation of sinapyl aldehyde alone has not been reported in *cad* mutants of other species, where sinapyl and coniferyl aldehyde levels accumulate similarly. This suggests that in *B. distachyon*, altered BdCAD1 activity not only reduces the incorporation of S units, but also promotes the β-*O*-4 coupling of sinapyl aldehyde. This is not altogether surprising considering that loss of BdCAD1 would interrupt the conversion of cinnamyl aldehydes to cinnamyl alcohol lignin monomers; however, the specific accumulation of sinapyl aldehydes and not coniferyl aldehydes is thus far distinct to *B. distachyon*.

While there was no difference in biomass between the mutants and wild type lines, *Bdcad1* mutant saccharification levels were significantly higher than the wild type (Trabucco et al. 2013; Bouvier et al. 2013). This increase in available sugars was also reflected in a higher ethanol yield (9–17 %) from *C. phytofermentans* digestion of mutant lines (Trabucco et al. 2013). Reduction in total lignin content, altered cross linking with hemicellulose, and modified branching degrees within the lignin polymer are all likely contributors to the observed increase in saccharification and ethanol yield in these mutants, indicating that modulation of *BdCAD1* could play a major role in the engineering of bioenergy crops (Bouvier et al. 2013).

Hydroxycinnamate Cross Linking

The lignin polymer is cross-linked to hemicelluloses via non-lignified hydroxycinnamates. The addition of H lignin to the G/S backbone has been shown to result from acylation of the coniferyl and sinapyl monolignols with *p*-coumaric acid at the γ-OH position prior to their incorporation into the lignin polymer. Enzymatically, this is accomplished by *p*-coumaroyl-CoA:monolignol transferase (PMT), which was recently characterized in *B. distachyon* (Petrik et al. 2014). *BdPMT* RNAi and the sodium-azide-generated *Bdpmt-1* loss-of-function mutant had significantly reduced *p*-coumaric acid levels.

Since *p*-coumaric acid can acylate both lignin and hemicelluloses in grasses, an important distinction was made between the total *p*-coumaric acid released through mild alkaline hydrolysis and the *p*-coumaric acid specifically from lignin. In wild type plants, *p*-coumaric acid was found almost exclusively on S lignin subunits, with no detectable levels found in the *Bdpmt-1* mutant. No difference was found in the levels of *p*-coumaric acid or ferulic acid linked to arabinose between *Bdpmt-1* and the wild type, showing that *BdPMT* is exclusively responsible for monolignol acylation, with another unknown enzyme(s) decorating hemicelluloses with *p*-coumaric acid and ferulic acid (Petrik et al. 2014).

Over-expression of *BdPMT* resulted in reduced overall lignin, but almost two-fold increased *p*-coumaric acid levels compared to wild type plants (Petrik et al. 2014). The loss of lignin in the over-expression line is potentially due to a redirection of *p*-coumaroyl-CoA from the general monolignol synthesis pathway to the enhanced acylation activity of excess BdPMT. Reduced lignin content also implies a higher ratio of arabinoxylan to lignin, a metric which may explain the increased levels of ferulic acid in the over-expression lines. Over-expression lines were more digestible than wild type, while *Bdpmt-1* showed no difference in digestibility (Petrik et al. 2014). The increase in digestibility in these over-expression lines is likely due not only to modified lignin content, but also composition. As with *cad* mutants, lines over-expressing *BdPMT* with an abundance of phenolic terminal units on their polymer chains are more susceptible to alkaline pretreatment degradation.

Laccases

Incorporation of the monolignol subunits into polymer chains has long been known to be dependent on hydrogen peroxide radicals (Vanholme et al. 2010). Peroxidases and laccases have recently been characterized as apoplastic enzymes involved in catalyzing and mediating this polymerization. In *A. thaliana*, two laccases (AtLAC4 and AtLAC17) have been shown to directly regulate lignification in the protoxylem tracheary elements (Schuetz et al. 2014). Orthology with these *A. thaliana* laccases was used to identify 29 genes in *B. distachyon* with putative laccase activity, numbered *BdLAC1-29* based on sequential genome location (Wang et al. 2015). The BdLACs cluster into four large clades when compared with laccases involved with phenolic oxidation in other species, with clade IV including genes known to be involved with lignification. BdLAC6 (Bradi1g74320) is most similar to AtLAC4, while BdLAC5 (Bradi1g66720), BdLAC8 (Bradi2g23370), and BdLAC12 (Bradi2g54740) are most similar to AtLAC17. *BdLAC5* was also orthologous to sugarcane *SofLAC* which is implicated in lignification (Wang et al. 2015; Cesarino et al. 2013). As with characterized laccases in *A. thaliana, BdLAC5* and *BdLAC6* exhibit an expression pattern similar to other genes that encode secondary-wall-related enzymes. Both BdLAC5 and BdLAC6 were also found to localize in the apoplast adjacent to sclerenchyma cells (Schuetz et al. 2014).

Genetic evidence for *BdLAC5* and *BdLAC6* function was provided from the analysis of TILLING mutants with defects at the respective loci. *Bdlac* mutants were smaller in stem height and diameter than wild type, with a modest reduction in lignin levels. More specifically, the disruption of *BdLAC5* resulted in decreased G lignin. Ferulate levels were also affected in these lines, with more ferulic acid found in ester linkages to xylans and less ferulic acid ether linked to lignin (Wang et al. 2015). This shift towards ferulation of xylans is likely due to reduced availability of integrated lignin monomers on which to form ether bonds due to the loss of laccase activity.

Additional Lignin Biosynthetic Enzymes

While two enzymes in the monolignol biosynthetic pathways, CAD and COMT, and two involved with incorporation of monolignols into the polymer backbone, LAC and PMT, have been characterized in *B. distachyon*, many more genes related to lignin biosynthesis remain to be characterized. Several candidates have been identified by screening a population of gamma-irradiated mutants for reduced lignin content as determined by histochemical staining of internode cross sections (Lee et al. 2014). Twenty-five lines were analyzed for polymorphisms in loci of genes with homology to known lignin biosynthetic enzymes, including putative genes for *PAL1*, *C4H1*, *CCR3*, *CAD4*, *COMT1*, *CCoACOMT1*, *CCoACOMT2*, and *CCoACOMT3* (Table 1) (Lee et al. 2014). Multiple mutations were found throughout these genes, with several predicted to be missense or nonsense mutations. Although these mutants were not characterized further, this study demonstrates the potential for using *B. distachyon* mutant populations for meaningful forward-genetic screens for cell wall traits.

Cellulose

Cellulose comprises the major fraction of cell wall polymers. It is synthesized from glucose monomers connected by β-1-4 glucan linkages to form long microfibrils that are extruded from the plasma membrane and aligned along microtubule tracks. The synthesis and deposition of the cellulose microfibrils is a long-studied area of plant cell biology that has been extensively elucidated in the model eudicot *A. thaliana* (Somerville 2006; McFarlane et al. 2014). The cellulose synthesis complex is comprised of CELLULOSE SYNTHASE A proteins arranged in a rosette formation on the plasma membrane. These processive glycosyltransferases typically represent a small gene family of about ten members in many plants. To an extent, these members have been shown to act differentially in the synthesis of either the primary or secondary cell wall.

The *B. distachyon* CESA family has ten members: *BdCESA1* (Bradi2g34240), *BdCESA2* (Bradi1g04597), *BdCESA3* (Bradi1g54250), *BdCESA4* (Bradi3g28350), *BdCESA5* (Bradi1g29060), *BdCESA6* (Bradi1g53207), *BdCESA7* (Bradi4g30540), *BdCESA8* (Bradi2g49912), *BdCESA9* (Bradi1g02510), and *BdCESA10* (Bradi1g36740) (Table 1) (Handakumbura et al. 2013). All ten members, with the exception of BdCESA10, contain the canonical eight transmembrane domains and the hyper-variable regions, as well as the UDP-glucose binding motif (D,D,D, QxxRW). Additionally, all the BdCESAs, with the exception of BdCESA5 and 10, have the RING-type zinc finger motif involved in protein dimerization. BdCESA1, 3, 6, and 9 are considered to be primary-wall-related CESAs, and exhibit high expression in both root and stem tissue (Handakumbura et al. 2013). The secondary-wall-associated CESAs are BdCESA4, 7, and 8, and these genes are expressed at higher levels in the stem compared to root tissue (Handakumbura et al. 2013).

The secondary CESAs in *B. distachyon* (BdCESA4, 7, 8) are numbered in accordance with their orthology to the *A. thaliana* secondary CESAs. BdCESA4 and 7 have been further characterized *in planta*. *In situ* hybridization was used to identify the expression patterns of these genes, and probe staining suggested high expression in vascular bundles, epidermis, and interfasicular fiber regions, with little expression detected in the pith parenchyma (Handakumbura et al. 2013). This expression pattern is consistent with tissue types that undergo secondary wall deposition. Knockdown of *BdCESA4* and *BdCESA7* resulted in delayed emergence of the inflorescence and smaller stature. Stem cross-section analysis showed thinner cell walls and reduced crystalline cellulose content in the knockdown lines (Handakumbura et al. 2013).

Mixed-Linkage Glucans

Mixed-linkage glucans are unbranched glucose chains with both β1,3- and β1,4 linkages. These polysaccharides are unique among land plants to the Poaceae, including the grasses, and the monilophyte *Equisetum* spp. These carbohydrates are also found in some pathogenic fungi and lichen-forming ascomycete species (Burton and Fincher 2009; Fry et al. 2008). The synthesis of these polymers is attributed in part to members of the cellulose synthase-like (CSL) families CSLF and CSLH. *CSLF4*, *CSLF6*, and *CSLH* from barley, as well as rice *CSLF2*, have been characterized *in planta* for the production of MLG (Burton et al. 2006, 2008, 2011; Doblin et al. 2009).

While manipulation of these *CSLF* and *CSLH* genes from barley and rice has been shown to modulate MLG content, parts of the mechanism behind MLG biosynthesis have been uncovered in *B. distachyon*. BdCSLF6 was shown to localize to the Golgi membrane, with the proposed catalytic domain on the cytosolic side (Kim et al. 2015). This is in line with the established paradigm of hemicelluloses and pectins being synthesized and modified at the Golgi before export to the cell wall; however, it also suggests that substrates do not need to be imported into the Golgi body, but rather synthesis occurs in the cytosol.

Furthermore, BdCSLF6 has been proposed to be capable of forming both β1,3 and β1,4 bonds. Transgenic yeast were found to have both bond types after expression of the single *BdCSLF6* transgene; previous studies in *A. thaliana* using rice and barley *CSLF6* orthologs also generated β1,3 and β1,4-containing MLG (Burton et al. 2006; Doblin et al. 2009; Kim et al. 2015). Although these observations do not rule out the role of endogenous genes in contributing to these multiple linkage types, similar dual functionality has been observed in the hyaluronic acid biosynthetic pathway where a related glycosyltransferase generates both linkages types (May et al. 2012; Weigel and DeAngelis 2007).

Transcriptional Regulation of the Secondary Cell Wall

The genetic control of secondary wall polymer synthesis has best been described through the elucidation of a transcription factor regulatory network. This network is most clearly defined in *A. thaliana*, with many transcription factors characterized as functional activators or repressors of secondary-wall-related genes. In this network there are a series of feed-forward loops containing NAC proteins that regulate a suite of other transcription factors, mostly MYB family proteins that also directly influence genes related to biosynthetic pathways (Taylor-Teeples et al. 2014; Zhong et al. 2008, 2010a). The cell wall regulatory network in grasses is far less well defined (Handakumbura and Hazen 2012). However, recent work in *B. distachyon* is beginning to assemble a framework similar to the *A. thaliana* network that is representative of cell wall transcriptional regulation in the grasses.

As in *A .thaliana*, several NAC proteins have been identified as regulators of secondary wall biosynthesis in *B. distachyon*. Eight such genes, the SECONDARY WALL NACs (SWNs), were identified by phylogenetic analysis (Valdivia et al. 2013). The first six members of the family, BdSWN1-6, are orthologs of the *A. thaliana* VASCULAR NAC DOMAIN (VND) transcription factors (Ohashi-Ito et al. 2010; Yamaguchi et al. 2011), while BdSWN7 and 8 are orthologous to the SND1, NST1, and NST2 transcription factors in *A. thaliana* (Mitsuda et al. 2005, 2007). All of the BdSWNs were found to be expressed in the stem internode, with particularly high expression of *BdSWN3*, *4*, *6*, and *7*. *BdSWN5* and *8* were found to be expressed in the leaves as well, while expression of all other *BdSWN* genes was very low in those organs. Likewise, with the exception of *BdSWN1* and *2*, expression in the roots was also low (Valdivia et al. 2013).

Transient expression of the *BdSWN*s in tobacco leaves via *Agrobacterium* infiltration resulted in ectopic deposition of secondary walls throughout the affected mesophyll and epidermal cells. This phenotype was particularly strong in the case of *BdSWN1*, *3*, *4*, and *7* (Valdivia et al. 2013). Furthermore, extensive cell death resulted from expression of *BdSWN1*, *2*, *3*, and *4*, sometimes even before secondary wall deposition began. Induced over-expression of *BdSWN5* resulted in ectopic cell wall deposition in the coleoptile parenchyma, as well as in the roots. In the roots, this deposition was only observed in cells close to the root tip, implying that only cells in the differentiation zone were responsive to *BdSWN5* (Valdivia et al. 2013).

Inducible expression of BdSWN5 in *B. distachyon* strongly upregulated *BdCESA4*, while the primary wall *BdCESA1* was not affected (Valdivia et al. 2013). Furthermore, *BdSWN5* expression was also shown to upregulate the representative secondary-wall-related genes encoding the xylem-specific cystine protease BdXCP1 and the MYB transcription factor BdMYB1. *BdXCP1* is the ortholog of *A. thaliana AtXCP1* and *2*, which are activated as part of the programmed cell death pathway that is associated with secondary wall deposition and senescence (Avci et al. 2008). Likewise, *BdMYB1* is the ortholog of *A. thaliana AtMYB46/83*, known transcriptional activators of secondary-wall-related genes (Zhong et al. 2007; Zhong and Ye 2012; Ko et al. 2012). The mechanism of BdSWN5-mediated activation was shown to be dependent on the presence of the highly conserved Secondary wall NAC Binding Element (SNBE) (Zhong et al. 2010a, b) in the promoters of *BdMYB1* and *BdXPC1*. Mutation of the SNBE site resulted in 93–95 % loss of BdSWN5 binding, indicating that this motif is necessary for activation by the BdSWNs (Valdivia et al. 2013).

Mutant Screens for Biofuel Feedstock Quality

With regard to the application of grasses as feedstocks for bioenergy production, the efficient extraction of fermentable sugars is paramount. Due to the recalcitrant nature of cell walls, these sugars are inaccessible without some measure of pretreatment. Force, heat, acid, enzymatic treatments, and other pretreatments have been employed to extract cell wall polysaccharides for fermentation with varying degrees of success. In developing analytical assays for investigating differences in conversion, total sugar extraction is not always the goal, but rather sensitive detection of changes in saccharification potential. One such assay employs a modest acid treatment combined with heat followed by enzymatic digestion (Gomez et al. 2008). This was found to release approximately 25 % of the total wall polysaccharides in *B. distachyon*, particularly xylose and xylose oligomers along with arabinose, glucose and small amounts of mannose (Gomez et al. 2008). This is typical of grasses, where the more labile hemicelluloses require the least treatment to extract. Heat- and acid-based pretreatment of *B. distachyon* resulted in saccharification profiles similar to those of other grasses. Compared to wheat, *M. x giganteus*, maize, and barley, *B. distachyon* had a similar pretreatment-generated profile of arabinose and xylose (Gomez et al. 2008).

Feedstock quality has also been assessed with a simultaneous saccharification and fermentation assay using the forest soil bacterium *Clostridium phytofermentans*. This anaerobic bacterium can not only degrade a variety of cell wall polymers but also metabolize both five- and six-carbon sugars directly to ethanol. *C. phytofermentans* has been shown to efficiently ferment cellulose, hemicelluloses, pectins, and starch from a variety of lignocellulosic feedstocks (Lee et al. 2012; Warnick et al. 2002). Because of these properties, *C. phytofermentans* has been used to investigate natural genetic variation in *B. distachyon* feedstock conversion efficiency. Five lines representing diverse genetic backgrounds were tested. The highest ethanol yield came from the accessions Bd2-1 and Bd3-1, which are relatively similar to each other, while the least ethanol came from the Spanish Bd30-1 accession (Lee et al. 2012).

In addition, mutant *B. distachyon* populations have been used to identify candidate genes related to saccharification efficiency. Chemically mutagenized plants were screened for saccharide release following alkaline pretreatment and enzymatic digestion. Twelve lines with altered saccharification values were identified and named *saccharification1-12* (*sac1-12*) (Marriott et al. 2014). These mutants exhibited a range of cell-wall-related phenotypes, including reduced lignin, altered S:G lignin ratios, increased H lignin, reduced crystalline cellulose, and decreased esterified ferulates. The causative mutations in these lines were mapped to a 2.6 Mbp region containing 14 polymorphisms, two of which were in exons and resulted in altered amino acid sequences. The glycine-to-serine substitution in Bradi2g01480 is predicted to disrupt the function of glycosyltransferase 61 (GT61) (Marriott et al. 2014). Bradi2g01480 is a member of the grass-specific clade C of glycosytransferases. Members of this clade, such as the wheat *XYLAN ARABINOTRANSFERASES* (*XAT*s) and the rice *XYLOSYL ARABINOSYL SUBSTITUTION OF XYLAN* (*XAX*), have been shown to mediate arabinoxylan substitution (Anders et al. 2012; Chiniquy et al. 2012). Rice *xax1* and *B. distachyon sac1* have similar phenotypes, including increased saccharification and decreased ferulate content, indicating that *Bdsac1* phenotypes may also be the result of a glycosyltransferase defect, such as that detected in Bradi2g01480.

Mutant screens for altered saccharification also identified the *spaghetti1* (*spa1*) mutant, which has brittle stems and is twofold more digestible than wild type lines (Timpano et al. 2015). These mutant phenotypes are believed to be a result of not only altered polymer levels, but also compromised structure in the cells walls. The *spa1* mutant has thinner secondary cell walls with less cellulose but higher hemicellulose and lignin levels than wild type. The cellulose also exhibited reduced crystallinity. Arabinoxylan levels were increased, as was the level of *p*-coumaric acid ester linked to these hemicelluloses. Lignin from the *spa1* plants had a higher frequency of inter-unit resistant bonds, as well as increased H monomer content (Timpano et al. 2015). Together, these changes are reminiscent of stress lignin deposited in compression wood (Timell 1986), suggesting that *spa1* modulates some aspect of cell wall formation related to mechanical stress. Digestibility of *spa1* is attributed to the high friability of its biomass, giving rise to significantly smaller particles than wild type plants after grinding. Smaller particle size favors saccharification efficiency, and it is thought that the altered wall structure of *spa1* gives rise to brittle and thus more friable biomass (Timpano et al. 2015). The causal mutation behind the *spa1* phenotype has yet to be identified.

Conclusion

Our understanding of grass cell wall biosynthesis is rapidly growing, in part due to the expansion of available genetic resources and more relevant model systems such as *B. distachyon*. Characterizations of previously identified cell wall biogenesis components have established *B. distachyon* as a viable platform for cell wall research, while compositional studies have shown this species to be comparable to many grass crops in both content and anatomy. Outstanding questions regarding components of hemicellulose synthesis and deposition, lignin polymerization, expanded transcriptional regulatory networks, and saccharification traits all remain to be answered. Ongoing research in *B. distachyon* will further our knowledge in these areas and in doing so will inform our understanding of biomass accumulation in grasses.

References

Anders N, Wilkinson MD, Lovegrove A, Freeman J, Tryfona T, Pellny TK, et al. Glycosyl transferases in family 61 mediate arabinofuranosyl transfer onto xylan in grasses. Proc Natl Acad Sci U S A. 2012;109(3):989–93.

Avci U, Petzold HE, Ismail IO, Beers EP, Haigler CH. Cysteine proteases XCP1 and XCP2 aid micro-autolysis within the intact central vacuole during xylogenesis in Arabidopsis roots. Plant J. 2008;56(2):303–15.

Bouvier d'Yvoire M, Bouchabke-Coussa O, Voorend W, Antelme S, Cézard L, Legée F, et al. Disrupting the *cinnamyl alcohol dehydrogenase 1* gene (*BdCAD1*) leads to altered lignification and improved saccharification in *Brachypodium distachyon*. Plant J. 2013;73(3):496–508.

Bukh C, Nord-Larsen PH, Rasmussen SK. Phylogeny and structure of the cinnamyl alcohol dehydrogenase gene family in *Brachypodium distachyon*. J Exp Bot. 2012;63(17):6223–36.

Burton RA, Fincher GB. (1,3;1,4)-beta-D-glucans in cell walls of the *Poaceae*, lower plants, and fungi: a tale of two linkages. Mol Plant. 2009;2(5):873–82.

Burton RA, Wilson SM, Hrmova M, Harvey AJ. Cellulose Synthase-Like CslF Genes Mediate the Synthesis of Cell Wall (1,3;1,4)-beta-D-Glucans. Science. 2006;311(5769):1940–2.

Burton RA, Jobling SA, Harvey AJ, Shirley NJ, Mather DE, Bacic A, et al. The genetics and transcriptional profiles of the cellulose synthase-like HvCslF gene family in barley. Plant Physiol. 2008;146(4):1821–33.

Burton RA, Collins HM, Kibble NA, Smith JA, Shirley NJ, Jobling SA, et al. Over-expression of specific *HvCslF* cellulose synthase-like genes in transgenic barley increases the levels of cell wall (1,3;1,4)-b-D-glucans. Plant Biotechnol J. 2011;9(2):117–35.

Carpita NC. Structure and biogenesis of the grass cell wall. Annu Rev Plant Physiol Plant Mol Biol. 1996;47(1):445–76.

Cesarino I, Araújo P, Sampaio Mayer JL, Vicentini R, Berthel S, Demedts B, et al. Expression of SofLAC, a new laccase in sugarcane, restores lignin content but not S:G ratio of Arabidopsis lac17 mutant. J Exp Bot. 2013;64(6):1769–81.

Chen L, Auh C, Chen F, Cheng X, Alijoe H, Dixon RA, et al. Lignin deposition and associated changes in anatomy, enzyme activity, gene expression, and ruminal degradability in stems of tall fescue at different developmental stages. J Agric Food Chem. 2002;50(20):5558–65.

Chen L, Auh C, Powling P, Bell J, Chen F, Hopkins A, et al. Improved forage digestibility of tall fescue (*Festuca arundinacea*) by transgenic down-regulation of cinnamyl alcohol dehydrogenase. Plant Biotechnol J. 2003;1(6):437–49.

Chiniquy D, SHarma V, Schultink A, Baidoo EE, Rautengarten C, Cheng K, et al. XAX1 from glycosyltransferase family 61 mediates xylosyltransfer to rice xylan. Proc Natl Acad Sci U S A. 2012;109(42):17117–22.

Christensen U, Harholt J, Alonso-Simon A, Willats WGT, Scheller HV. Characterization of the primary cell walls of seedlings of *Brachypodium distachyon* - a potential model plant for temperate grasses. Phytochemistry. 2010;71(1):62–9.

Collazo P, Montoliu L, Puigdomènech P, Rigau J. Structure and expression of the lignin O-methyltransferase gene from *Zea mays*. Plant Mol Biol. 1992;20(5):857–67.

Dalmais M, Antelme S, Ho-Yue-Kuang S, Wang Y, Darrac O, Bouvier d'Yvoire M, et al. A TILLING platform for functional genomics in *Brachypodium distachyon*. PLoS One. 2013;8(6):e65503.

Doblin MS, Pettolino FA, Wilson SM, Campbell R, Burton RA, Fincher GB, et al. A barley cellulose synthase-like CSLH gene mediates (1,3;1,4)-b-D-glucan synthesis in transgenic Arabidopsis. Proc Natl Acad Sci U S A. 2009;106(14):5996.

Fincher GB. Revolutionary times in our understanding of cell wall biosynthesis and remodeling in the grasses. Plant Physiol. 2009;149(1):27–37.

Fornalé S, Capellades M, Encina A, Wang K, Irar S, Lapierre C, et al. Altered lignin biosynthesis improves cellulosic bioethanol production in transgenic maize plants down-regulated for cinnamyl alcohol dehydrogenase. Mol Plant. 2011;5(4):817–30.

Fry SC, Nesselrode BHWA, Miller JG, Mewburn BR. Mixed-linkage (1,3;1,4)-b-D-glucans. New Phytol. 2008;179(1):104–15.

Gomez L, McQueen-Mason S, Statham E, Bristow J. Analysis of saccharification in Brachypodium distachyon stems under mild conditions of hydrolysis. Biotechnol Biofuels. 2008;1(1):1–12.

Grabber JH, Ralph J, Lapierre C, Barrière Y. Genetic and molecular basis of grass cell-wall degradability. I. Lignin-cell wall matrix interactions. C R Biol. 2004;327(5):455–65.

Guo DM, Ran JH, Wang XQ. Evolution of the cinnamyl/sinapyl alcohol dehydrogenase (CAD/SAD) gene family: the emergence of real lignin is associated with the origin of bona fide CAD. J Mol Evol. 2010;71(3):202–18.

Handakumbura PP, Hazen SP. Transcriptional regulation of grass secondary cell wall biosynthesis: playing catch-up with *Arabidopsis thaliana*. Front Plant Sci. 2012;3:74.

Handakumbura PP, Matos DA, Osmont KS, Harrington MJ, Heo K, Kafle K, et al. Perturbation of *Brachypodium distachyon CELLULOSE SYNTHASE A4* or *7* results in abnormal cell walls. BMC Plant Biol. 2013;13(1):131.

Harrington MJ, Mutwil M, Barriere Y, Sibout R. Molecular biology of lignification in grasses. In: Jouanin L, Lapierre C, Jacquot J, Gadal P, editors. Advances in botanical research lignins: biosynthesis, biodegradation and bioengineering, vol. 61. London: Academic Press; 2012. p. 75–105.

Hatfield RD, Marita JM, Frost K, Grabber J, Ralph J, Lu F, et al. Grass lignin acylation: p-coumaroyl transferase activity and cell wall characteristics of C3 and C4 grasses. Planta. 2009;229(6):1253–67.

Hirano K, Aya K, Kondo M, Okuno A, Morinaka Y, Matsuoka M. OsCAD2 is the major CAD gene responsible for monolignol biosynthesis in rice culm. Plant Cell Rep. 2012;31(1):91–101.

Kiesselbach TA. The structure and reproduction of corn. Lincoln: University of Nebraska College of Agriculture, Agricultural Experiment Station; 1949.

Kim S-J, Zemelis S, Keegstra K, Brandizzi F. The cytoplasmic localization of the catalytic site of CSLF6 supports a channeling model for the biosynthesis of mixed-linkage glucan. Plant J. 2015;81(4):537–47.

Ko JH, Kim WC, Kim JY, Ahn SJ, Han KH. MYB46-mediated transcriptional regulation of secondary wall biosynthesis. Mol Plant. 2012;5(5):961–3.

Lee SJ, Warnick TA, Pattathil S, Alvelo-Maurosa JG, Serapiglia MJ, McCormick H, et al. Biological conversion assay using *Clostridium phytofermentans* to estimate plant feedstock quality. Biotechnol Biofuels. 2012;5:1–14.

Lee MB, Kim DY, Hong MJ, Lee YJ, Seo YW. Identification of gamma irradiated *Brachypodium* mutants with altered genes responsible for lignin biosynthesis. Genes Genomics. 2014;36 (1):65–76.

Li H, Rotter D, Hartman T, Pak F, Havkin-Frenkel D, Belanger F. Evolution of novel *O*-methyltransferases from the *Vanilla planifolia* caffeic acid *O*-methyltransferase. Plant Mol Biol. 2006;61(3):537–52.

Ma QH. Functional analysis of a cinnamyl alcohol dehydrogenase involved in lignin biosynthesis in wheat. J Exp Bot. 2010;61(10):2735–44.

Marriott PE, Sibout R, Lapierre C, Fangel JU, Willats WG, Hofte H, et al. Range of cell-wall alterations enhance saccharification in *Brachypodium distachyon* mutants. Proc Natl Acad Sci U S A. 2014;111(40):14601–6.

Matos DA, Whitney IP, Harrington MJ, Hazen SP. Cell walls and the developmental anatomy of the *Brachypodium distachyon* stem internode. PLoS One. 2013;8(11):e80640.

May JF, Levengood MR, Splain RA, Brown CD, Kiessling LL. A processive carbohydrate polymerase that mediates bifunctional catalysis using a single active site. Biochemistry. 2012;51(6):1148–59.

McFarlane HE, Döring A, Persson S. The cell biology of cellulose synthesis. Annu Rev Plant Biol. 2014;65:69–94.

Meineke T, Manisseri C, Voigt CA. Phylogeny in defining model plants for lignocellulosic ethanol production: a comparative study of *Brachypodium distachyon*, wheat, maize, and *Miscanthus x giganteus* leaf and stem biomass. PLoS One. 2014;9(8):e103580.

Mitsuda N, Seki M, Shinozaki K, Ohme-Takagi M. The NAC transcription factors NST1 and NST2 of Arabidopsis regulate secondary wall thickenings and are required for anther dehiscence. Plant Cell. 2005;17(11):2993–3006.

Mitsuda N, Iwase A, Yamamoto H, Yoshida M, Seik M, Shinozaki K, et al. NAC transcription factors, NST1 and NST3, are key regulators of the formation of secondary walls in woody tissues of Arabidopsis. Plant Cell. 2007;19(1):270–80.

Ohashi-Ito K, Oda Y, Fukuda H. Arabidopsis VASCULAR-RELATED NAC-DOMAIN6 directly regulates the genes that govern programmed cell death and secondary wall formation during xylem differentiation. Plant Cell. 2010;22(10):3461–73.

Patrick JW. Vascular system of the stem of the wheat plant. I. Mature state. Aust J Bot. 1972;20 (1):49–63.

Petrik DL, Karlen SD, Cass CL, Padmakshan D, Lu F, Liu S, et al. *p*Coumaroyl-CoA:monolignol transferase (PMT) acts specifically in the lignin biosynthetic pathway in *Brachypodium distachyon*. Plant J. 2014;77(5):713–26.

Ralph J. Hydroxycinnamates in lignification. Phytochem Rev. 2010;9(1):65–83.

Rancour DM, Marita JM, Hatfield RD. Cell wall composition throughout development for the model grass *Brachypodium distachyon*. Front in Plant Sci. 2012;3:266.

Saballos A, Ejeta G, Sanchez E, Kang C, Vermerris W. A genomewide analysis of the cinnamyl alcohol dehydrogenase family in sorghum [*Sorghum bicolor* (*L.*) *Moench*] identifies SbCAD2 as the *brown midrib6* gene. Genetics. 2009;181(2):783–95.

Sattler SE, Saathoff AJ, Haas EJ, Palmer NA, Funnell-Harris DL, Sarath G, et al. A nonsense mutation in a cinnamyl alcohol dehydrogenase gene is responsible for the *Sorghum brown midrib6* phenotype. Plant Physiol. 2009;150(2):584–95.

Schuetz M, Benske A, Smith RA, Watanabe Y, Tobimatsu Y, Ralph J, et al. Laccases direct lignification in the discrete secondary cell wall domains of protoxylem. Plant Physiol. 2014;166(2):798–807.

Somerville C. Cellulose synthesis in higher plants. Annu Rev Cell Dev Biol. 2006;22(1):53–78.

Taylor-Teeples M, Lin L, de Lucas M, Turgo G, Toal TW, Gaudinier A, et al. An Arabidopsis gene regulatory network for secondary cell wall synthesis. Nature. 2014;517(7536):571–5.

Timell TE. Compression wood in gymnosperms. Berlin: Springer-Verlag; 1986.

Timpano H, Sibout R, Devaux MF, Alvarado C, Looten R, Falourd X, et al. Brachypodium cell wall mutant with enhanced saccharification potential despite increased lignin content. Bioenerg Res. 2015;8(1):53–67.

Trabucco GM, Matos DA, Lee SJ, Saathoff AJ, Priest HD, Mockler TC, et al. Functional characterization of cinnamyl alcohol dehydrogenase and caffeic acid O-methyltransferase in *Brachypodium distachyon*. BMC Biotechnol. 2013;13:61.

Tu Y, Rochfort S, Liu Z, Ran Y, Griffith M, Badenhorst P, et al. Functional analyses of caffeic acid *O*-methyltransferase and cinnamoyl-CoA-reductase genes from perennial ryegrass (*Lolium perenne*). Plant Cell. 2010;22(10):3357–73.

Tufarelli C, Stanley JA, Garrick D, Sharpe JA, Ayyub H, Wood WG, et al. Transcription of antisense RNA leading to gene silencing and methylation as a novel cause of human genetic disease. Nat Genet. 2003;34:157–65.

Urahara T, Tsuchiya K, Konishi T, Kotake T, Tohno-oka T, Komae K, et al. A beta-(1;4)-xylosyltransferase involved in the synthesis of arabinoxylans in developing barley endosperms. Physiol Plant. 2004;122(2):169–80.

Valdivia ER, Herrera MT, Gianzo C, Fidalgo J, Revilla G, Zarra I, et al. Regulation of secondary wall synthesis and cell death by NAC transcription factors in the monocot Brachypodium distachyon. J Exp Bot. 2013;64(5):1333–43.

Vanholme R, Demedts B, Morreel K, Ralph J, Boerjan W. Lignin biosynthesis and structure. Plant Physiol. 2010;153(3):895–905.

Vignols F, Rigau J, Torres MA, Capellades M, Puigdomènech P. The *brown midrib3* (*bm3*) mutation in maize occurs in the gene encoding caffeic acid O-methyltransferase. Plant Cell. 1995;7(4):407–16.

Vogel J. Unique aspects of the grass cell wall. Curr Opin Plant Biol. 2008;11(3):301–7.

Wang Y, Bouchabke-Coussa O, Lebris P, Antelme S, Soulhat C, Gineau E, et al. LACCASE 5 is required for lignification of the *Brachypodium distachyon* culm. Plant Physiol. 2015;168 (1):192–204.

Warnick TA, Methé BA, Leschine SB. Clostridium phytofermentans sp. nov., a cellulolytic mesophile from forest soil. Int J Syst Evol Microbiol. 2002;52:1155–60.

Weigel PH, DeAngelis PL. Hyaluronan synthases: a decade-plus of novel glycosyltransferases. J Biol Chem. 2007;282(51):36777–81.

Wilson JR, Mertens DR, Hatfield RD. Isolates of cell types from sorghum stems: digestion, cell wall and anatomical characteristics. J Sci Food Agric. 1993;63(4):407–17.

Wu X, Wu J, Luo Y, Bragg J, Anderson O, Vogel J, et al. Phylogenetic, molecular, and biochemical characterization of caffeic acid O-methyltransferase gene family in *Brachypodium distachyon*. Int J Plant Genomics. 2013;2013:12.

Yamaguchi M, Mitsuda N, Ohtani M, Ohme-Takagi M, Kato K, Demura T. VASCULAR-RELATED NAC-DOMAIN7 directly regulates the expression of a broad range of genes for xylem vessel formation. Plant J. 2011;66(4):579–90.

Zhong R, Ye ZH. MYB46 and MYB83 bind to the SMRE sites and directly activate a suite of transcription factors and secondary wall biosynthetic genes. Plant Cell Physiol. 2012;53(2):368–80.

Zhong R, Richardson EA, Ye ZH. The MYB46 transcription factor is a direct target of SND1 and regulates secondary wall biosynthesis in Arabidopsis. Plant Cell. 2007;19(9):2776–92.

Zhong R, Lee C, Zhou J, McCarthy RL, Ye Z-H. A battery of transcription factors involved in the regulation of secondary cell wall biosynthesis in *Arabidopsis*. Plant Cell. 2008;20(10).

Zhong R, Lee C, Ye ZH. Global analysis of direct targets of secondary wall NAC master switches in Arabidopsis. Mol Plant. 2010a;3(6):1087–103.

Zhong R, Lee C, Ye ZH. Evolutionary conservation of the transcriptional network regulating secondary cell wall biosynthesis. Trends Plant Sci. 2010b;15(11):625–32.

Zubieta C, Kota P, Ferrer JL, Dixon RA, Noel JP. Structural basis for the modulation of lignin monomer methylation by caffeic acid/5-hydroxyferulic acid 3/5-O-methyltransferase. Plant Cell. 2002;14(6):1265–77.

Brachypodium Seed: A Potential Model for Studying Grain Development of Cereal Crops

Jiajie Wu, Roger Thilmony, and Yong Gu

Abstract Seeds of small grains are important resources for human and animal food. The understanding of seed biology is essential for crop improvement by increasing grain yields and nutritional value. In the last decade, *Brachypodium distachyon* has been developed as a model plant for temperate cereal grasses. Recently, several studies have been published that compare Brachypodium seed anatomy and grain development to those of wheat and barley. While seeds of these three species share many properties, distinct features were identified in Brachypodium, including relatively smaller endosperms with thick cell walls and irregularities in cell sizes in the aleurone layer. Brachypodium seeds have lower starch and higher (1,3;1,4)-β-glucan content, and lower prolamin and higher globulin protein contents, compared to its domesticated relatives. The sequences and expression of genes involved in starch biosynthesis are conserved between Brachypodium and domesticated cereals, but the expression of certain genes in Brachypodium seeds is earlier and at much lower levels, providing a possible explanation for their relatively low starch content. Proteomics analyses indicated that the predominant globulins in the storage proteins were the 11S type and that they are encoded by a multi-gene family. Less than 12 % of the storage proteins were of the prolamin class. Annotation of the Brachypodium genome revealed it contains much fewer prolamin genes than the genomes of wheat and maize, suggesting that the high levels of seed globulins in the former are at the expense of prolamins. The demonstration that expression of a wheat High-Molecular-Weight glutenin gene promoter was endosperm-specific in Brachypodium transgenic plants opens the way for analyses of other gene promoters from cereal crop species that are difficult to transform. The ease of obtaining transgenic Brachypodium plants and the relatively large size of its seeds compared to its small stature make it an ideal system for research of seed properties including dormancy, germination and maturation.

J. Wu, Ph.D.
State Key Laboratory of Crop Biology, Shandong Key Laboratory of Crop Biology, Shandong Agricultural University, Tai'an 271018, China

R. Thilmony, Ph.D. • Y. Gu, Ph.D. (✉)
USDA-ARS, Crop Improvement and Genetics Research Unit, Western Regional Research Center, 800 Buchanan Street, Albany, CA 94710, USA
e-mail: yong.gu@ars.usda.gov

J. Vogel (ed.), *Genetics and Genomics of Brachypodium*,
Plant Genetics and Genomics: Crops Models, DOI 10.1007/7397_2015_12
Published online: 02 October 2015

Keywords Brachypodium seed • Grain development • Storage protein • Starch synthesis • β-glucan

Development of the Brachypodium Seed

Seed Anatomy

The Brachypodium seed is a typical caryopsis with its anatomy resembling that of wheat and barley grains. Considering the small stature of Brachypodium plants, its seeds are relatively large in size. It is approximately 17 % of the size of a wheat seed, and 35–40 % of a polished rice grain (Garvin et al. 2008). The length of Brachypodium seeds is similar to those of cereal grains, mostly longer than 6 mm (Opanowicz et al. 2011; Tyler et al. 2014). The seed width ranges from 1.3 to 1.6 mm and the depth is less than 1 mm, thus a width: depth ratio of 1.7 makes a narrow and flat profile for Brachypodium seeds (Opanowicz et al. 2011; Tyler et al. 2014). The seed mass ranges from 3.5 to 6.0 mg per seed (Trafford et al. 2013; Tyler et al. 2014). The Bd21-3 genotype which was used to generate a large collection of T-DNA mutants (Bragg et al. 2012) has the biggest seeds compared to other accessions investigated (Tyler et al. 2014). As the main part of a caryopsis, the endosperm comprises around 75 % of the total de-hulled grain weight in Brachypodium (Guillon et al. 2011). Different from wheat, barley and oat which have closed crease on the adaxial side of their grains, the Brachypodium endosperm forms a crescent-shaped structure making the seed more flat. This open crease appearance is also present in *Bromus*, *Elymus repens*, *Lolium perenne* and *Festuca pratensis* within the core pooids (Hands et al. 2012). The Brachypodium endosperm forms a convex indentation at the adaxial region facing the nucellar projection region. The outer layers of Brachypodium grain consist of the pericarp, seed coat and the nucellus layer that developed from inner and outer nucellar integuments, and persistent nucellar epidermis comprising squashed cells (Fig. 1; Guillon et al. 2011).

Seed Composition

The protein content in Brachypodium grain is reported between 17 and 28 %, higher than that of wheat, barley, oat, or rice (Guillon et al. 2011; Guillon et al. 2012; Tanackovic et al. 2014). The globulin is the prominent protein stored in the endosperm cells of Brachypodium. While starch is the dominant carbohydrate in most domesticated cereals and accounts for approximately 50–70 % of the grain weight, it only represents about 6–12 % of the Brachypodium grains (Guillon et al. 2011; Tanackovic et al. 2014). At the possible expense of starch, more (1,3;1,4)-β-glucan is deposited in Brachypodium grain and comprises approximately 45 % of the grain weight and 50 % of the endosperm weight at maturity. The (1,3;1,4)-β-glucan is stored in the thick cell walls of endosperm (Guillon et al. 2011; Guillon et al. 2012). Based on the alcohol-insoluble materials prepared

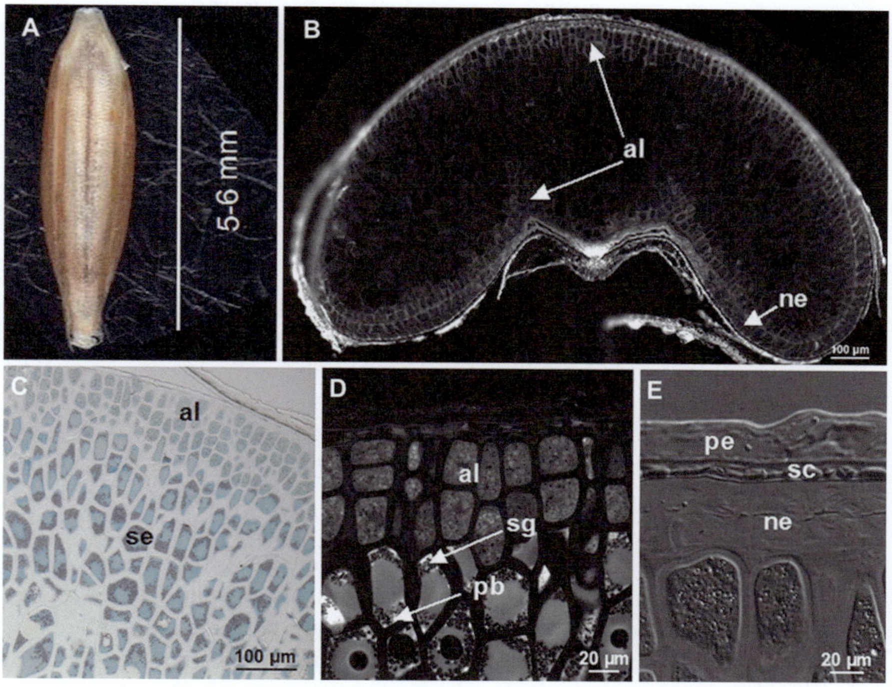

Fig. 1 Image of a whole dehulled *B. distachyon* grain (**a**) and micrographs of dehulled grain cross-sections (**b**–**e**). (**b**) Primary fluorescence of a paraffin-embedded grain. Excitation light, 340–380 nm band-pass filter; emission light, >425 nm. (**c** and **d**) A resin-embedded grain stained with fast green and iodine, showing the aleurone layer (al) and storage endosperm (se). (**c**) Bright field micrograph; (**d**) fluorescence micrograph, higher magnification showing the localization of proteins bodies (pb) around starch granules (sg). (**e**) An interferential contrast micrograph of the outer layers of a paraffin-embedded grain. The outer layer consists of the nucellus epidermis (ne), the seed coat (sc), and the pericarp (pe). Reproduced with permission from Guillon et al. (2011) and *Journal of Experimental Botany*

from whole grain, polysaccharides account for about 60 % of the Brachypodium grain. Glucose (91 %) is the main sugar detected in endosperm and outer layers. These high glucose levels were primarily derived from (1,3;1,4)-β-glucan in the cell walls and a small portion came from starch. Only a small amount of arabinose (3.7 %), xylose (4.2 %), uronic acid (1.5 %) and trace of mannose and galactose were present in the grain (Guillon et al. 2011).

Similar to most cereal grains, the phenolic acid of Brachypodium are mainly composed of ferulic acid, dehydrodimers of ferulic acid and para-coumaric acid (Guillon et al. 2011). Higher ratio of ferulic acid to arabinoxylan was observed in Brachypodium compared to other cereals, suggesting a high degree of arabinoxylan cross-linking which may be involved in cell-cell adhesion (Guillon et al. 2011). The content of total fatty acids in Brachypodium is about 1.64 %, being close to that of *A. fatua* (1.4 %), but much lower than that of wheat (10 %) and rice (16 %) (Guillon et al. 2012). The fatty acids in Brachypodium grain are mainly composed of C18:2 (36.97 %), C18:1 (35.71 %) and C16:0 (15.92 %). A small amount of C12:0, C14:0,

C18:3, C14:1, C16:1, C18:0 and very long chain fatty acids are also present in Brachypodium grain. Of these fatty acids, acylated molecular of PL and TAG account for 13.6 % and 57.4 %, respectively. The composition of fatty acid in Brachypodium is similar to that of oats (Guillon et al. 2012).

Seed Development

Brachypodium has a short generation time, one of important attributes for being an ideal model system for temperate cereals. However, growth conditions including light intensity, temperature, and day length could significantly affect the process of developmental stages (Garvin et al. 2008; Matos et al. 2014). For grain development, the reference accession Bd21 takes around 24 days from anthesis to fully filled ripe grains under growth condition of 22 h of light and 22 °C temperature (Opanowicz et al. 2011). The caryopsis elongates rapidly after fertilization between 2 and 6 DAF (days after flowering), and reaches its maximum length by 6 DAF (Opanowicz et al. 2011). The fresh weight of the grain also increases dramatically during this period (Guillon et al. 2012). After this fast elongating stage, the grain width, fresh weight and dry weight gradually increase until maturity. The water content decreases linearly through grain development from nearly 80 % to 45 %, and when the seeds desiccate, it sharply decreases to 5 % (Guillon et al. 2012), which is much lower than the water content of matured cereal grains (13–15 %).

Embryo and Endosperm Development

Embryo and endosperm are the products of double fertilization during the seed development. The mature embryo of Brachypodium shows dorso-ventral differentiation surrounded by an asymmetrical scutellum, a morphological pattern similar to the mature embryo of wheat and barley. The process of embryo development has been finely reconstructed using a modified pseudo-Schiff propidium iodide staining technique (Guillon et al. 2012). The general shape of the Brachypodium embryo is developed at 18 DAF and all the structures of the mature grain are formed at 26 DAF (Guillon et al. 2012).

The development of endosperm experiences a serial of key stages, such as syncytial endosperm, cellularization, aleurone cell differentiation, and endosperm cell expanding and filling. Similar observations on the timing of Brachypodium endosperm development have been obtained by several studies, with little differences from grains being developed under different growth conditions. In the cytological analysis by Opanowicz et al (2011), endosperm cellularization begins at 3–5 DAF. At the same time, endosperm becomes flat, and deep adaxial indentations of endosperm flanking the nucellar projection starts developing. At 8 DAF, endosperm is fully cellularized and presumptive aleurone cells can be observed around the periphery of the endosperm. Small square aleurone cells distributed

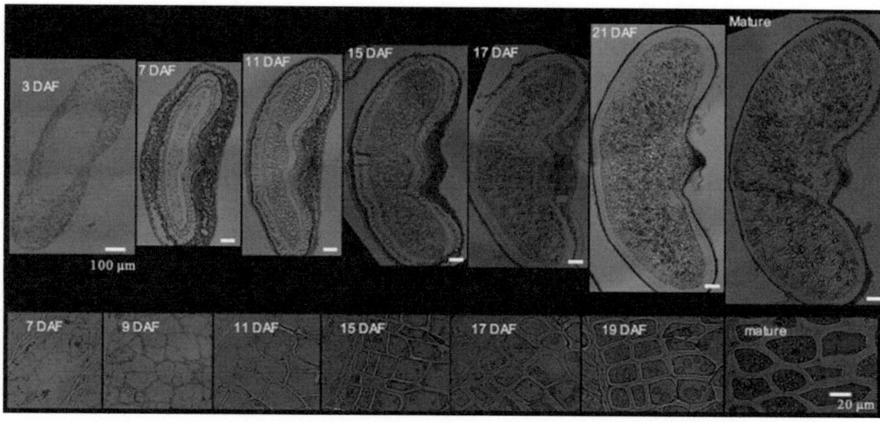

Fig. 2 *B. distachyon* grain cross-sections from 3 DAF to mature stages. Bright-field micrographs of toluidine blue-stained sections. *Top*, morphological development of the grain; *Bottom*, Aleurone differentiation and development. Bar = 100 μm (*Top*), 20 μm (*Bottom*). Reproduced with permission from Guillon et al. (2012) and *Journal of Experimental Botany*

around the edge of endosperm are clearly visible at 15 DAF (Opanowicz et al. 2011). Similarly, Guillon et al (2012) observed fully cellularization at 7 DAF and apparent aleurone layers at 15 DAF. The types of proteins accumulated and their amount changed significantly at 15 DAF (Fig. 2), indicating the change in direction in protein synthesis and function from metabolic activity to storage (Guillon et al. 2011). The active phase of reserve accumulation starts from 22 to 24 DAF when embryo morphogenesis and endosperm expansion are complete. Consistently, from 22 to 24 DAF, the hexose:sucrose ratio begins decreasing and some globulin genes are maximally expressed (Guillon et al. 2012).

Features of the Brachypodium Seed

Cell Differentiation and Maternal Tissue Organization

For endosperm, Brachypodium and other cereals differ in their cell proliferation and wall thickness of cells. In Brachypodium, the endosperm cells continue to divide for several days after cellularization, followed by cell expansion and thickening of cell walls (Guillon et al. 2012). The cell division and expansion in Brachypodium are limited. From cellularization (8 DAF) to maximum fresh weight, which can be reached 19 DAF, the number of cells across the endosperm lobe varies little and the endosperm area only increased 2.5-fold on the mid-grain cross-section of Brachypodium. In comparison, over the same developmental period in barley, the cell number increased 2-fold and the area increased 15-fold (Trafford et al. 2013). Much reduced expression of cell-cycle-related genes cyclin A3 and

CDKB1 indicated that the low rate of cell proliferation in Brachypodium endosperm is probably due to the blocking of mitosis (Trafford et al. 2013).

The endosperm of Brachypodium is very distinctive for its small cell size and thick cell walls (Guillon et al. 2011; Hands et al. 2012). Its cell size is about 21.4–27.5 μm, much smaller than that in other species investigated in Pooideae (Hands et al. 2012; Opanowicz et al. 2011). However, the cell walls of the central endosperm in Brachypodium seeds (2.7–4.4 μm) are several times thicker than that of other species, and are about twice as thick as that of wheat (Hands et al. 2012; Opanowicz et al. 2011). The thickness of endosperm cell walls are generally less than half of that of aleurone cells in members of core pooids, however, they are comparable or slightly greater than aleurone cell walls in Brachypodium (Hands et al. 2012; Opanowicz et al. 2011).

Aleurone, the outermost cell layer of the endosperm is a general feature of the plant endosperm, represents the only endosperm tissue that is alive at maturity. The aleurone plays an important role in seed germination. The endosperm is surrounded by one layer of aleurone cells in wheat and three layers of cells in barley. In temperate small grain cereals, the aleurone layer at the crease region facing the nucellar projection differentiates into modified aleurone as a transfer tissue. The Brachypodium *BdPPDK*, orthologue of the wheat aleurone specific gene encoding pyruvate orthophosphate dikinase (PPDK), expressed weak and later in the peripheral endosperm, suggesting that the aleurone differentiates after endosperm cellularization and is less distinct in Brachypodium (Opanowicz et al. 2011). The Brachypodium 7S globulin encoding gene *BdGLO1* is specifically accumulated in aleurone and embryo. Its expression pattern indicated that Brachypodium aleurone is irregularly distributed with 1 to 4 cell layers in the abaxial region and single cell layer in the adaxial region (Opanowicz et al. 2011). Continuous presence of *BdGLO1* transcripts surrounds the endosperm periphery further suggested another big difference between Brachypodium and cereals, that the modified aleurone is absent in Brachypodium. Consistently, Evans Blue staining showed Brachypodium endosperm was surrounded by continuous layer of living aleurone cells (Opanowicz et al. 2011). This was also confirmed by staining with tetrazolium chloride as a marker of living tissues. Interestingly, the *Bromus* and some *Triticeae* species of *speltoides*, *uratu* and *dicoccoides* also showed living cells in the expected modified aleurone region (Hands et al. 2012). The control of aleurone differentiation in cereal grains was found very complex and many factors are involved, such as the hormones ABA and GA, transcription factors and other important genes (Becraft and Yi 2011). Orthologues of the key genes controlling aleurone development in maize and barley have been identified in Brachypodium, and no significant differences in gene expression were observed between Brachypodium and barley (Hands et al. 2012).

The nucellar epidermis in Brachypodium, like those in rice, persists during grain development, while it is greatly reduced and compressed by maturity in *Triticeae* (wheat, barley, rye), *Aveneae* (oat) and *Poeae* (*Lolium*) (Hands et al. 2012; Opanowicz et al. 2011). The Brachypodium nucellar epidermis is thicker at the adaxial region (29.9 μm) than at the abaxial region (16 μm), and at the lobes it can reach 50 μm which is around five times that the thickness of wheat (Opanowicz et al. 2011). Similar to that of rice, the Brachypodium nucellar epidermal cells were found rich in plasmodesmata, therefore they possibly function in nutrient transportation at the absence of modified aleurone (Opanowicz et al. 2011).

The nucellar projection developed from nucellus tissue facing the main vascular bundle is an important tissue for maternal nutritional supplies during grain development. The size of the nucellar projection region in Brachypodium is vastly reduced compared to that of the small grains (Opanowicz et al. 2011). The robust nucellar projection is believed to be important for the size and shape of grains (Hands et al. 2012), and the reduced development of nucellar projection in barley compromised transfer of signals and assimilates from maternal to the filial grain and caused a shrunken endosperm (Weier et al. 2014). Lack of the robust nucellar projection in Brachypodium might explain its flat and starch-poor grains (Hands et al. 2012).

Low Starch Content

The starch which dominates the endosperms of cereal grains, only accounts for 6–12 % of the Brachypodium grain weight. During Brachypodium grain development, the starch content remained stable (~100 μg per grain) from 4 DAF to 20 DAF and increased to 300 μg per grain by maturity. However, when expressed as μg mg^{-1} of grain dry weight, its concentration decreased from 160 to 60 μg mg^{-1} dry weight (Guillon et al. 2012), indicating that starch accumulation plays a minor role in the increase of dry weight during Brachypodium seed development. It has been observed that the starch is distributed along a gradient in the Brachypodium seed, being more abundant in the proximal end of the endosperm than at the distal end which is away from embryo, and is restricted in the central zone of the endosperm at the distal end (Trafford et al. 2013). The starch transiently accumulated in the pericarp at an early developmental stage, and started to accumulate in the endosperm after cellularization (Chen et al. 2014a; Guillon et al. 2012; Trafford et al. 2013). Starch granules appeared at 15 DAF and were apparently distributed around the periphery of endosperm cells at 17 DAF, and afterwards took over all the spaces excluding the fused vacuoles in mature seeds (Guillon et al. 2012).

The Brachypodium granules are mostly flattened, appearing as concave disks with depressions in the centre like doughnuts (Tanackovic et al. 2014). Several studies have shown the Brachypodium starch granules are small in size and are not bimodally distributed (Chen et al. 2014a; Guillon et al. 2012; Hands et al. 2012; Opanowicz et al. 2011), although Tanackovic et al (2014) reported a bimodal population of Brachypodium granules consisting of small B-type (2.5–10 μm) and very small C-type (0.5–2.5 μm). The large A type granules, a typical type in cereals, are absent in Brachypodium. In addition, the Brachypodium starch granule is smooth at its surface, similar to the Barley granule, but its internal part does not have obvious pores, channels or growth rings (Tanackovic et al. 2014). Starch is made of two polysaccharides: amylose and amylopectin. The content and branch profile of amylose and amylopectin in Brachypodium starch granules are generally similar to those in barley. However, starch-bound phosphate content, important for starch degradation, is about half of that of barley starch, suggesting a difference in starch mobilization (Tanackovic et al. 2014).

Analysis of genes involved in starch biosynthesis and metabolism suggested that the low content of starch in Brachypodium endosperm was possibly due to reduced expression of starch synthesis genes and the reduced catalytic activity compared to that of barley (Chen et al. 2014a; Trafford et al. 2013). It was further hypothesized that the grain filling with less starch granules represents one possible reason for the small sizes of Brachypodium endosperm cells (Trafford et al. 2013).

High β-Glucan Content in the Cell Walls

The thick cell walls of the storage endosperm of Brachypodium makes its grains very distinct to other cereals, resulting in its high content of cell wall polysaccharides. They account for ~52 % (w/w) of the endosperm in comparison with 2–7 % in other cereals (Guillon et al. 2011). In cereals, the endosperm cell wall is mainly composed of arabinoxylan and (1,3;1,4)-β-glucan, with low amounts of cellulose and structural proteins (Burton and Fincher 2014; Fincher and Stone 1986). The relative ratio of arabinoxylan and (1,3;1,4)-β-glucan varies significantly among the cereals. For instance, barley endosperm cell wall contains 20 % of arabinoxylan and 70 % of β-glucan (Shewry et al. 2003), while the wheat endosperm cell walls comprise about 70 % of arabinoxylan and 15 % of β-glucan (Feng et al. 2009).

Similar to barley, (1,3;1,4)-β-glucan represents the most abundant polysaccharides in the cell walls of Brachypodium grains (Burton and Fincher 2014; Guillon et al. 2011; Trafford et al. 2013). The β-glucan accounts for about 80 % (w/w) of the endosperm cell walls and 19–43 % (% FWT) of the Brachypodium grain (Guillon et al. 2011; Trafford et al. 2013). It is enriched in both the endosperm and aleurone cell walls, but was absent in the middle lamella and cell junction zones (Guillon et al. 2011). (1,3;1,4)-β-Glucan first appeared in the outer layer at early stage, and then started to accumulate in the endosperm cell walls when cellularization begins. During grain development, the speed of (1,3;1,4)-β-glucan deposition were estimated seven times faster than that of starch (Guillon et al. 2012). Digestion of the (1,3;1,4)-β-glucan of Brachypodium grain using lichenase yielded gluco-oligosaccharides comprising 3-O-β-cellobiosyl-D-glucose (BG_3) and 3-O-β-cellotrissyl-D-glucose (BG_4) with a ratio of 5.8–8:1 (Guillon et al. 2011; Guillon et al. 2012), which is significantly higher than the 2–3:1 observed previously from Poaceae (Burton and Fincher 2014). Besides (1,3;1,4)-β-glucan, arabinoxylan and cellulose are also present in cell walls of Brachypodium grain, and accounted for about 2.7 % and 7.3 % of the endosperm, respectively (Guillon et al. 2011). As in barley (Jääskeläinena et al. 2013), the arabinoxylan was mainly distributed in the middle lamella and junction zones while cellulose was not present in these areas (Guillon et al. 2011).

The observation of high (1,3;1,4)-β-glucan and low starch content in the endosperm leads to the hypothesis that (1,3;1,4)-β-glucan could function as a storage compound in Brachypodium. Although the evidence obtained from seed germination experiments is limited, several facts are supporting this hypothesis (Burton and Fincher 2012; 2014). For example, cell wall polysaccharides have been found to be used as principal storage compounds in many plant species (Buckeridge 2010), and compared to starch, the (1,3;1,4)-β-glucan can be synthesized and deposited more easily, as well as the depolymerization requires less enzymes. It has been further argued that β-glucan might have broader roles than previously considered given its accumulation pattern in senescing and mature tissues (Vega-Sanchez et al. 2013).

Starch Synthesis in Brachypodium

Genes Are Conserved Between Brachypodium and Cereals

The synthesis of starch requires a series of reactions catalyzed by multiple enzymes, including ADP-glucose pyrophosphorylase (AGPase), soluble starch synthases (SSs), granule-bound starch synthase (GBSSs), starch-branching enzymes (SBEs), debranching enzymes of isoamylase (ISA) and limit dextrinase (PUL), and α-glucan phosphorylase (PHOs). A total of 28 genes involved in starch biosynthesis and degradation have been identified and annotated from the Brachypodium genome (Tanackovic et al. 2014). The result indicated that the genetic capacity of starch metabolism is highly conserved in the grass genomes. However, a few differences were also observed among species. Orthologues of all of the rice starch metabolism genes could be identified in Brachypodium except one *SSIV* and one *AGPase* subunit (Trafford et al. 2013). These two genes were also absent in wheat and barley, suggesting their lineage specific evolution. One *GBSSIb* (Bradi4g00650) and one *SBEIII* (Bradi1g41970) were recently identified as starch synthase genes (Tanackovic et al. 2014). The *GBSSIb* (Bradi4g00650) was possibly a pseudogene and was grouped as an outlier in the phylogenetic tree (Tanackovic et al. 2014). The presence of only two bands in SDS-PAGE of granule-bound synthase further indicated that this *GBSSIb* might not be functional in Brachypodium (Chen et al. 2014a). For the *SBEIII* (Bradi1g41970), no orthologues were found in rice and barley, however, it was highly conserved with the *SBEIII* (JQ346193) of wheat (Chen et al. 2014a).

The starch genes of Brachypodium are randomly distributed along the five chromosomes. Gene clusters or tandem duplications of these genes were not observed. The phylogenetic tree analysis of the starch synthase genes of Brachypodium and *Poaceae* species was performed, resulting in individual clades consisting of clusters with specific enzymatic functions (Chen et al. 2014a; Tanackovic et al. 2014; Trafford et al. 2013). The close phylogenetic relationships and the conserved content of starch synthase genes suggested starch metabolism of Brachypodium might be overall conserved with those in cereal grasses. In addition, the sequences of Brachypodium starch synthase genes were more conserved to wheat than to rice or maize (Chen et al. 2014a; Tanackovic et al. 2014), confirming the closer relatedness of Brachypodium with *Triticeae* species.

Starch synthases including SS, GBSS, SBE and DBE are mainly composed of catalytic domain and non-catalytic carbohydrate-binding module (CBM) with affinity to starch and plastid-targeting signal peptides. Four different CBM domains (CBM 28, CBM53, CBM20, and CBM45) were identified and positioned in the Brachypodium gene sequences. There are 3 tandem repeats of CBM53 in the middle region of *SSIIIs*, one CBM48 located at the N-termini of *SBEs*, *ISOs* and *PUL* (Chen et al. 2014a; Tanackovic et al. 2014). CBM20 and CBM45 were identified in the starch degradation related genes of glucan water dikinases (*WGDs*) of Brachypodium (Tanackovic et al. 2014).

Gene Expression and Enzyme Activities Were Decreased in Brachypodium

Basically, the starch synthase genes of Brachypodium are highly expressed at 5 DAF before the endosperm cellularization, and then undergo declining expression towards seed maturity (Tanackovic et al. 2014). For example, the *AGP-L1*, *AGP-S2*, *GBSS1a*, *SSI*, *SSIIa*, *SBEIIa*, *SBEIIa*, *SBEIIb*, *ISA1* and *PHO1* showed maximum expression at 5 DAF, and after that the expression level declined (Tanackovic et al. 2014). Exceptions come from the *SBE1*, *SSIIIa* and *ISA2* genes, of which expression were constant during seed development or were slightly increased at late stage (Tanackovic et al. 2014). Consistently, similar expression pattern were observed by qRT-PCR analysis, showing that the expression of *GBSSI*, *SSIIa*, *SSIIb*, *SBEII*, *ISAI*, and *PUL* declined from 4 DAF to 25 DAF (Chen et al. 2014a). However, at 30 DAF when seeds mature, some genes like *SSI*, *SSIIb* and *SSIIc* were clearly increasingly expressed (Chen et al. 2014a). In comparison, the expression of barley starch synthase genes were low and induced at the pre-cellularization stage, and then increase to peaks at 10–16 DAF and 26 DAF (Tanackovic et al. 2014). As for the abundance of transcripts of starch synthase genes, significant differences were observed between Brachypodium and barley. By comparing the gene orthologues, lower levels of the transcripts accumulated in Brachypodium than in barley, and the maximum transcript abundance of seven Brachypodium genes was less than 1/10 of the abundance of the corresponding barley genes. The *SSI* and *BEI* transcript levels were significantly lower in Brachypodium compared with barley where the transcript levels were 2000- and 580-fold higher, respectively (Trafford et al. 2013).

Consistent with the low expression of the Brachypodium starch synthase genes, the enzymes' catalytic activities were also relatively low when compared to barley. The Brachypodium SS activity was less than half of barley SS, and the AGPase activity was 18 times lower in Brachypodium than in barley (Trafford et al. 2013). Phosphorylation of some starch synthase enzymes is an important modification related to protein-protein interactions. The starch synthase catalytic domain and relevant amino acids of *GBSSI* were well conserved between Brachypodium and wheat. Two phosphorylation sites were located between the third and fourth alpha-helix in wheat *GBSSI*, however, the conserved sites were not phosphorylated in Brachypodium (Chen et al. 2014a). The missing of phosphorylation of *GBSS1* might cause a reduction of its glucan-substrate recognition and affinity (Chen et al. 2014a). The low starch content Brachypodium grains could be directly due to the low expression of the starch synthase genes, however, the impact of the phosphorylation sites on the catalytic activity requires further investigation.

Brachypodium Seed Storage Proteins

Storage proteins serve as a reservoir of carbon, nitrogen and sulphur for post-germination growth of seedlings. In cereal grains such as wheat, barley, rice and maize, storage proteins account for 7–15 % of the grain dry weight and 60–80 % of the total proteins, and therefore, are an important source of protein for human and animal food. Based on their solubility in different agents, storage proteins can be classified into four classes, including albumins soluble in water, globulins soluble in dilute saline, gliadins soluble in alcohol solutions and glutenins soluble in dilute acid or alkali solutions. Since a mixture of gliadins and glutenins is present in protein extractions under reducing condition, and they both contain high percentage of proline and glutamine amino acids, therefore, they were also classified as prolamin (Payne 1987). Prolamins constitute the main storage proteins in wheat, barley, and rye. In wheat, the content and composition of the prolamins plays important roles in determining the properties of the flour dough (Shewry and Halford 2002). Among the wheat prolamins, gliadins are monomers and determine the viscosity and extensibility of wheat dough, while glutenins are polymers and determine its elasticity (Payne 1987). In rice and oat, globulins are the main components of the storage proteins (Shewry and Halford 2002).

Storage Proteins of Cereals Are Present in Brachypodium

In the recent years, a number of studies have been performed to investigate the seed storage proteins of Brachypodium. Larre et al (2010) quantified the total amino acid composition of mature grains by reversed-phase high-performance liquid chromatography (RP-HPLC), and found the total nitrogen content of the Bd21-3 grain was nearly 3 %. Similarly, the total nitrogen content of Bd21 grain was reported to be about 3.7 % (Guillon et al. 2012). Using a conversion factor of 5.7 for cereals, the protein content of Brachypodium grains was estimated to be 16–21 % of the dry weight, which was higher than that of the small grains (10–12 %) (Shewry and Halford 2002). In addition, an extremely high 28 % protein content was reported by

Tanackovic et al (2014), in Bd21 and Bd21-3 compared to the 12 % present in barley cultivars. Amino acid composition analysis in Brachypodium seeds showed that Glx (20 %) was the most abundant, followed by Asx (10 %), Pro (8.6 %), and Gly (8.5 %) (Larre et al. 2010).

Different types of cereal storage proteins were identified in Brachypodium grains, including albumins (Wang et al. 2010), globulins (Laudencia-Chingcuanco and Vensel 2008) and prolamins (Larre et al. 2010). Based on solubility, the majority of the Brachypodium grain proteins were soluble in urea buffer in reducing conditions, 10 % were soluble in salt buffer and very few were soluble in alcoholic solutions (Larre et al. 2010). Brachypodium storage proteins are predominately comprised of globulins, while gliadins and glutenins only accounted for minor percentage (Larre et al. 2010). This composition is closer to that of rice and oat, and much lower than in wheat.

Using Fast Green staining and deep ultraviolet fluorescence imaging, the accumulation of Brachypodium storage proteins was visualized to be present at the cell periphery around starch granules (Guillon et al. 2012; Larre et al. 2010). Two types of protein bodies were observed in Brachypodium endosperm cells and the sizes ranged from 0.5 to 2 μm. The Fast Green staining and fluorescence imaging also showed that the large 'vacuoles' in endosperm cells are filled with proteins, but anti-11S globulin labeling was not detected (Guillon et al. 2012). In another study, the immunolabeling of anti-11S globulin indicated there were big irregular shaped protein matrixes in the endosperm cells (Larre et al. 2010).

Globulins

Laudencia-Chingcuanco and Vensel (2008) first reported the storage proteins of Brachypodium seed. By SDS-PAGE of Brachypodium total seed proteins, seven dominant bands ranging from 57 to 20 kDa in size were revealed. Each of the major bands were excised from the gel and were identified by electron spray ionization mass spectrometry. Six of them were shown to be globulin proteins as these peptide sequences had high similarity to the known globulin proteins, including maize Globulin-1 (GenBank accession AAC31462), oat 12S globulin protein (1515394A) and rice globulin proteins. These globulins all represent the main storage proteins in the different species. Larre et al (2010) further investigated different components of Brachypodium storage proteins by sequential extraction followed by 1D and 2D electrophoresis and mass spectrometry identification. Similarly, eight major bands ranging from 70 to 10 kDa were revealed on 1D SDS-PAGE. Furthermore, both the salt-soluble (50 mM sodium phosphate buffer pH 8 and 0.5 M NaCl) and UCT-soluble (8 M urea, 2 % CHAPS, 2 M thiourea and 18 mM DTT) fractions were separated on 2D gel for proteomic analyses of the storage proteins. For the salt-soluble fraction, major intense spots were parts of spots in three strings on the gel. The protein spots from the same string usually belong to the same family. Four spots in one string located around 72 kDa and four spots in the second string around 57 kDa were excised and subjected to mass spectrometry for protein identification. The result indicated that the proteins in the two strings were annotated as homologues of wheat globulin 3 and maize globulin 2, respectively, which both belong to the 7S globulin family (Larre et al. 2010). These 7S globulins comprised at least 30 % of the salt extract. For the urea extract, a total of 52 (43 %) spots on the 2D gel were identified by mass spectrometry, and 23 of them were annotated as 11S globulins. These 11S globulins were found varying from 18 to 70 kDa in molecular weight, and accounted for 72 % of the UCT-soluble proteins. In addition, Wang et al (2010) also reported identification of 12S seed storage globulin 1 precursor, 11S globulin and seed storage globulins in Brachypodium seed proteins. These studies demonstrated that globulins are the major components of storage proteins in Brachypodium, and are mainly comprised of 11S globulins.

Glutenins

In wheat, the high molecular weight glutenin (HMW-glutenin) and low molecular weight glutenins (LMW-glutenin) subunits are predominant components of seed storage proteins, and contribute to the biochemical properties of wheat dough and its baking quality. Therefore, investigation of glutenins in Brachypodium seeds will help evaluate the utility of Brachypodium as a model system for studying wheat quality attributes. A number of studies reported the characterization of Brachypodium seed storage proteins (Guillon et al. 2012; Larre et al. 2010; Laudencia-Chingcuanco and Vensel 2008). However, unlike the globulins, the reports on the glutenins have been inconsistent. The work from Laudencia-Chingcuanco and Vensel (2008) suggested that prolamins are not the major storage proteins in Brachypodium seed. A more detailed proteomic analysis of Brachypodium seed storage proteins by Larre et al (2010) detected both globulins and prolamins, but no albumins. However, none of the detected prolamins was closely related to HMW-glutenin and LMW-glutenin. Recently, Wang et al (2010, 2012) reported characterization of seed storage proteins in Brachypodium. In the SDS-PAGE analysis of different protein fractions, a similar band pattern in the LMW-GS regions of common wheat could be found in Brachypodium, suggesting the presence of protein products similar to LMW-glutenin. However, no bands within the size range of the wheat HMW-glutenin regions were detected, although the MALDI-TOF identified a clear peak of 70 kDa which is similar to the size of HMW y-type subunit of wheat. Based on the protein analysis data, the 70 kDa protein is similar to the wheat HMW glutenin y-type subunit (Wang et al. 2010). The presence of LMW-GS in Brachypodium was further confirmed by Western blotting using a polyclonal antibody (CIPGLERPWQQQPL) specific to wheat LMW-GS (Wang et al. 2012). Approximately ten bands of Brachypodium LMW-GS were revealed by SDS-PAGE, and subsequent MALDI-TOF identifications showed their molecular mass spectra were mostly ranging from 5 to 50 kDa (Wang et al. 2010;

Wang et al. 2012). Based on the protein separation data, the authors stated that Brachypodium has no LMW-D type (~60 kDa), less LMW-B type (40–45 kDa), and that the LMW-C type subunits were more abundant than the equivalent proteins in common wheat (Wang et al. 2012).

Gliadins

The alcohol-soluble fractions of seed storage proteins are classified as gliadins which are encoded by a multigene family and are comprised of highly diverse monomer proteins with molecular weight ranging from 28 to 75 kDa. Recently, Chen et al (2014b) reported that no gliadin protein was expressed in Brachypodium seeds based on A-PAGE, SDS-PAGE, RP-HPLC and western blotting analysis of the alcohol-soluble protein extract. Similarly, only traces of proteins could be recovered with alcohol solutions and very few bands were detectable on the SDS-PAGE gel (Larre et al. 2010; Wang et al. 2012). However, with the subsequent urea extract, four spots on the 2-DE gel were identified as products of Brachypodium prolamin (Bradi3g17070), and three spots of 19 kDa were identified as products of Bradi1g50300 which is similar to wheat avenin-like protein (CAJ32654) (Larre et al. 2010). They both could be annotated as gliadin proteins. On the 2D gel, these gliadins represent about only 5 % of expressed proteins found in the urea extract (Larre et al. 2010). Taken together, gliadins or gliadin-related prolamins represent a very small proportion of Brachypodium seed proteins.

Annotation of Brachypodium Storage Protein Genes

The completion of whole genome sequencing of Bd21 and the re-sequencing of more accessions provided an excellent platform for the structural and functional studies of storage protein encoding genes of Brachypodium. A total of 24 genes have been annotated as storage protein genes by the sequencing project, including 12 globulin genes and 12 prolamin genes, respectively (Brachypodium annotation release version 1.2). The Brachypodium proteome was also explored by individual studies to facilitate the identification of storage proteins. Larre et al (2010), using combined proteomic and genome annotated approaches, identified a total of 15 globulin genes and 12 prolamin genes (Table 1). Of the 15 globulin genes, 12 encode 11S globulins and 3 encode 7S globulins. For the 11S globulin genes, conserved NG (Asparagine-Glycine) cleavage sites were present in the deduced amino acids of most genes, especially of those expressed globulin genes (Larre et al. 2010). Previously, another two globulin genes, *Bd.glo1* homologue to maize 7S globulin, and *Bd.glo2* homologue to oat 12S globulin, were cloned from EST library of Bd21 (Laudencia-Chingcuanco and Vensel 2008). Sequence alignments showed that these two genes are identical to the annotated globulin genes, *Bd.glo1*

Table 1 The storage protein encoding genes annotated from Brachypodium accession Bd21

Brachypodium genes	CDS length (bp)	Homologue and protein name	Expression	Reference	Note
Type 11S globulin					
Bradi2g20860.1	699	Q0Q5E3, globulin 1 (*Triticum aestivum*)		Larre et al. (2010), Gu et al. (2010)	
Bradi2g37470.1	2132	B6SLE7, legumin-like protein (*Zea mays*)		Larre et al. (2010)	
Bradi2g37470.2	2132	B6SLE7, legumin-like protein (*Zea mays*)		Larre et al. (2010)	
Bradi2g37860.1	1764	O49258, 12S globulin (*Avena sativa*)		Larre et al. (2010)	
Bradi2g38050.1		Q38780, 11S globulin (*Avena sativa*)	Yes	Larre et al. (2010)	Absent in MIPS_v1.2
Bradi2g38060.1	1924	Q38780, 11S globulin (*Avena sativa*)	Yes	Larre et al. (2010), Trafford et al. (2013)	
Bd.glo2a	1482	Q38780, 11S globulin (*Avena sativa*)	Yes	Laudencia-Chingcuanco and Vensel (2008)	Identical to *Bradi2g38060.1*
Bd.glo2b	1482	Q38780, 11S globulin (*Avena sativa*)	Yes	Laudencia-Chingcuanco and Vensel (2008)	Identical to *Bradi2g38060.1*
Bradi2g38070.1	1769	Q38780, 11S globulin (*Avena sativa*)		Larre et al. (2010)	
Bradi2g40840.1	1828	Q38780, 11S globulin (*Avena sativa*)	Yes	Larre et al. (2010), Trafford et al. (2013)	
Bradi2g62590.1		B6TDD3, legumin-like protein (*Zea mays*)		Larre et al. (2010)	Absent in MIPS_v1.2
Bradi2g62590.2		B6TDD3, legumin-like protein (*Zea mays*)		Larre et al. (2010)	Absent in MIPS_v1.2
Bradi4g28220.1	1853	P12615, 12S seed storage globulin (*Avena sativa*)	Yes	Larre et al. (2010), Trafford et al. (2013)	

(continued)

Table 1 (continued)

Brachypodium genes	CDS length (bp)	Homologue and protein name	Expression	Reference	Note
Bradi4g29130.1	1826	O49258, 12S globulin (*Avena sativa*)	Yes	Larre et al. (2010)	
Bradi1g05940	987	Q8L8I0, globulin-like protein (*Oryza sativa*)		MIPS_v1.2	*BdGLO1B*
Bradi4g37310	1128	W5FJB1, uncharacterized protein (*Triticum aestivum*)		MIPS_v1.2	*BdGLO1C*
Type 7S					
Bradi1g05910.1	1886	Q7M1Z8, globulin-2 (*Zea mays*)	Yes	Larre et al. (2010), Trafford et al. (2013)	
Bradi1g13040.1	2404	B7U6L4, globulin 3 (*Triticum aestivum*)	Yes	Larre et al. (2010), Trafford et al. (2013)	
Bd.glo1	1683	Q03865, vicilin-like embryo storage protein (*Zea mays*)	Yes	Laudencia-Chingcuanco and Vensel (2008)	Identical to *Bradi1g13040*
Bradi1g63180.1	2178	B6SK46, Cupin family protein (*Zea mays*)		Larre et al. (2010)	
Prolamin					
Type S-rich					
Bradi1g50200.1	778	Q2A784, putative avenin-like a1 (*Triticum aestivum*)		Larre et al. (2010)	
Bradi1g50290.1	739	Q2A781, putative avenin-like a1 (*Aegilops tauschii*)		Larre et al. (2010)	
Bradi1g50300.1	685	Q2A784, putative avenin-like a1 (*Triticum aestivum*)	Yes	Larre et al. (2010)	
Bradi2g33280.1	530	C3SAE7, prolamine (*Brachypodium distachyon*)		Larre et al. (2010)	

(continued)

Table 1 (continued)

Brachypodium genes	CDS length (bp)	Homologue and protein name	Expression	Reference	Note
Bradi2g38530.1	785	C3SAE7, prolamine (*Brachypodium distachyon*)		Larre et al. (2010)	
Bradi2g39910	315	J7I1U6, gamma-gliadin (*Aegilops speltoides*)		MIPS_v1.2	
Bradi2g39920	378	Q09071, avenin (*Avena sativa*)		MIPS_v1.2	
Bradi2g39940.1	826	C3SAE7, prolamine (*Brachypodium distachyon*)		Larre et al. 2010	
Bradi3g17070.1	1056	C3SAE7, prolamine (*Brachypodium distachyon*)	Yes	Larre et al. (2010)	
JN628948	876	K7XRA1, alpha-gliadin (*Triticum aestivum*)		Chen et al. (2014b)	Pseudogene
JN628949	900	Q1WA39, alpha-gliadin (*Triticum aestivum*)		Chen et al. (2014b)	Pseudogene
JN628950	849	U3PVE0, alpha-gliadin (*Aegilops tauschii*)		Chen et al. (2014b)	Pseudogene
JN628951	852	R9XW63, alpha-gliadin (*Aegilops tauschii*)		Chen et al. (2014b)	Pseudogene
JN628952	876	K7XRA1, alpha-gliadin (*Triticum aestivum*)		Chen et al. (2014b)	Pseudogene
JN628953	876	K7XRA1, alpha-gliadin (*Triticum aestivum*)		Chen et al. (2014b)	Pseudogene
JN628966	876	K7XRA1, alpha-gliadin (*Triticum aestivum*)		Chen et al. (2014b)	Able to be expressed in *E. coli* with T7 promoter
JN628967	876	K7XRA1, alpha-gliadin (*Triticum aestivum*)		Chen et al. (2014b)	Able to be expressed in *E. coli* with T7 promoter

(continued)

Table 1 (continued)

Brachypodium genes	CDS length (bp)	Homologue and protein name	Expression	Reference	Note
HQ220191	1164	D0ES15, LMW glutenin subunit (*Aegilops uniaristata*)	Yes	Wang et al. (2012)	LMW-i
HQ220193	1134	V5NEC7, LMW glutenin subunit (*Triticum dicoccoides*)		Wang et al. (2012)	LMW-i
HMW					
Bradi2g20870.1	300	Q8LKI8, HMW glutenin subunit (*Aegilops speltoides*)		Larre et al. (2010), Gu et al. (2010)	Pseudogene
Bradi2g20910.1	885	Q670Q5, HMW glutenin subunit 1Dy10.1 (*Triticum aestivum*)		Larre et al. (2010)	
JX915838	1977	Q6RX92, HMW glutenin subunit (*Triticum aestivum*)	Yes	Subburaj et al. (2014)	Pseudogene

Table was modified from Larre et al (2010).

corresponds to Bradi1g13040, and *Bd.glo2* corresponds to Bradi2g38060. The globulin proteins encoded by both *Bd.glo1* and *Bd.glo2* have been purified and identified in two separate studies (Larre et al. 2010; Laudencia-Chingcuanco and Vensel 2008) and demonstrated that they are highly expressed compared to other storage protein genes of Brachypodium.

The annotated prolamin genes in Brachypodium included seven genes as S-rich prolamin due to their cysteine content (3–6.6 %) and HMW glutenin related genes (Bradi2g20870 and Bradi2g20910) (Table 1). These two HMW glutenin related genes are likely pseudogenes since they are much shorter in length and missing the central repeat domain and 3′ end regions compared to HMW-GS gene of other species. The sequence alignment showed that the deduced peptide sequence of Bradi2g20870 shared good sequence similarity (47 % identities) with the y-type HMW glutenin only in the 5′ translated region with five conserved cysteine residues present in both genes (Gu et al. 2010). Furthermore, Bradi2g20870 was located in the genomic region that showed significant collinearity with the wheat HMW glutenin region. Taken together, Bradi2g20870 is orthologous to the wheat HMW glutenin genes. The orthologs of wheat HMW glutenin genes were not present in rice and maize, suggesting that the ancestral gene of the wheat HMW glutenin gene evolved after the divergence of the Brachypodium and wheat lineages from that of rice and maize. Interestingly, a globulin gene (Bradi2g20860) is next to the Brachypodium HMW glutenin-like gene (Bradi2g20870), similar to what is found in wheat and other *Triticeae* species. Protein sequence alignment showed that this globulin shared sequence similarity (~50 %) with the wheat HMW glutenin at the 5′ translated region. Given the sequence similarity, close location, and seed specific expression, it was hypothesized that the ancestral HMW glutenin gene was derived from a duplication of the globulin gene (Gu et al. 2010). In the wheat lineage, one of the duplicated genes further evolved to gain the central repeat domain sequences, resulting in the change of its biochemical properties including solubility.

In addition to these prolamin genes confirmed through genome annotation, other approaches have been taken to search for other prolamin-like genes. Southern blot analyses using the probes of wheat prolamin genes suggested more than 10 copies of gliadin genes and 4–5 copies of LMW-GS are present in Brachypodium (Chen et al. 2014b; Wang et al. 2012). To further characterize the Brachypodium prolamin genes, allele specific primer pairs used for amplifying wheat prolamin genes were employed in PCR reactions using Brachypodium genomic DNAs as template (Chen et al. 2014b; Subburaj et al. 2014; Wang et al. 2012). A total of 30 gliadin genes were cloned from 9 Brachypodium accessions, and 8 genes were obtained from Bd21 (Table 1) (Chen et al. 2014b). The isolated gliadin genes are highly conserved both in nucleotide sequences and deduced amino acids. Compared to wheat gliadin genes, common structures of alpha-gliadins, including signal peptide, repetitive domain, polyglutamine domain and unique domains, were also conserved in these deduced proteins (Chen et al. 2014b). Consistent with the finding that few gliadin proteins were deposited in Brachypodium endosperm, most of the isolated gliadin genes were pseudogenes and only four of them have full length ORF (Chen et al. 2014b). The same strategy was used to identify a total of 18 LMW-GS genes from Brachypodium, 2 from diploid accession Bd21 (Table 1) and 16 from five polyploidy accessions (Wang et al. 2012). All of these genes had typical structural characteristics identified in the LMW glutenin genes in wheat such as conserved 20 aa signal peptides, variable repetitive domain and three subregions of the C-terminal domain (Wang et al. 2012).

In addition, Subburaj et al (2014) recently reported isolation of 13 HMW-GS genes from several Brachypodium accessions. All of these genes have typical structural features of a y-type HMW-GS gene of wheat. Premature stop codons were present in all of these Brachypodium HMW-GS genes. Gene expression analysis showed the HMW-GS gene of Bd21 (Table 1) were expressed but at a much lower level compared to common wheat (Subburaj et al. 2014). However, these genes share unexpectedly high levels of similarity (99 %) with the wheat *1Dy12* gene (JN255520).

Phylogenetic analyses were performed using gliadin and LMW-GS genes from Brachypodium and *Triticae* species. Based on this analysis, wheat was more closely related to Brachypodium than to barley and rye (Chen et al. 2014b; Wang et al. 2012). Also, surprisingly, all these PCR amplified Brachypodium prolamin gene sequences were not present in the existing Brachypodium genome sequence database. One explanation could be that the absence of these gene sequences reflects gap regions in the Brachypodium genome assembly as suggested by the group (Chen et al. 2014b). However, screening Brachypodium BAC clones with the prolamin probes and subsequent sequencing of the positive clones can not only provide further validation of these Brachypodium prolamin genes, but also help fill in these gap regions in the genome, thereby providing insights into the structure, organization, and evolution of these prolamin genes in the Brachypodium lineage.

Genome Distribution and Gene Expression of Storage Protein Genes

The storage protein encoding genes appear to not be randomly distributed in the Brachypodium genome. They are mainly clustered in chr. 2 and chr. 1 (Table 1). Only 3 storage protein genes are located on chr. 4 and 1 gene on chr. 3, and no storage protein genes were found on chr. 5. A total of 15 storage protein genes on chr. 2 were clustered in two regions. The first region is 43 kb long and contains one globulin gene (Bradi2g20860) and the two HMW-glutenin genes (Bradi2g20870 and Bradi2g20910). The other region spans 3.37 Mb with 12 genes, including 7 genes of 11S globulin and 5 genes of S-rich prolamin. Gene duplication might play important roles in this non-random gene distribution. It has been reported that Bradi2g38060 (*Bd.glo2*) belongs to a gene family with evidence of multiple bands in Southern blotting (Laudencia-Chingcuanco and Vensel 2008). The hybridized bands of Bradi2g38060 are likely resulted from the duplicated genes in this region. Clustering arrangement of highly related genes also indicates a local gene duplication event. Tandem duplication was found for these storage protein genes, including Bradi1g50200, Bradi1g50290 and Bradi1g50300 on chr.1, and Bradi2g39910, Bradi2g39920 and Bradi2g39940 on chr2. However, compared with wheat and maize (Anderson et al. 2013; Feng et al. 2009; Shewry et al. 2003), fewer prolamin genes can be identified in Brachypodium. It is not clear if this resulted in less amount of accumulated prolamin content. Nevertheless, understanding prolamin gene arrangement in different grass species is important for comparative analysis. In cereals, seed storage protein genes often belong to large gene families and have a unique evolutionary path (Shewry and Halford 2002). Study of the maize prolamin gene organization has provided insights into the evolution of genome and gene duplications in grass species (Xu and Messing 2008). Syntenic alignment of the wheat *Glu-1* region with Brachypodium and other grass species revealed that the wheat HMW-glutenin gene is originated from a duplication of a globulin gene (Gu et al. 2010). Therefore, Brachypodium will be useful for better understanding of the origin and evolution of storage protein genes in cereals.

Until now, few studies reported on investigating the gene expression feature of storage protein encoding genes of Brachypodium. Many of these genes are likely pseudogenes due to the presence of premature stop codons. For example, six of the 8 alpha-gliadin genes isolated from Bd21 were pseudogenes (Table 1). The transcriptional expression of gliadin genes have been studied using qRT-PCR, and the results showed they were expressed at a significantly lower lever than that of common wheat (Chen et al. 2014b). However, similar expression patterns during grain developmental stages were observed between the two species. The gliadin genes were rapidly up-regulated from 9 DAP (days after pollination) to 18 DAP and then were down-regulated through the mature stage (Chen et al. 2014b).

Gene expression can also be evaluated based on the accumulation of the protein. Based on the analysis of SDS-PAGE and 2-DE gels, a total of six globulin genes and three prolamin genes are expressed and could be translated into storage proteins in Brachypodium seeds (Table 1). Larre et al (2010) estimated that the total products of 7S globulin genes Bradi1g13040 and Bradi1g05910 might represent about 1/3 of the salt-soluble fraction of the total protein, and the products of four 11S globulin genes (Bradi2g38060, Bradi2g40840, Bradi4g28220, Bradi4g29130) represent more than 2/3 of the urea extract. In the same study, prolamins expressed by two genes (Bradi1g50300, Bradi3g17070) only accounted for 5 % of the urea extract of the total seed protein. Therefore, the expression level of globulin genes must be much higher than that of prolamin genes. Of the globulin genes, the expression of 7S globulin (Bradi1g13040) is restricted to the aleurone layers and the embryo of the mature Brachypodium grain (Opanowicz et al. 2011), as is observed in barley.

Brachypodium as a Model for Studying Seed Biology of Cereal Crops

In the last decade, Brachypodium research tool boxes have been developed, including germplasm resources, the complete genome sequence as well as re-sequencing data, a high resolution genetic map, efficient transformation methods, genome editing protocols with TALENs or the CRISPR/Cas system, large populations of T-DNA mutants, sequenced EMS mutants, and detailed investigations of plant or tissue development (Mur et al. 2011; Shan et al. 2013). These resources have laid a solid basis for research using Brachypodium as a model system. Numerous studies have been performed using Brachypodium as a model to understand cereal crop biology. Examples include the use of Brachypodium to study the interactions between Brachypodium and fungal pathogens of cereals (Ayliffe et al. 2013; Zhong et al. 2014), root development of wheat (Chochois et al. 2012), copper transport (Jung et al. 2014), and traits of importance in biofuel crops (Douche et al. 2013). This research has shown that quite similar traits and common underlying mechanisms are present in both Brachypodium and cereal crops. The advantages of using Brachypodium as a model system becomes even more apparent when the natural attributes of the plant are assessed.

In addition to the similarity in its grain development with other cereal crops, the large gain size, relative to the small stature of Brachypodium plants, is another useful feature in studying seed biology. Brachypodium has also been exploited to study the expression and function of genes from other cereal species. For example, a T-DNA construct that included the *uidA* reporter gene under control of a wheat

1Dy10 gene promoter and the *1Dx5* transcription terminator was transformed into Brachypodium plants to examine the heterologous expression of this wheat HMW-GS endosperm-specific expression cassette (Thilmony et al. 2014). For comparison, the same promoter-reporter gene construct was also introduced into wheat. Histochemical staining for β-glucuronidase activity showed that the reporter gene was highly and specifically expressed in the seed endosperm of both the Brachypodium and wheat transgenic plants (Fig. 3; Thilmony et al. 2014). This result demonstrated the overall faithfulness of expression of the wheat HMW-GS expression cassette in Brachypodium, showing that this system is a good model for

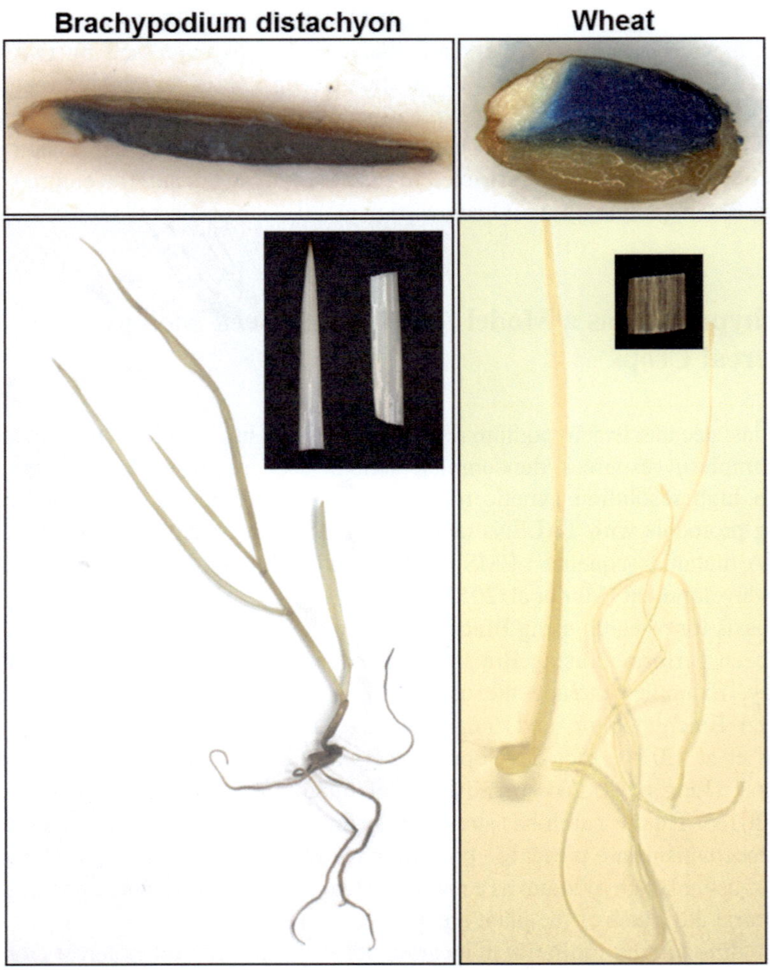

Fig. 3 Function of the wheat HMW-GS expression cassette in transgenic Brachypodium and wheat plants. Transgenic 1Dy10-GUS-1Dx5 seedlings and mature seeds were histochemically stained to detect β-glucuronidase reporter gene activity. Both species exhibited an endosperm-specific expression pattern (Thilmony et al. 2014)

studying seed specific expression components from wheat and potentially other cereal grain crops. Taken together with the ease of obtaining transgenic Brachypodium plants, this model system will not only be useful for studies of other regulatory elements from cereal crop species that are difficult to transform, but can also be employed for other research related to seed biology, such as seed dormancy, germination and maturation.

References

Anderson OD, Huo N, Gu YQ. The gene space in wheat: the complete gamma-gliadin gene family from the wheat cultivar Chinese Spring. Funct Integr Genomics. 2013;13(2):261–73.

Ayliffe M, Singh D, Park R, Moscou M, Pryor T. Infection of *Brachypodium distachyon* with selected grass rust pathogens. Mol Plant Microbe Interact. 2013;26(8):946–57.

Becraft PW, Yi G. Regulation of aleurone development in cereal grains. J Exp Bot. 2011;62 (5):1669–75.

Bragg JN, Wu J, Gordon SP, Guttman ME, Thilmony R, Lazo GR, et al. Generation and characterization of the Western Regional Research Center *Brachypodium* T-DNA insertional mutant collection. PLoS One. 2012;7(9), e41916.

Buckeridge MS. Seed cell wall storage polysaccharides: models to understand cell wall biosynthesis and degradation. Plant Physiol. 2010;154(3):1017–23.

Burton RA, Fincher GB. Current challenges in cell wall biology in the cereals and grasses. Front Plant Sci. 2012;3:130.

Burton RA, Fincher GB. Evolution and development of cell walls in cereal grains. Front Plant Sci. 2014;5:456.

Chen G, Zhu J, Zhou J, Subburaj S, Zhang M, Han C, et al. Dynamic development of starch granules and the regulation of starch biosynthesis in *Brachypodium distachyon*: comparison with common wheat and *Aegilops peregrina*. BMC Plant Biol. 2014a;14:198.

Chen GX, Lv DW, Li WD, Subburaj S, Yu ZT, Wang YJ, et al. The alpha-gliadin genes from *Brachypodium distachyon* L. provide evidence for a significant gap in the current genome assembly. Funct Integr Genomics. 2014b;14(1):149–60.

Chochois V, Vogel JP, Watt M. Application of Brachypodium to the genetic improvement of wheat roots. J Exp Bot. 2012;63(9):3467–74.

Douche T, San Clemente H, Burlat V, Roujol D, Valot B, Zivy M, et al. *Brachypodium distachyon* as a model plant toward improved biofuel crops: search for secreted proteins involved in biogenesis and disassembly of cell wall polymers. Proteomics. 2013;13(16):2438–54.

Feng L, Zhu J, Wang G, Tang Y, Chen H, Jin W, et al. Expressional profiling study revealed unique expressional patterns and dramatic expressional divergence of maize alpha-zein super gene family. Plant Mol Biol. 2009;69(6):649–59.

Fincher G, Stone B. Cell walls and their components in cereal grain technology. In: Pomeranz Y, editor. Advances in cereal science and technology, vol. 8. St. Paul: American Association of Cereal Chemists; 1986. p. 207–95.

Garvin DF, Gu Y-Q, Hasterok R, Hazen SP, Jenkins G, Mockler TC, et al. Development of genetic and genomic research resources for *Brachypodium distachyon*, a new model system for grass crop research. Crop Sci. 2008;48 Suppl 1:S-69.

Gu YQ, Wanjugi H, Coleman-Derr D, Kong X, Anderson OD. Conserved globulin gene across eight grass genomes identify fundamental units of the loci encoding seed storage proteins. Funct Integr Genomics. 2010;10(1):111–22.

Guillon F, Bouchet B, Jamme F, Robert P, Quemener B, Barron C, et al. *Brachypodium distachyon* grain: characterization of endosperm cell walls. J Exp Bot. 2011;62(3):1001–15.

Guillon F, Larre C, Petipas F, Berger A, Moussawi J, Rogniaux H, et al. A comprehensive overview of grain development in *Brachypodium distachyon* variety Bd21. J Exp Bot. 2012;63(2):739–55.

Hands P, Kourmpetli S, Sharples D, Harris RG, Drea S. Analysis of grain characters in temperate grasses reveals distinctive patterns of endosperm organization associated with grain shape. J Exp Bot. 2012;63(17):6253–66.

Jääskeläinena A-S, Holopainen-Mantilaa U, Tamminena T, Vuorinenb T. Endosperm and aleurone cell structure in barley and wheat as studied by optical and Raman microscopy. J Cereal Sci. 2013;57(3):8.

Jung HI, Gayomba SR, Yan J, Vatamaniuk OK. *Brachypodium distachyon* as a model system for studies of copper transport in cereal crops. Front Plant Sci. 2014;5:236.

Larre C, Penninck S, Bouchet B, Lollier V, Tranquet O, Denery-Papini S, et al. *Brachypodium distachyon* grain: identification and subcellular localization of storage proteins. J Exp Bot. 2010;61(6):1771–83.

Laudencia-Chingcuanco DL, Vensel WH. Globulins are the main seed storage proteins in *Brachypodium distachyon*. Theor Appl Genet. 2008;117(4):555–63.

Matos DA, Cole BJ, Whitney IP, MacKinnon KJ, Kay SA, Hazen SP. Daily changes in temperature, not the circadian clock, regulate growth rate in *Brachypodium distachyon*. PLoS One. 2014;9(6), e100072.

Mur LA, Allainguillaume J, Catalan P, Hasterok R, Jenkins G, Lesniewska K, et al. Exploiting the Brachypodium Tool Box in cereal and grass research. New Phytol. 2011;191(2):334–47.

Opanowicz M, Hands P, Betts D, Parker ML, Toole GA, Mills EN, et al. Endosperm development in *Brachypodium distachyon*. J Exp Bot. 2011;62(2):735–48.

Payne PI. Genetics of wheat storage proteins and the effect of allelic variation on bread-making quality. Annu Rev Plant Physiol. 1987;38:14.

Shan Q, Wang Y, Chen K, Liang Z, Li J, Zhang Y, et al. Rapid and efficient gene modification in rice and Brachypodium using TALENs. Mol Plant. 2013;6(4):1365–8.

Shewry PR, Halford NG. Cereal seed storage proteins: structures, properties and role in grain utilization. J Exp Bot. 2002;53(370):947–58.

Shewry PR, Halford NG, Lafiandra D. Genetics of wheat gluten proteins. Adv Genet. 2003;49:111–84.

Subburaj S, Chen G, Han C, Lv D, Li X, Zeller FJ, et al. Molecular characterisation and evolution of HMW glutenin subunit genes in *Brachypodium distachyon* L. J Appl Genet. 2014;55 (1):27–42.

Tanackovic V, Svensson JT, Jensen SL, Buleon A, Blennow A. The deposition and characterization of starch in *Brachypodium distachyon*. J Exp Bot. 2014;65(18):5179–92.

Thilmony R, Guttman ME, Lin JW, Blechl AE. The wheat HMW-glutenin 1Dy10 gene promoter controls endosperm expression in *Brachypodium distachyon*. GM Crops Food. 2014;5 (1):36–43.

Trafford K, Haleux P, Henderson M, Parker M, Shirley NJ, Tucker MR, et al. Grain development in Brachypodium and other grasses: possible interactions between cell expansion, starch deposition, and cell-wall synthesis. J Exp Bot. 2013;64(16):5033–47.

Tyler L, Fangel JU, Fagerstrom AD, Steinwand MA, Raab TK, Willats WG, et al. Selection and phenotypic characterization of a core collection of *Brachypodium distachyon* inbred lines. BMC Plant Biol. 2014;14:25.

Vega-Sanchez ME, Verhertbruggen Y, Scheller HV, Ronald PC. Abundance of mixed linkage glucan in mature tissues and secondary cell walls of grasses. Plant Signal Behav. 2013;8(2): e23143.

Wang K, Han X, Dong K, Gao L, Li H, Ma W, et al. Characterization of seed proteome in *Brachypodium distachyon*. J Cereal Sci. 2010;52(2):177–86.

Wang S, Wang K, Chen G, Lv D, Han X, Yu Z, et al. Molecular characterization of LMW-GS genes in *Brachypodium distachyon* L. reveals highly conserved Glu-3 loci in Triticum and related species. BMC Plant Biol. 2012;12:221.

Weier D, Thiel J, Kohl S, Tarkowska D, Strnad M, Schaarschmidt S, et al. Gibberellin-to-abscisic acid balances govern development and differentiation of the nucellar projection of barley grains. J Exp Bot. 2014;65(18):5291–304.

Xu JH, Messing J. Organization of the prolamin gene family provides insight into the evolution of the maize genome and gene duplications in grass species. Proc Natl Acad Sci U S A. 2008;105 (38):14330–5.

Zhong S, Ali S, Leng Y, Wang R, Garvin DF. *Brachypodium distachyon-Cochliobolus sativus* pathosystem is a new model for studying plant-fungal interactions in cereal crops. Phytopathology. 2014. doi:10.1094/PHYTO-08-14-0214-R.

The *Brachypodium distachyon* Root System: A Tractable Model to Investigate Grass Roots

Christian S. Hardtke and David Pacheco-Villalobos

Abstract The root systems of land plants mine the soil for water and essential edaphic nutrients that are needed for the vegetative and reproductive phases of shoot growth. Different root system architectures exist across the angiosperms, and while there are many variants, two principal layouts are associated with the monocotyledon-dicotyledon divide: whereas a primary taproot and its branch roots typically dominate dicotyledon root systems, monocotyledon root systems appear overall more complex and are typically dominated by post-embryonic shoot-borne roots. *Brachypodium distachyon* (Brachypodium) displays all the characteristics of a monocotyledon root system; however its complexity is minimal as compared to many other monocotyledon species, notably crops. Together with its relatively small size, this makes the Brachypodium root system a tractable model for monocotyledon root development that can be easily investigated in tissue culture but also in soil. First molecular genetic and physiological studies already point to distinct regulatory mechanisms and environmental responses in Brachypodium as compared to well-characterized dicotyledon model species. These results highlight the worthwhileness of studying the Brachypodium root system and its value as a credible model to decipher major evolutionary-developmental facets of angiosperm root system diversity. Moreover, the fact that Brachypodium is a wild plant that has not undergone human selection contrasts with the crops that serve as key monocotyledon models so far. Therefore, analysis of Brachypodium can be instructive with respect to root traits that have been modified or lost during crop domestication, especially in the closely related temperate cereals, barley, rye and wheat. Combined with natural germplasm collections, Brachypodium is thus an ideal model to investigate ecological, evolutionary and developmental aspects of monocotyledon root systems and their relation to crop performance.

Keywords Brachypodium • Root • Seminal root • Coleoptile node root • Leaf node root • Monocotyledon • Auxin • Brassinosteroid • Ethylene • Arabidopsis

---

C.S. Hardtke (✉) • D. Pacheco-Villalobos
Department of Plant Molecular Biology, University of Lausanne,
Biophore Building, 1015 Lausanne, Switzerland
e-mail: christian.hardtke@unil.ch

J. Vogel (ed.), *Genetics and Genomics of Brachypodium*,
Plant Genetics and Genomics: Crops Models, DOI 10.1007/7397_2015_6
Published online: 02 October 2015

The Intrinsic Value of Model Systems

The concentration of research on model organisms with the aim to decipher the molecular-genetic basis of life has been tremendously successful across all disciplines of biology (Brenner 2009). In the Molecular Plant Sciences, these efforts have mainly targeted *Arabidopsis thaliana* (Arabidopsis) from the Brassicaceae family of dicotyledons, a small, annual and selfing weed with a small genome and a rapid life cycle (Meyerowitz 1989). The Arabidopsis genome was the first plant genome to be sequenced and has served as a blueprint for functional plant genome annotation ever since (AGI 2000; McCourt and Benning 2010). Subsequent genome sequences from both monocotyledons and dicotyledons revealed roughly similar gene content across angiosperms with substantial fractions of new, species-specific genes, yet also a remarkable conservation of the majority of genes across all species. Thus, the genetic toolkit of angiosperms is very similar, yet their morphology and physiology can be very different. To what degree this is due to species-specific gene functions or variations on the genetic 'core' toolkit remains to be determined. Clearly, analysis of additional model species can advance our understanding of this issue, maybe with an "Arabidopsis of monocotyledons" as a foremost priority. Brachypodium can fulfil this role, because it displays many of the key features that made Arabidopsis such a successful model, i.e. it is small, annual and selfing and has a relatively compact genome and rapid life cycle (Draper et al. 2001; Girin et al. 2014). Together with the released genome sequence and established transformation protocols, Brachypodium has become a tractable experimental system (International Brachypodium Initiative 2010; Thole and Vain 2012; Vain et al. 2008). Brachypodium is also particularly interesting for the analysis of monocotyledon root system traits because unlike other extensively researched monocotyledon crops, Brachypodium is a wild plant; and because its root system is relatively simple yet comprises all the hallmarks of monocotyledon roots.

Monocotyledon Root Systems: An Overview

The emergence of root systems was a key step in plant evolution that allowed their large-scale conquest of land. From the soil, roots absorb water, macro- and micronutrients that are essential for both vegetative and reproductive shoot growth. In turn, photosynthates delivered to the root system from the shoot are required for root growth, which is driven by apical meristems located at the root tips. Especially for annual plants, continuous soil exploration and thus root growth is necessary to sustain plant growth as the rhizosphere nutrient content of a soil patch is exhausted.

Various root system architectures are observed across the angiosperms, yet while there are many variants, two principal layouts can be distinguished and are associated with the monocotyledon-dicotyledon divide (Osmont et al. 2007). Dicotyledon root systems, such as in Arabidopsis, typically display a dominant primary taproot, which is formed during embryogenesis and persists throughout the plant's life cycle. The root system enlarges through branching of this primary root, i.e. the formation of lateral roots, and can be complemented by adventitious roots that typically emerge from the hypocotyl (the embryonic stem), or less common from other shoot structures. Lateral roots can emerge from all initial root types and branch again to give rise to higher order lateral roots. Finally, the formation of root hairs on all root types contributes substantially to increase the root-soil interaction surface area (Grierson and Schiefelbein 2002).

Monocotyledons also form an axial primary root during embryogenesis, but in addition they often develop multiple secondary roots that arise from the mesocotyl (Osmont et al. 2007). Collectively, they are referred to as seminal roots, a term that is sometimes also applied to a sole primary root, as in Brachypodium. Depending on the species and on environmental conditions, seminal roots can persist over a portion of, or over the total plant life cycle, or they can degenerate and die off as post-embryonic, shoot-borne roots gradually dominate the system. The latter can emerge from the first node above the mesocotyl, the coleoptile, or from later developed stem nodes and are therefore termed coleoptile node roots or leaf node roots, respectively. Similar to dicotyledons, any of the root types described can branch through the formation of lateral roots, and all roots form root hairs. Finally, since secondary growth does not occur in monocotyledons, their roots do not form a lateral meristem, a cambium, or a periderm, unlike dicotyledon roots.

The Brachypodium Root System: Minimal, Yet Complete

The Brachypodium root system displays all the characteristics of a typical monocotyledon root system (Fig. 1), yet its complexity is minimal as compared to many other monocotyledon species, notably crops (Chochois et al. 2012; Pacheco-Villalobos and Hardtke 2012). To start with, Brachypodium develops a single axial primary root during embryogenesis and no other seminal roots. All other roots are formed post-embryonically, starting with up to three, but typically two coleoptile node roots. At later stages a variable number of leaf node roots can be formed, which might depend on nutrient conditions (Ingram et al. 2012; Poire et al. 2014). In fact, when grown in culture room conditions, leaf node roots might be rarely if at all observed in many accessions (Poire et al. 2014), especially if planting density is high or stems are fixed in an upright position. Brachypodium can complete its life cycle without leaf node roots, since both the primary root and the coleoptile node roots persist throughout it. Interestingly, leaf node roots are more frequent on tillers that acquire a nearly horizontal position, eventually anchoring the node distant from the main root system and contributing to the spread of Brachypodium plants in a stolon-like fashion (Alves et al. 2009). *Brachypodium distachyon* therefore displays a somewhat intermediate phenotype as compared to other Brachypodium species that form proper stolons, like *Brachypodium kotschyi* or *Brachypodium madagascariense* (Fig. 2).

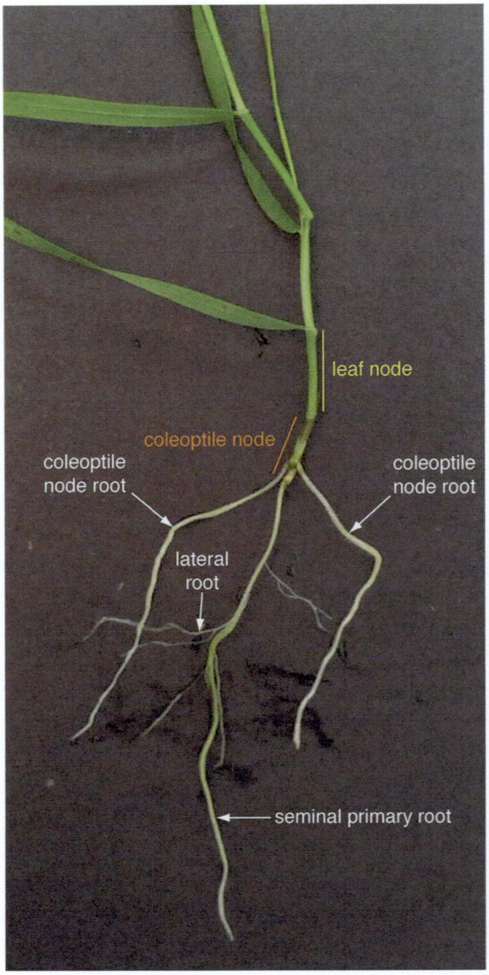

Fig. 1 Essential features of the Brachypodium root system, illustrated with a 25-day-old plant that was grown on standard media in a glass cylinder

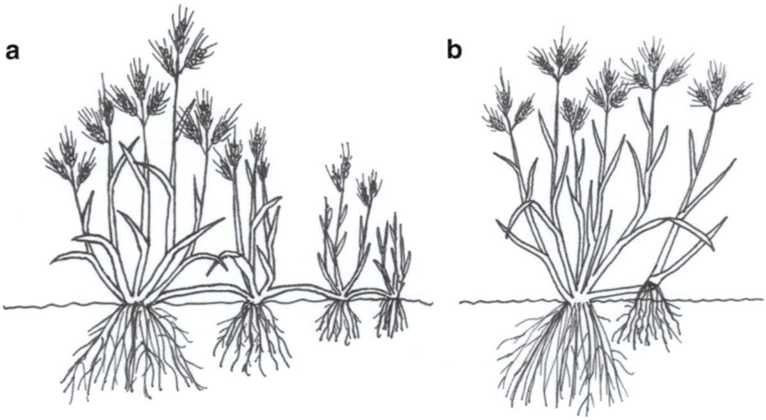

Fig. 2 Whereas some Brachypodium species (e.g. *Brachypodium kotschyi, Brachypodium madagascariense*) develop proper stolons (**a**), *Brachypodium distachyon* does not, but can develop leaf node roots that help the plant to spread in a stolon-like fashion (**b**)

Cellular Features of Brachypodium Roots

At the cellular level, the Brachypodium primary root displays a radial tissue organization that is laid down in the meristem and does not undergo any subsequent rearrangements (Pacheco-Villalobos and Hardtke 2012; Pacheco-Villalobos et al. 2013). Thus, unlike in the Arabidopsis primary root, there is no gradual switch in the vascular pattern along the root from bilateral to radial. The general organization of the primary root comprises an epidermis layer, followed by ground tissue that consists of typically five cortex layers and a single endodermis layer (Fig. 3a, b) (Pacheco-Villalobos et al. 2013; Watt et al. 2009). In the stele, which is delimited by a single surrounding pericycle layer, the vascular tissues are arranged in a radial pattern of alternating xylem and phloem poles (Fig. 3c). Each xylem pole initially consists of several protoxylem cell files that are in contact with the pericycle to the outside and a metaxylem cell file to the inside. Likewise, in phloem poles a single protophloem cell file is in contact with the pericycle. To the inside, the protophloem is neighboured by two flanking companion cell files, which themselves are next to a single metaphloem cell file that is located further inside. Finally, the ring of xylem and phloem poles surrounds the central metaxylem, a large xylem cell file that is right at the centre of the root. This pattern is representative for all root types, although some variation can be observed, even within one root type, for instance with respect to the number of xylem and phloem poles (typically eight, but sometimes one to two more or less) and the central metaxylem (typically one cell file, but sometimes two or three) (Fig. 3a, b) (Pacheco-Villalobos and Hardtke 2012; Watt et al. 2009).

In the longitudinal dimension, the Brachypodium primary root displays an equally clean layout, which is particularly useful for investigation of the root meristem (Pacheco-Villalobos and Hardtke 2012). Unlike for instance in maize, where the quiescent centre of the root meristem can comprise up to several hundred cells and in which cell divisions are frequently not oriented and thus oblique (Kerk and Feldman 1995), the Brachypodium root tip has a quiescent centre of at most two dozen cells and an overall clear structure of cell divisions that allow the identification of, e.g., stem cells and their derivatives (Fig. 3d). Distal of the meristem centre the root forms a simple columella with typically a dozen cell layers and a lateral root cap of a single cell layer (Pacheco-Villalobos and Hardtke 2012).

Just like most other flowering plants, Brachypodium develops root hairs from a subset of epidermal cells in a regular pattern, with root hair-bearing cells reaching a shorter mature cell length than non-hair cells (Marzec et al. 2014). In Arabidopsis, files of short hair cells alternate with files of longer non-hair cells around the root circumference (Grierson and Schiefelbein 2002). By contrast, in Brachypodium a longitudinal rather than circumferential alternation of hair cells and non-hair cells is found (Fig. 3e). The emergence of Brachypodium root hairs is preceded by an asymmetric cell division, with the smaller daughter cell adopting hair fate and the larger daughter cell adopting non-hair fate (Kim and Dolan 2011). This is different

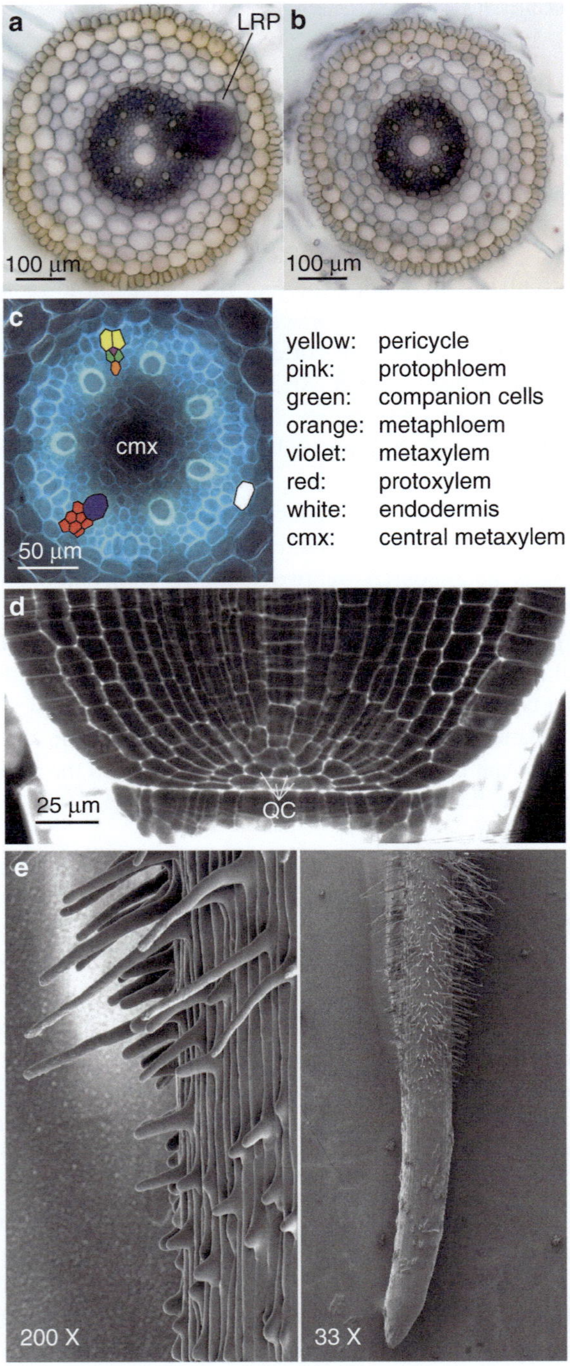

Fig. 3 Histological cross sections through Brachypodium primary roots (**a**, **b**) illustrate the arrangement of concentric tissue layers and the vasculature, and their variation. *LRP* lateral root primordium. A close up of the stele shows the different vascular tissues (**c**). Confocal imaging of the primary root meristem's center plane (**d**) reveals an ordered arrangement of cells in which the quiescent center and the stem cells can be easily identified. Root hairs are forming in the maturation zone of the root meristem, in a regular longitudinal pattern of alternating hair and non-hair cells (**e**). Drs. Chulmin Kim and Liam Dolan generously provided the scanning electron microscopy pictures of root hair formation in (**e**)

from, for instance, rice, where root hair formation is preceded by a symmetric cell division and subsequently reduced elongation of the hair as compared to the non-hair daughter cell. In general, the overall root hair formation process and pattern in Brachypodium appears to be representative for the Pooideae, which include barley, rye and wheat (Kim and Dolan 2011).

The Brachypodium Root Is Amenable to Full-Scale Experimental Investigation

An important advantage of Brachypodium as compared to crops is that its roots are amenable to full-scale experimental analysis in a standard plant science laboratory. Numerous physiological assays and other manipulations can be performed in tissue culture, and on large 24 × 24 cm petri dishes plants up to 3 weeks old can be investigated (Pacheco-Villalobos and Hardtke 2012). In soil, the root system will eventually reach the bottom of standard pots, but growing the plants, e.g., in cylinders of 50 cm height in greenhouse conditions allows monitoring of the entire root system (Chochois et al. 2012). At the microscopic level, Brachypodium root tips can be imaged across all cell layers with state-of-the-art confocal microscopes (Pacheco-Villalobos and Hardtke 2012; Pacheco-Villalobos et al. 2013), although they are clearly much thicker (typically ~500 μm in diameter) than Arabidopsis root tips (typically ~200 μm in diameter), the gold standard model system for the analysis of root development. Together with adapted protocols for the detection of chromogenic markers, such as beta-glucuronidase staining, and routine detection of fluorescent protein reporters that do not require any specific settings, such as green fluorescent protein and derivatives, Brachypodium can thus serve as a full-fledged model for monocotyledon root system analysis across all scales.

Conserved Gene Functions in Brachypodium Root Development and Their Variants

The precedence of functional gene analysis in Arabidopsis can guide investigations of Brachypodium, and in turn Arabidopsis mutants can serve as a test case for principal conservation of gene function. For instance, with respect to root development, it was shown that most homologous proteins from the small Brachypodium *BREVIS RADIX-LIKE (BdBRXL)* gene family could complement the impaired protophloem development and the associated root growth defects observed in Arabidopsis *brevis radix* loss-of-function mutants (Beuchat et al. 2010; Mouchel et al. 2004). Likewise, the Brachypodium homolog of *SHORT ROOT (BdSHR)* can fully rescue the Arabidopsis *shr* mutant phenotype (Wu et al. 2014). In Arabidopsis, *SHR* is required for the periclinal division of the endodermis-cortex initial cell as well as for the specification of endodermis fate (Helariutta et al. 2000). Consequently, instead of the single endodermis cell layer and the single cortex cell layer found in Arabidopsis roots, *shr* loss-of-function mutants only form a single cell layer that displays cortex identity. Interestingly, the Arabidopsis *SHR* gene is expressed in the stele and its propensity to induce periclinal division of the endodermis-cortex initial as well as endodermis fate depends on cell-to-cell movement of SHR protein via plasmodesmata (Nakajima et al. 2001). Feed-forward regulated capture and nuclear trapping of the SHR protein in the endodermis is thought to restrict its mobility, and consistently it has been shown that ectopic SHR expression, for instance under control of an endodermis-specific promoter, induces additional cell layers (Nakajima et al. 2001). The same is true when *BdSHR* is expressed under the native Arabidopsis *SHR* promoter, because although BdSHR protein displays all the features of its Arabidopsis counterpart in functional assays, its mobility is higher (Wu et al. 2014). Thus, the further reach of BdSHR induces additional periclinal cell divisions in Arabidopsis that give rise to additional cortex cell layers. This particularity of BdSHR observed in Arabidopsis-based assays connects directly to Brachypodium root development, since it suggests that it is responsible for the formation of multiple cortex layers.

Variations on Hormonal Responses in Brachypodium Roots

Similar to developmental genes like *BdBRXL* or *BdSHR*, genes that constitute central components of hormone signalling or biosynthesis pathways in Arabidopsis are also conserved in Brachypodium. While they likely have similar molecular function, first studies already indicate important variations in their regulatory and physiological wiring that could be responsible for key characteristics of the Brachypodium root system. For example, Arabidopsis loss-of-function mutants in the brassinosteroid receptor BRASSINOSTEROID INSENSITIVE 1 (BRI1) display a pleiotropic dwarf phenotype, including reduced root growth. The Arabidopsis *bri1* mutant phenotype is largely recapitulated in a mutant of a Brachypodium homolog. However, the root phenotype appears comparatively mild (Thole et al. 2012; Goddard et al. 2014), which is possibly due to redundancy because just as in Arabidopsis, treatment of Brachypodium wild type with the brassinosteroid biosynthesis inhibitor brassinazole strongly inhibits root growth. Brachypodium primary roots also appear to be much more sensitive to brassinosteroid dosage effects than Arabidopsis. For instance, in Arabidopsis, the conversion of brassinosteroid precursors into active brassinosteroids requires cytochrome P450 (CYP) proteins such as CYP85A1. Whereas Arabidopsis *cyp85A1* loss-of-function mutants display no notable phenotype due to redundancy (Nomura et al. 2005), corresponding mutants in rice or maize, named *brassinosteroid-dependent 1* or *brassinosteroid-dependent dwarf 1 (brd1)*, are severe dwarfs (Makarevitch et al. 2012; Mori et al. 2002). The latter is also the case for the Brachypodium *brd1* mutant, however, an intermediate phenotype, including

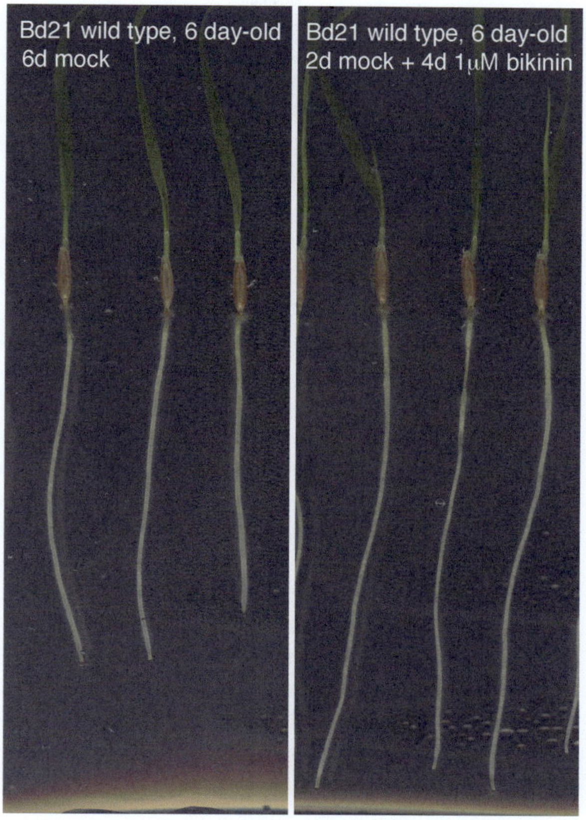

Fig. 4 Response of Brachypodium primary roots to bikinin treatment, applied to 2-day-old seedlings for 4 days as compared to mock control

already strongly reduced root growth, can already be observed in hemizygous plants (Xu et al. 2015).

A different response of Brachypodium to brassinosteroids as compared to Arabidopsis is also suggested by pharmacological treatments. Typically, Arabidopsis roots react to a reduction as well as an increase of a particular hormonal stimulus by external applications in the same way, that is with a reduction in growth (Depuydt and Hardtke 2011). This also applies to brassinosteroids, for instance both inhibition of their biosynthesis through brassinazole or stimulation of their signalling through bikinin induces a short root phenotype in Arabidopsis (Cho et al. 2014; De Rybel et al. 2009). However, while brassinazole also suppresses root growth in Brachypodium, bikinin treatment dramatically stimulates root elongation (Fig. 4), likely by stimulating cell elongation. Thus, mutant phenotypes could be substantially different from what could be expected from their Arabidopsis counterparts.

Examples in this respect are hormonal Brachypodium mutants in the auxin and ethylene pathways. Loss-of-function mutants in both the auxin biosynthesis gene *TRYPTOPHAN AMINOTRANSFERASE OF ARABIDOPSIS 1-RELATED 2-LIKE (BdTAR2L)* and the ethylene signaling gene *ETHYLENE INSENSITIVE*

2-LIKE 1 (BdEIN2L1) display strongly enhanced primary root elongation (Pacheco-Villalobos et al. 2013). In both cases, this phenotype is caused by strongly enhanced cell elongation of the root cells, which is also associated with an increase in cellular anisotropy. Given that the respective Arabidopsis mutants display either no phenotype or a short root phenotype (depending on redundancy levels) (Stepanova et al. 2008; Tao et al. 2008), the strongly elongated roots of *Bdtar2l* and *Bdein2l1* mutants are highly counterintuitive. TRYPTOPHAN AMINO-TRANSFERASE OF ARABIDOPSIS 1 (TAA1) and homologous proteins are involved in a developmentally regulated two-step auxin biosynthesis pathway, in which they catalyze the conversion of tryptophan into indole-3-pyruvic acid (IPA). IPA is then converted into indole-3-acetic acid (IAA), the principal active auxin, by another class of enzymes that are encoded by the *YUCCA* genes (Stepanova et al. 2011; Won et al. 2011). However, *Bdtar2l* mutants contain more rather than less auxin in the root elongation zone, as do *Bdein2l* mutants (Pacheco-Villalobos et al. 2013). These paradoxical observations could be resolved by the finding that in Brachypodium, ethylene signaling represses the second, rate-limiting step of the pathway, i.e. the expression of *YUCCA* genes (Pacheco-Villalobos et al. 2013; Stepanova et al. 2011). Because auxin and ethylene biosynthesis are metabolically connected through IPA (Zheng et al. 2013), this means that lower IPA levels presumably also result in lower ethylene levels. Thus, as IPA levels are reduced in *Bdtar2l* mutants, the ethylene inhibition of YUCCA expression is relieved, resulting in an overall higher flux of tryptophan to IPA to IAA (Pacheco-Villalobos et al. 2013). This is unlike the Arabidopsis primary root, where the ethylene pathway positively regulates the expression of TAA1 and its homologues as well as of YUCCA genes (Stepanova et al. 2008; Liang et al. 2012). Thus, altered gene redundancy relations and alternative wiring of one regulatory step leads to very different outcomes in the root morphology and physiology of these Brachypodium mutants as compared to their Arabidopsis counterparts.

Variations of Brachypodium Root Response to Nutrient Supply

First insight has also been gained into the Brachypodium root system response to macronutrients, notably nitrogen and phosphate. In Arabidopsis, the availability of either nutrient has a strong impact on the formation, emergence and growth of lateral roots and thereby total root system size (Svistoonoff et al. 2007; Lopez-Bucio et al. 2003). For instance, limited phosphate supply results in reduced primary root growth due to reduced meristematic activity and enhanced root branching. Similarly, the Brachypodium root system reacts to variable nitrogen and phosphate supply with altered growth rates and branching patterns (Ingram et al. 2012; Poire et al. 2014). However, the absolute contribution of the primary root and its branches remains remarkably robust across a range of concentrations

(Poire et al. 2014). Rather, it is the number and growth rate of leaf node roots that changes most prominently across the gradient, and who dominate the root system in both absolute and relative terms in optimal conditions when neither phosphate nor nitrogen are limiting. By contrast, coleoptile node roots are the least responsive and contribute little to overall root system size. The differential behavior of different root types points to a complex regulation of the root system's nutrient response at the systemic level, and it will be interesting to decipher these distinct responses at the level of individual root types in further analyses. Moreover, it will be interesting to determine to what degree corresponding responses of root hair formation and growth contribute to change the root-soil interaction surface in Brachypodium.

Natural Variation in Root System Traits

Given that Brachypodium is a wild plant that has not undergone human selection, one could expect substantial plasticity in root traits due to natural variation across Brachypodium accessions. Indeed, individual accessions can display quite variable root system architectures, and first analyses suggest that this can be functionally relevant (Pacheco-Villalobos and Hardtke 2012; Ingram et al. 2012). For instance, one of the studies of Brachypodium response to varying nutrient levels found that the root architecture differences between two commonly used accessions, Bd21 and Bd3-1, translate into distinct qualitative as well as quantitative responses to limiting phosphate or nitrogen conditions (Ingram et al. 2012). Many of these features, especially in response to low phosphate levels, displayed high heritability, suggesting that root system adaptation to different soil types exists among Brachypodium accessions and can be traced genetically, possibly down to the causative gene variants.

It appears possible that among those genes one will not only find allelic variation, but even accession-specific genes. Plant genome sequences reveal a remarkable conservation of the majority of genes across both monocotyledon and dicotyledon species, pointing to widespread gene orthology. Yet, large fractions of new genes, reaching from the hundreds to the thousands, can be found. To what degree the latter are species-specific remains to be determined, since substantial differences in gene content can already be observed between accessions of the same species. Genome re-sequencing of many accessions will enable us to define the Brachypodium pan-genome, and initial efforts already give a glimpse of the magnitude that can be expected (Gordon et al. 2014). The recently reported genome sequences of only six accessions identified hundreds of genes that are not present in the reference genome accession Bd21, and dozens of genes that are specific to individual accessions. Moreover, genotype-specific gene expression profiles were observed in response to drought stress (Gordon et al. 2014), suggesting the existence of important trans- but also cis-regulatory variation across accessions.

Outlook: This Is Just the Beginning

The examples cited above illustrate root system-related research in Brachypodium, which, however, is still in its infancy as this chapter is being written. As projects are progressing, tools are becoming increasingly standardized and more genetic resources, such as mapped T-DNA mutant collections or TILLING lines (Bragg et al. 2012; Catalan et al. 2014; Dalmais et al. 2013; Thole et al. 2010), become available, an exponential increase in publications revolving around Brachypodium root development and physiology can be expected. These studies will likely also take advantage of the Brachypodium root system as a model for biotic interactions, including genuine plant-pathogen interactions (Goddard et al. 2014; Parker et al. 2008; Routledge et al. 2004) as well as association with mycorrhizae (Hong et al. 2012). Finally, with the advent of genome editing tools such as TALEN or CRISPR/Cas9 technology (Baltes and Voytas 2014; de Lange et al. 2014), a wide range of functional studies becomes possible to further advance the use of Brachypodium as a model for the ecology, evolution and development of monocotyledon root systems.

References

AGI. Analysis of the genome sequence of the flowering plant Arabidopsis thaliana. Nature. 2000;408(6814):796–815.

Alves SC, Worland B, Thole V, Snape JW, Bevan MW, Vain P. A protocol for Agrobacterium-mediated transformation of Brachypodium distachyon community standard line Bd21. Nat Protoc. 2009;4(5):638–49.

Baltes NJ, Voytas DF. Enabling plant synthetic biology through genome engineering. Trends Biotechnol. 2014;33(2):120–31.

Beuchat J, Li S, Ragni L, Shindo C, Kohn MH, Hardtke CS. A hyperactive quantitative trait locus allele of Arabidopsis BRX contributes to natural variation in root growth vigor. Proc Natl Acad Sci U S A. 2010;107(18):8475–80.

Bragg JN, Wu J, Gordon SP, Guttman ME, Thilmony R, Lazo GR, et al. Generation and characterization of the Western Regional Research Center Brachypodium T-DNA insertional mutant collection. PLoS One. 2012;7(9):e41916.

Brenner S. In the beginning was the worm. Genetics. 2009;182(2):413–5.

Catalan P, Chalhoub B, Chochois V, Garvin DF, Hasterok R, Manzaneda AJ, et al. Update on the genomics and basic biology of Brachypodium: International Brachypodium Initiative (IBI). Trends Plant Sci. 2014;19(7):414–8.

Cho H, Ryu H, Rho S, Hill K, Smith S, Audenaert D, et al. A secreted peptide acts on BIN2-mediated phosphorylation of ARFs to potentiate auxin response during lateral root development. Nat Cell Biol. 2014;16(1):66–76.

Chochois V, Vogel JP, Watt M. Application of Brachypodium to the genetic improvement of wheat roots. J Exp Bot. 2012;63(9):3467–74.

Dalmais M, Antelme S, Ho-Yue-Kuang S, Wang Y, Darracq O, d'Yvoire MB, et al. A TILLING platform for functional genomics in. PLoS One. 2013;8(6):e65503.

de Lange O, Binder A, Lahaye T. From dead leaf, to new life: TAL effectors as tools for synthetic biology. Plant J. 2014;78(5):753–71.

De Rybel B, Audenaert D, Vert G, Rozhon W, Mayerhofer J, Peelman F, et al. Chemical inhibition of a subset of Arabidopsis thaliana GSK3-like kinases activates brassinosteroid signaling. Chem Biol. 2009;16(6):594–604.

Depuydt S, Hardtke CS. Hormone signalling crosstalk in plant growth regulation. Curr Biol. 2011;21(9):R365–73.

Draper J, Mur LA, Jenkins G, Ghosh-Biswas GC, Bablak P, Hasterok R, et al. Brachypodium distachyon. A new model system for functional genomics in grasses. Plant Physiol. 2001;127 (4):1539–55.

Girin T, David LC, Chardin C, Sibout R, Krapp A, Ferrario-Mery S, et al. Brachypodium: a promising hub between model species and cereals. J Exp Bot. 2014;65(19):5683–96.

Goddard R, Peraldi A, Ridout C, Nicholson P. Enhanced disease resistance caused by BRI1 mutation is conserved between Brachypodium distachyon and barley (Hordeum vulgare). Mol Plant-Microbe Interact. 2014;27(10):1095–106.

Gordon SP, Priest H, Des Marais DL, Schackwitz W, Figueroa M, Martin J, et al. Genome diversity in Brachypodium distachyon: deep sequencing of highly diverse inbred lines. Plant J. 2014;79(3):361–74.

Grierson C, Schiefelbein J. Root hairs. Arabidopsis Book. 2002;1:e0060.

Helariutta Y, Fukaki H, Wysocka-Diller J, Nakajima K, Jung J, Sena G, et al. The SHORT-ROOT gene controls radial patterning of the Arabidopsis root through radial signaling. Cell. 2000;101 (5):555–67.

Hong JJ, Park YS, Bravo A, Bhattarai KK, Daniels DA, Harrison MJ. Diversity of morphology and function in arbuscular mycorrhizal symbioses in Brachypodium distachyon. Planta. 2012;236 (3):851–65.

Ingram PA, Zhu J, Shariff A, Davis IW, Benfey PN, Elich T. High-throughput imaging and analysis of root system architecture in Brachypodium distachyon under differential nutrient availability. Philos Trans R Soc Lond B Biol Sci. 2012;367(1595):1559–69.

International Brachypodium Initiative. Genome sequencing and analysis of the model grass Brachypodium distachyon. Nature. 2010;463(7282):763–8.

Kerk NM, Feldman LJ. A biochemical-model for the initiation and maintenance of the quiescent center - implications for organization of root-meristems. Development. 1995;121(9):2825–33.

Kim CM, Dolan L. Root hair development involves asymmetric cell division in Brachypodium distachyon and symmetric division in Oryza sativa. New Phytol. 2011;192(3):601–10.

Liang X, Wang H, Mao L, Hu Y, Dong T, Zhang Y, et al. Involvement of COP1 in ethylene- and light-regulated hypocotyl elongation. Planta. 2012;236(6):1791–802.

Lopez-Bucio J, Cruz-Ramirez A, Herrera-Estrella L. The role of nutrient availability in regulating root architecture. Curr Opin Plant Biol. 2003;6(3):280–7.

Makarevitch I, Thompson A, Muehlbauer GJ, Springer NM. Brd1 gene in maize encodes a brassinosteroid C-6 oxidase. PLoS One. 2012;7(1):e30798.

Marzec M, Melzer M, Szarejko I. The evolutionary context of root epidermis cell patterning in grasses (Poaceae). Plant Signal Behav. 2014;9(1), e27972.

McCourt P, Benning C. Arabidopsis: a rich harvest 10 years after completion of the genome sequence. Plant J. 2010;61(6):905–8.

Meyerowitz EM. Arabidopsis, a useful weed. Cell. 1989;56(2):263–9.

Mori M, Nomura T, Ooka H, Ishizaka M, Yokota T, Sugimoto K, et al. Isolation and characterization of a rice dwarf mutant with a defect in brassinosteroid biosynthesis. Plant Physiol. 2002;130(3):1152–61.

Mouchel CF, Briggs GC, Hardtke CS. Natural genetic variation in Arabidopsis identifies BREVIS RADIX, a novel regulator of cell proliferation and elongation in the root. Genes Dev. 2004;18 (6):700–14.

Nakajima K, Sena G, Nawy T, Benfey PN. Intercellular movement of the putative transcription factor SHR in root patterning. Nature. 2001;413(6853):307–11.

Nomura T, Kushiro T, Yokota T, Kamiya Y, Bishop GJ, Yamaguchi S. The last reaction producing brassinolide is catalyzed by cytochrome P-450s, CYP85A3 in tomato and CYP85A2 in Arabidopsis. J Biol Chem. 2005;280(18):17873–9.

Osmont KS, Sibout R, Hardtke CS. Hidden branches: developments in root system architecture. Annu Rev Plant Biol. 2007;58:93–113.

Pacheco-Villalobos D, Hardtke CS. Natural genetic variation of root system architecture from Arabidopsis to Brachypodium: towards adaptive value. Philos Trans R Soc Lond B Biol Sci. 2012;367(1595):1552–8.

Pacheco-Villalobos D, Sankar M, Ljung K, Hardtke CS. Disturbed local auxin homeostasis enhances cellular anisotropy and reveals alternative wiring of auxin-ethylene crosstalk in Brachypodium distachyon seminal roots. PLoS Genet. 2013;9(6):e1003564.

Parker D, Beckmann M, Enot DP, Overy DP, Rios ZC, Gilbert M, et al. Rice blast infection of Brachypodium distachyon as a model system to study dynamic host/pathogen interactions. Nat Protoc. 2008;3(3):435–45.

Poire R, Chochois V, Sirault XR, Vogel JP, Watt M, Furbank RT. Digital imaging approaches for phenotyping whole plant nitrogen and phosphorus response in Brachypodium distachyon. J Integr Plant Biol. 2014;56(8):781–96.

Routledge AP, Shelley G, Smith JV, Talbot NJ, Draper J, Mur LA. Magnaporthe grisea interactions with the model grass Brachypodium distachyon closely resemble those with rice (Oryza sativa). Mol Plant Pathol. 2004;5(4):253–65.

Stepanova AN, Robertson-Hoyt J, Yun J, Benavente LM, Xie DY, Dolezal K, et al. TAA1-mediated auxin biosynthesis is essential for hormone crosstalk and plant development. Cell. 2008;133(1):177–91.

Stepanova AN, Yun J, Robles LM, Novak O, He W, Guo H, et al. The Arabidopsis YUCCA1 flavin monooxygenase functions in the indole-3-pyruvic acid branch of auxin biosynthesis. Plant Cell. 2011;23(11):3961–73.

Svistoonoff S, Creff A, Reymond M, Sigoillot-Claude C, Ricaud L, Blanchet A, et al. Root tip contact with low-phosphate media reprograms plant root architecture. Nat Genet. 2007;39 (6):792–6.

Tao Y, Ferrer JL, Ljung K, Pojer F, Hong F, Long JA, et al. Rapid synthesis of auxin via a new tryptophan-dependent pathway is required for shade avoidance in plants. Cell. 2008;133 (1):164–76.

Thole V, Vain P. Agrobacterium-mediated transformation of Brachypodium distachyon. Methods Mol Biol. 2012;847:137–49.

Thole V, Worland B, Wright J, Bevan MW, Vain P. Distribution and characterization of more than 1000 T-DNA tags in the genome of Brachypodium distachyon community standard line Bd21. Plant Biotechnol J. 2010;8(6):734–47.

Thole V, Peraldi A, Worland B, Nicholson P, Doonan JH, Vain P. T-DNA mutagenesis in Brachypodium distachyon. J Exp Bot. 2012;63(2):567–76.

Vain P, Worland B, Thole V, McKenzie N, Alves SC, Opanowicz M, et al. Agrobacterium-mediated transformation of the temperate grass Brachypodium distachyon (genotype Bd21) for T-DNA insertional mutagenesis. Plant Biotechnol J. 2008;6(3):236–45.

Watt M, Schneebeli K, Dong P, Wilson IW. The shoot and root growth of Brachypodium and its potential as a model for wheat and other cereal crops. Funct Plant Biol. 2009;36(10-11):960–9.

Won C, Shen X, Mashiguchi K, Zheng Z, Dai X, Cheng Y, et al. Conversion of tryptophan to indole-3-acetic acid by TRYPTOPHAN AMINOTRANSFERASES OF ARABIDOPSIS and YUCCAs in Arabidopsis. Proc Natl Acad Sci U S A. 2011;108(45):18518–23.

Wu S, Lee CM, Hayashi T, Price S, Divol F, Henry S, et al. A plausible mechanism, based upon Short-Root movement, for regulating the number of cortex cell layers in roots. Proc Natl Acad Sci U S A. 2014;111(45):16184–9.

Xu Y, Zhang X, Li Q, Cheng Z, Lou H, Ge L, et al. BdBRD1, a brassinosteroid C-6 oxidase homolog in Brachypodium distachyon L., is required for multiple organ development. Plant Physiol Biochem. 2015;86:91–9.

Zheng Z, Guo Y, Novak O, Dai X, Zhao Y, Ljung K, et al. Coordination of auxin and ethylene biosynthesis by the aminotransferase VAS1. Nat Chem Biol. 2013;9(4):244–6.

Dissecting the Control of Flowering Time in Grasses Using *Brachypodium distachyon*

Daniel P. Woods and Richard M. Amasino

Abstract The timing of flowering is a critical life history trait that has been shaped over evolutionary time to maximize the ability to flower at a time that optimizes reproductive success. Furthermore, timing of flowering is one of many traits that has been manipulated by humans for increased crop productivity. It can be difficult to determine the molecular underpinnings controlling flowering in cereals due to their large complex genomes and larger stature. However, many attributes of Brachypodium distachyon makes it a useful model grass system to accelerate understanding of the genetic basis of flowering time in grasses. Here we will first discuss what is currently known about flowering in temperate grasses, which largely comes from studies of natural variation for flowering in wheat and barley, followed by some of the progress made in B. distachyon. We will then discuss practical considerations of flowering behavior when growing different accessions of B. distachyon for studies of other traits of interest.

Keywords Flowering • Vernalization • Photoperiod • Grasses • Pooideae • Brachypodium

Photoperiodic Flowering in Temperate Grasses

In many plant species, flowering takes place during a particular time of year in response to perception of seasonal cues, such as changes in day length (photoperiod). The general concept of photoperiodic induction of flowering and

---

D.P. Woods
Laboratory of Genetics, University of Wisconsin, 425-G Henry Mall, Madison, WI, USA

United States Department of Energy Great Lakes Bioenergy Research Center, University of Wisconsin-Madison, Madison, WI, USA

R.M. Amasino (✉)
Department of Biochemistry, University of Wisconsin, Madison, WI, USA

United States Department of Energy Great Lakes Bioenergy Research Center, University of Wisconsin-Madison, Madison, WI, USA

433 Babcock Drive, 215 Biochemistry Addition, Madison, WI 53706, USA
e-mail: amasino@biochem.wisc.edu

J. Vogel (ed.), *Genetics and Genomics of Brachypodium*,
Plant Genetics and Genomics: Crops Models, DOI 10.1007/7397_2015_10
Published online: 20 October 2015

the range of response types found among plant species was first well established by Garner and Allard (1920). For example, short day plants such as rice flower when the night length exceeds a critical amount of time, whereas long day plants such as Brachypodium and wheat flower as day length increases. There are of course other response types found in plants; however, for the purpose of this chapter we will focus on what is known about flowering in temperate grasses in which long days (LD) are inductive for flowering.

An early key finding in photoperiodism is that day length is perceived in leaves (Knott 1934), which indicates there must be a signal that travels from leaves to the shoot apical meristem for flowering to occur. This mobile floral signal was called florigen (Chailakhyan 1936). Decades later, florigen was discovered to be a small mobile protein called FT (Corbesier et al. 2007; Jaeger and Wigge 2007). Timing of flowering depends largely on the seasonal changes in the expression of *FT*. In temperate grasses, *FT* is activated by exposure of plants to LD (Song et al. 2014; Yan et al. 2006). FT is synthesized in leaves and travels from the leaves to the shoot apical meristem where it interacts with the B-zip transcription factor FD to activate floral homeotic genes and transform the vegetative meristem into a floral meristem (Abe 2005; Wigge et al. 2005; Li and Dubcovsky 2008). The role of *FT* in initiating flowering appears to be conserved across flowering plants (Ballerini and Kramer 2011). Indeed, allelic variation at the *FT* locus in both barley and wheat is responsible for lines of barley and wheat that flower rapidly (Yan et al. 2006).

Variation in the LD promotion of flowering in wheat and barley can also be due to allelic variation at *Photoperiod-H1* (*PPD1*) (Turner et al. 2005). Allelic variation at the *PPD1* locus results in two types of spring varieties that are either sensitive to photoperiod and early flowering (*PPD1*) or insensitive to photoperiod and later flowering (*ppd1*) (Turner et al. 2005). The early flowering *PPD1* allele is associated with elevated *FT* levels whereas the delayed flowering *ppd1* allele is associated with reduced *FT* expression (Turner et al 2005; Faure et al 2007; Kitagawa et al. 2012). *PPD1* is a member of the pseudoresponse regulator gene family; members of this family contain two conserved domains: a pseudo receiver domain and a CCT DNA binding domain (Turner et al. 2005). *PPD1* may regulate *FT* expression directly, but this has yet to be tested. Additionally, allelic variation at *PPD1* is correlated with altered expression of the transcription factor *CONSTANS* (*CO*), which in Arabidopsis and rice has been shown to directly regulate *FT* expression (Putterill et al. 1995; Griffiths et al. 2003; Turner et al. 2005; Kitagawa et al. 2012). Neither mutations or natural allelic variation in *CO* has been reported in temperate grasses; however, consistent with the role of *CO* and light-dependent activation of *FT* in Arabidopsis (Putterill et al. 1995), overexpression of *CO* in barley results in activation of *HvFT1* in LD and rapid flowering (Campoli et al 2012).

Vernalization in Temperate Grasses: A Block in the Photoperiod Pathway

In addition to the photoperiod pathway, a key adaptation for optimal timing of flowering in temperate climates is the evolution of a vernalization requirement. Vernalizaton is the process by which plants become competent to flower only after prolonged exposure to the cold temperatures of winter (Chouard 1960). Cold exposure alone is typically not enough to induce flowering; rather, it often must be followed by an inductive photoperiod such as the increasing day lengths during spring and summer months. The combination of a requirement for a prolonged period of cold followed by a requirement for increasing day lengths has the adaptive value of preventing flowering in the fall season prior to the onset of winter which would likely compromise reproductive success (Amasino and Michaels 2010).

To date, information about the vernalization response in grasses has largely been the result of studies of existing allelic variation in wheat and barley. Wheat and barley varieties have been broadly classified into spring and winter varieties. Spring varieties do not require vernalization in order to flower rapidly in inductive LD, whereas winter varieties have a vernalization requirement that must be met before LD can promote flowering (Greenup et al. 2009; Distelfeld et al. 2009). A current molecular model of vernalization in grasses, which is based on studies in wheat and barley, consists of the genes, *VERNALIZATION1* (*VRN1*), *VERNALIZATION2* (*VRN2*) and *VERNALIZATION3* (*VRN3*), forming a regulatory loop (Fig. 1, Dennis and Peacock 2009; Distelfeld et al. 2009; Greenup et al. 2009). *VRN3* is orthologous to *FT* and from here on will be referred to as *FT* (Yan et al. 2006). As illustrated in Fig. 1, the regulatory loop is thought to exist in leaves where all three genes are expressed (Yan et al. 2004, 2006; Sasani et al. 2009). In the fall, prior to cold exposure, high levels of VRN2 repress *FT* either directly or indirectly to prevent flowering (Fig. 1; Yan et al. 2004). VRN2 is a CCT domain containing transcription factor and is part of the type VI *CO-like* family of genes (Yan et al. 2004; Griffiths et al. 2003). Prior to cold exposure, both *VRN1* and *FT* are expressed at low levels (Greenup et al. 2009). However, during the cold of winter, *VRN1* is up-regulated proportionately to the length of cold experienced, and in turn, higher *VRN1* levels are thought to cause a decrease of *VRN2* expression (Sasani et al. 2009; Dubcovsky et al. 2006). The grass *VRN1* is related to the *APETALA1*/*FRUITFULL* (*AP1*/FUL)-like class of genes in Arabidopsis (Yan et al. 2003). *VRN1* arose from a duplication event at the base of the grass family (Preston and Kellogg 2006) In crucifers such as Arabidopsis, the *AP1*/*FUL* family of genes is expressed in reproductive tissues; however, in grasses *VRN1* is expressed in leaves as well as in reproductive tissues (Yan et al. 2003; Fig. 1). Furthermore, *VRN1* is up-regulated by cold in the leaves of grasses, but there is no evidence for cold affecting expression of *AP1* or *FUL* in crucifers (Yan et al. 2003). In the shoot apex, *VRN1* appears to act downstream of *FT* as in Arabidopsis; thus, an increase in *FT* expression results in an increase in *VRN1* expression (Fig. 1). Furthermore, FT forms a complex with TaFDL2 (a homolog of FD in Arabidopsis) and binds to the

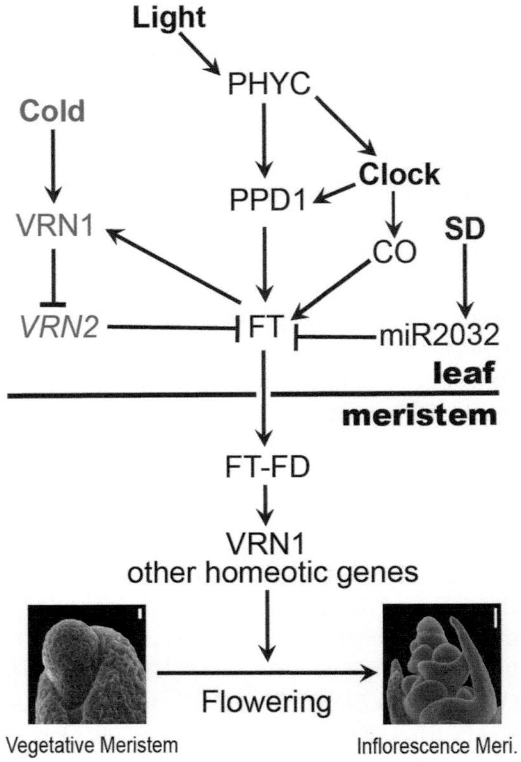

Fig. 1 Overview of the flowering time pathways in temperate grasses from studies in wheat, barley and Brachypodium. *Gray text* represents genes involved in the vernalization pathway. *In Brachypodium exposure to cold does not result in repression of VRN2 and VRN1 does not appear to regulate VRN2*

promoter of *VRN1 in vitro* (Li and Dubcovsky 2008). Thus, *VRN1* has been co-opted to play a role upstream of *FT* in leaves, activating flowering by turning off *VRN2* either directly or indirectly in wheat and barley (Shimada et al. 2009; Chen and Dubcovsky 2012; Fig. 1). VRN1 likely has different targets when expressed in leaves versus meristem, but this has yet to be confirmed. Activation of *VRN1* by cold is accompanied by changes in chromatin in a presumed regulatory region of its first intron (Oliver et al. 2009). These chromatin modifications shift from the repressive H3K27 methylation to the active H3K4 methylation state during vernalization, but no changes in chromatin modifications have been observed around the *VRN2* locus during or after cold (Oliver et al. 2009; Fig. 1). Upon repression of *VRN2, FT* up-regulates *VRN1* expression in leaves creating a positive feedback loop (Distelfeld et al. 2009; Yan et al. 2006; Shimada et al. 2009; Fig. 1). Additionally *FT* likely travels to the meristem and, with *FD,* activates floral homeotic genes as in Arabidopsis (Fig. 1).

As mentioned above, the flowering model in wheat and barley is based primarily on the study of epistatic interactions of existing allelic variation that exists within wheat and barley accessions (Shimada et al. 2009; Dubcovsky et al. 2006; Karsai et al. 2005; Hemming et al. 2008; Trevaskis et al. 2003). Particular varieties of spring wheat and barley that flower rapidly without vernalization carry either deletions of the *VRN2* locus, or point mutations in the CCT domain of *VRN2* (Yan et al. 2004; Dubcovsky et al. 2005; Karsai et al. 2005; von Zitzewitz et al. 2005). Furthermore high *VRN2* expression results in delayed flowering as measured by heading date whereas knock down of *VRN2* expression results in rapid flowering (Yan et al. 2004). Such studies have led to the conclusion that VRN2 is a flowering repressor and that an active *VRN2* creates a vernalization requirement.

Other spring varieties have dominant alleles of *VRN1* or alleles of *FT* that are constitutively activated (Yan et al. 2006; von Zitzewitz et al. 2005; Fu et al. 2005; Loukoianov et al. 2005). Dominant *VRN1* alleles render plants able to flower without vernalization presumably by down regulating *VRN2* (Loukoianov et al. 2005). *VRN1* seems to maintain suppression of *VRN2* after vernalization because *VRN2* declines during cold exposure in both *vrn1* mutants and in wild type, but in *vrn1* mutants *VRN2* increases after cold exposure has ended resulting in plants that are delayed in flowering (Chen and Dubcovsky 2012). The delayed flowering phenotype observed in *vrn1* plants appear to be due to the presence of *VRN2* because *vrn1/vrn2* double mutants flower several months earlier than *vrn1* single mutants (Chen and Dubcovsky 2012). Dominant alleles of *FT* as well as transgenes overexpressing *FT* result in rapid flowering with low levels of *VRN2* and high levels of *VRN1* (Distelfeld et al. 2009).

Vernalization responses exist in a variety of plant lineages spanning angiosperm diversification (Preston and Sandve 2013). Vernalization has independently evolved multiple times. The three vernalization pathways studied to date at a molecular level in Arabidopsis, beet, and wheat/barley all contain different repressors consistent with these vernalization systems resulting from convergent evolution (Ream et al. 2012). However, all three groups of plants contain orthologous genes related to photoperiod perception, for example, *FT, CO* and several floral homeotic genes, indicating that photoperiod-mediated flowering occurs via an ancient pathway in plants that existed before flowering plants diversified (Amasino and Michaels 2010). This is not surprising given that the ability to respond to day length changes likely served as an important environmental cue in the progenitors of flowering plants. In contrast, the advantage of perceiving and adapting to cold climates is likely to have been a relatively more recent event in plant evolution: climates were much warmer 120–150 million years ago when flowering plants were diversifying during a period when there was only one major super-continent called Pangea (Speelman et al. 2009; Soltis et al. 2008). It wasn't until one continent became many and migrated to different locations while the climate was cooling that something like a vernalization response would be adaptive.

Are the Flowering Pathways Described in Wheat and Barley Conserved in Brachypodium?

Brachypodium contains orthologs of all of the major flowering-time genes described so far in wheat and barley. To a large extent these genes appear to be functionally conserved in Brachypodium with the possible exception of the role of *BdVRN2* in the vernalization process is less clear (Higgins et al. 2010; Ream et al. 2012, 2014; Lv et al. 2014). For example, consistent with studies from wheat and barley, *FT* and *VRN1* are promoters of flowering in Brachypodium (Ream et al. 2014). Overexpression of *FT* and *VRN1* results in rapid flowering in Brachypodium (Ream et al. 2014) and knockdown of *FT* results in non-flowering plants (Lv et al. 2014). To date no *Bdvrn1* mutants have been reported; however, *BdVRN1* expression is induced in leaves proportional to the amount of time in cold temperatures and expression remains high post cold treatment, which correlates with higher *FT* levels and rapid flowering (Ream et al. 2014). Also, we observe high expression levels of *FT* in lines overexpressing *VRN1* and high expression levels of *VRN1* in lines overexpressing *FT* consistent with the positive feedback loop suggested from flowering studies in wheat and barley (Ream et al. 2014; Fig. 1). No mutants of other key floral promoting genes such as *PPD1* and *CO* have been reported in Brachypodium.

The role of *BdVRN2* in the vernalization process is less clear. Unlike *VRN2* in barley and wheat, *BdVRN2* mRNA levels are the same before and after cold exposure, which is not consistent with a role of *BdVRN2* as a floral repressor suppressed by vernalization (Ream et al. 2014). Additionally *BdVRN2* mRNA levels in leaves do not always correlate with *VRN1* and *FT* expression levels in the absence of cold as they do in wheat and barley where high mRNA levels of *FT* and *VRN1* are associated with lower *VRN2* levels (Ream et al. 2014). However, overexpression of *FT* and *VRN1* does result in lower *VRN2* mRNA levels compared with wild type, but this may be due to an artifact of expressing these genes at a level not normally occurring. It will be informative to determine the expression level of *VRN2* in a *vrn1* mutant in Brachypodium. Even if VRN2 is not involved in vernalization, it may still retain an ancestral role in repressing flowering, as it does in wheat and barley, and as the closely related gene *Ghd7* does in rice. It will be interesting to determine if loss of *VRN2* results in more rapid flowering and if over-expression of *VRN2* results in a delay of flowering as it does in wheat and barley.

Brachypodium as a Model Plant for Flowering Time Gene Discovery in Temperate Grasses

Although there has been great progress in understanding photoperiodic flowering and vernalization at a molecular level in grasses, much remains much to be learned including identifying additional components of the vernalization response and the extent to which vernalization pathways are conserved throughout grasses. Brachypodium has many attributes that make it an attractive model to pursue flowering research. It is placed phylogenetically in the grass family within the Pooideae sub-family, sister to the agronomically important "crown pooids"—wheat, oat, barley and rye (Grass Phylogeny Working Group 2001; See Chapters "Phylogeny and Evolution of the Genus *Brachypodium*" and "*Brachypodium* Paleogenomics: From Genome Evolution to Translational Research in Grass

Crops"). Second, its relatively small both in stature and in genome size (272mb, diploid) compared to its crop relatives, which often are polyploid. (International Brachypodium Initiative et al. 2010; Brkljacic et al. 2011; See Chapter "The *Brachypodium distachyon* Reference Genome"). Third, there are hundreds of inbred accessions that have been collected throughout the Middle-East and Europe and display a wide range of flowering behaviors (Brkljacic et al. 2011; Schwartz et al. 2010; Ream et al. 2014; Tyler et al. 2014; See Chapter "Genomic Diversity and Climate Adaptation in Brachypodium"), Fourth, studies have shown that Brachypodium has the highest rate of recombination known in plants which makes it useful in mapping projects (Brkljacic et al. 2011). Lastly, a growing number of T-DNAs and TILLING populations have been created for use in reverse genetic approaches (Chapters "Transformation and T-DNA Mutagenesis" and "Mutagenesis: Chemical, Radiation, TILLING"). Thus, Brachypodium is an ideal genetic system to accelerate gene discovery in flowering research.

One route to gene discovery is the use of Brachypodium to conduct forward genetic screens to identify novel genes. We have streamlined mapping Brachypodium mutants and have identified the causative lesion of 11 EMS-induced flowering time mutants using a mapping by sequencing pipeline called Cloudmap (Minevich et al. 2012; Woods et al. 2014b). Interestingly, most of the genes we have mapped are in novel flowering-time genes that have not been described before in other plants (Woods and Amasino unpubl.).

Mutants were obtained by treating Bd21-3 with 0.4 and 0.6 % EMS. We then conducted screens to identify mutants that flower rapidly without vernalization and other screens to identify mutants that are delayed in flowering despite having been exposed to a saturating vernalization treatment (Woods et al. 2014b; Woods and Amasino unpubl.). To create mapping populations, homozygous mutant plants were crossed to the polymorphic accession, Bd3-1. Bd3-1 was chosen as a mapping partner because it has flowering behavior essentially identical to Bd21-3, and, most importantly, no flowering-time variation occurs in an F2 population derived from these two accessions, indicating no modifier loci of flowering behavior are present. Additionally, Bd3-1 has abundant polymorphisms relative to Bd21-3 that can be used to generate both indel and single-nucleotide based markers (Gordon et al. 2014; Woods et al. 2014b). For mapping, segregating mutants in the F2 mapping population are pooled and total DNA is extracted from leaves using a standard CTAB protocol with the addition of RNAse and Proteinase K steps (Woods et al. 2014b). We have pooled between 20 and 200 segregating mutant plants for fine mapping. Of course the more mutant plants pooled the greater the probability of resolving an interval further; however, we have found that 20 mutants is often sufficient to generate a mapping interval <500 kb in Cloudmap, which is small enough to identify candidate loci. We typically sequence bulked DNA using an Illumina HiSeq 2500 generating roughly 120 million single-end reads that average in 100 bp length. Mapping is done using the Cloudmap mapping pipeline optimized for mapping EMS-induced mutations in Brachypodium through the web based bioinformatic platform Galaxy (Blankenberg et al. 2010; Giardine et al. 2010; Goecks et al. 2010; Woods et al. 2014b). However, one can run Cloudmap locally if desired. The Cloudmap pipeline mapping and variant calling were performed with the Hawaiian Variant mapping workflow as previously described (Minevich et al. 2012). As part of the pipeline, Burrows-Wheeler Aligner was used for mapping the 100 bp illumina reads against the newly sequenced Bd21-3 genome (Li and Durbin 2010; Gordon et al. 2014), PICARD was used for removing duplicate reads (http://picard.sourceforge.net), Genome Analysis Toolkit was used for variant calling and subtraction (Depristo et al. 2011), and SnpEff was used for variant annotation (Cingolani et al. 2012). The workflow is modular within Galaxy so other bioinformatic tools can be used if desired, such as a different aligner or variant caller.

One of the first mutants isolated and mapped using the above-described pipeline was an extremely delayed flowering mutant that is due to a lesion in the light receptor *PHYTOCHROME C* (*PHYC*) (Woods et al. 2014b). We isolated three mutant alleles of *PHYC* all of which only flower sporadically after a year of growth in LD of 20 h, which is a highly inductive photoperiod for flowering (wild type Bd21-3 flowers around 30 days after germination in 20 h days). We found that *PHYC* is required for the transcriptional activation of all of the known genes within the photoperiod pathway including *FT*, *CO* and *PPD1* (Woods et al. 2014b). Additionally expression of genes that comprise the circadian clock were also altered in the *phyC* mutant background (Woods et al. 2014b). Lack of *FT* expression in the *phyC* mutant likely accounts for the severe delay in flowering because over-expression of *FT* rescues the delayed flowering phenotype; this is consistent with *FT* being the essential downstream component required for flowering (Woods et al. 2014b). Similar results were also obtained in studies of wheat and barley (Chen et al. 2014; Pankin et al. 2014). Thus, it appears that *phyC* is essential for photoperiodic flowering broadly across Pooideae. However, the essential role of PHYC in photoperiodic flowering is not universal as loss of *phyC* function in Arabidopsis and rice has only small effects on flowering (Monte et al. 2003; Takano et al. 2005). That PHYC evolved to be the dominant photoreceptor for photoperiodic flowering in pooid grasses may be due to the fact that it can form stable homodimers in pooid grasses whereas the Arabidopsis PHYC requires PHYB for stability (Chen et al. 2014). In summary, PHYC appears to be the main light receptor required for photoperiodic flowering in temperate grasses and may have played a role in the evolution of LD flowering in temperate grasses (Fig. 1).

In addition to the progress that can be made using genetics, transcriptome studies have already begun to shed light on additional layers of flowering-time regulation such as the post-transcriptional regulation of *FT* (Wu et al. 2013). Recently an endogenous microRNA, miR2032, has been shown to target FT, and the microRNA itself is regulated by photoperiod (Wu et al. 2013). miR2032 appears to be responsible for the SD repression of *FT* as miR2032 is highly expressed in SD but not LD (Wu et al. 2013; Fig. 1). Consistent with miR2032's ability to repress *FT*, miR2032 over-expressing plants have very low *FT* levels even in LD conditions and a delay of flowering by several months (in some cases plants did not flower) (Wu et al. 2013). Interestingly, miR2032 appears to be pooid specific as miR2032 is not detectable in other grasses or monocot species. However, miR2032's expression pattern is conserved across pooid grasses with miR2032 more highly expressed in SD than in LD (Wu et al. 2013). Given miR2032 is specific to the pooid lineage indicates that miR2032 represents a fairly recent evolutionary path to regulate *FT* expression.

Practical Considerations When Growing Different Accessions of Brachypodium

With respect to flowering and vernalization, there is a surprisingly rich amount of natural variation in Brachypodium for an inbreeding species (Schwartz et al. 2010; Ream et al. 2012, 2014; Tyler et al. 2014). This is quite promising as natural variation has the potential to reveal allelic variation upon which natural selection has acted. This is of particular significance in pooids because to date natural variation studies have been limited to domesticated backgrounds, and a wild species such as Brachypodium may contain additional variation that might not be present in the domesticated crops.

We have characterized the range of flowering behavior in many of accessions (Schwartz et al. 2010; Ream et al. 2014; Woods and Amasino unpubl.). Variation for the vernalization requirement ranges from accessions that flower rapidly in 20 h long days without a vernalizing cold treatment (no vernalization requirement; Bd21, Bd3-1 and Bd21-3) to accessions that have an obligate vernalization requirement (i.e., flowering rarely occurs without an extensive prior cold treatment) such as in Bd29-1, TR8n, TR7a (Ream et al. 2014). Furthermore, there is a wide range of responses between these extremes: there are accessions that require 2, 4, 6 or 8 weeks of cold exposure to saturate the vernalization response (Ream et al. 2014). There are also accessions with a conditional vernalization response; these respond to vernalizing cold when grown in 16 h day-lengths, but have a limited vernalization response in 20 h day-lengths (Bd21-3 and Bd3-1) (Ream et al. 2014). Unlike wheat and barley which can be spring or winter lines, all tested Brachypodium accessions are likely “winter types” as growth in photoperiods of 16 h or shorter require at least 2–3 weeks of cold to accelerate flowering (Ream et al. 2014; Tyler et al. 2014). Thus, if you plan to grow accessions in 16 h or less, it is critical to vernalize plants to obtain flowering in a reasonable time (see Fig. 2 for an example of the flowering difference between a fully vernalized and sub saturated vernalized plant).

We prefer to vernalize as imbibed seed either in soil or between moist paper towels in a petri plate that is sealed to prevent moisture loss. Before the vernalization treatment, it is useful to imbibe seed in water overnight in dark at 4 °C which aids in obtaining a uniform vernalization response. The photoperiod does not matter for seed vernalization and this is a major reason why we prefer seed vernalization to vernalizing germinated seedlings, which vernalize much more effectively in SD (Ream et al. 2014). If we need to vernalize for longer than 6 weeks, it is best to vernalize directly in soil in the pots in which the plants will be grown because the

Fig. 2 Developmental effect of vernalization. Plant on *left* received a saturating vernalization treatment and flowered rapidly whereas the plant on the *right* received a sub-saturating vernalization treatment taking several months to flower

shoot and roots will elongate rendering the plants difficult to transplant. Note that after 10 weeks of vernalization, the seedlings will be between the one and two leaf stage. Although a range of temperatures will satisfy the vernalization requirement, temperatures between 2 and 6 °C are optimal for vernalization, and we tend to vernalize around 5 °C. Vernalization does not occur robustly at temperatures below 2 °C (Ream et al. 2014).

For vernalization of plants >2 leaf stage it is important to be aware that vernalization under SD (8 h) is more effective than LD. Perhaps that vernalization is more effective in SD is related to the observation that when in the cold there are higher *VRN1* levels in SD than LD (Ream et al. 2014; Colton-Gagnon et al. 2014). It is possible to vernalize mature Brachypodium plants; however, it is more difficult to keep mature plants healthy although providing sufficient light can help to maintain plant health. Light intensity is not as critical at the seedling stage, perhaps because seedlings have stored reserves. The duration of cold required to saturate the vernalization requirement does not appear to change between younger and older plants in several accessions we studied (Woods and Amasino unpubl.).

One limitation of doing genetics in Brachypodium is that you only get one seed per cross. This can make it difficult to do experiments that require examining a large number of plants. When a large number of seeds are needed from a given cross, we first grow a "mother" plant in 8 h day lengths. When the plants reach the 4–5 leaf stage (about a month), we begin to make cuttings of the newly forming tillers being sure to cut just below the lowest node but being careful not to damage the remaining portion of the plant. The newly cut tillers are then placed in moist soil and moved into a 20 h photoperiod to induce flowering. We have found that the plants root and establish at a much higher frequency when covered with a plastic dome for a week. We continue making cuttings from the "mother" plant grown in SD until enough tiller cuttings are obtained to produce sufficient seed. We have kept "mother" plants alive for well over a year in SD and have been able to produce thousands of seeds from a single cross. If an F1 requires vernalization to flower (which is likely if both or even one of the parents has a vernalization requirement), we first vernalize the F1 for the required time before growing in SD. Even if vernalized extensively, Brachypodium will not flower in SD because Brachypodium has a strict requirement for LD in order to flower (Ream et al. 2014). However, the vernalization treatment of the F1 will be "remembered" in the tiller cuttings and thus the cuttings will readily flower when shifted into LD (Woods et al. 2014a). The vernalized state is mitotically stable for at least a year and perhaps longer in Brachypodium (Woods et al. 2014a). Finally, it is also useful to keep in mind that exposing many Brachypodium accessions to SD alone can accelerate flowering even without cold, a phenomenon commonly called short day vernalization (Purvis and Gregory 1937); thus, often one can skip the vernalization treatment altogether if plants have been growing in SD for at least 6 weeks. The short day substitution for vernalization occurs in approximately 80 % of the Brachypodium accessions we have tested (Woods and Amasino unpubl.). Exceptions in which short days do not substitute for vernalization once moved into LD include lines such as Bd29-1 which require >8 weeks of cold to saturate the vernalization requirement (Woods et al. 2014a).

Acknowledgments This work was funded in part by the National Science Foundation (grant no. IOS–1258126), the Great Lakes Bioenergy Research Center (Department of Energy Biological and Environmental Research Office of Science grant no. DE–FCO2–07ER64494), and a National Institutes of Health-sponsored pre-doctoral training fellowship to the University of Wisconsin Genetics Training Program to Daniel Woods.

References

Abe M, Kobayashi Y, Yamamoto S, Daimon Y, Yamaguchi A, Ikeda Y. FD, a bZIP protein mediating signals from the floral pathway integrator FT at the shoot apex. Science. 2005;309:1052–6. doi:10.1126/science.1115983.

Amasino RM, Michaels SD. The timing of flowering. Plant Physiol. 2010;154:516–20. doi:10.1104/pp. 110.161653.

Ballerini ES, Kramer EM. In the light of evolution: a reevaluation of conservation in the CO-FT regulon and its role in photoperiodic regulation of flowering time. Front Plant Sci. 2011;2:1–13. doi:10.3389/fpls.2011.00081.

Blankenberg D, Von Kuster G, Coraor N, Ananda G, Lazarus R, et al. Galaxy: a web-based genome analysis tool for experimentalists. Curr Prot Mol Biol. 2010;19:Unit 19.10.1–21.

Brkljacic J, Grotewold E, Scholl R, Mockler T, Garvin DF, Vain P, et al. Brachypodium as a model for the grasses: today and the future. Plant Physiol. 2011;157:3–13. doi:10.1104/pp. 111. 179531.

Campoli C, Drosse B, Searle I, Coupland G, von Korff M. Functional characterisation of HvCO1, the barley (Hordeum vulgare) flowering time ortholog of CONSTANS. Plant J. 2012;69:868–80. doi:10.1111/j.1365-313X.2011.04839.x.

Chailakhyan MK. New facts in support of the hormonal theory of plant development. C R (Dokl) Acad Sci USSR. 1936;13:79–83.

Chen A, Dubcovsky J. Wheat TILLING mutants show that the vernalization gene VRN1 down-regulates the flowering repressor VRN2 in leaves but is not essential for flowering. PLoS Genet. 2012;8, e1003134. doi:10.1371/journal.pgen.1003134.s009.

Chen A, Li C, Hu W, Lau MY, Lin H, Rockwell NC, et al. PHYTOCHROME C plays a major role in the acceleration of wheat flowering under long-day photoperiod. Proc Natl Acad Sci U S A. 2014. doi:10.1073/pnas.1409795111.

Chouard P. Vernalization and its relations to dormancy. Annu Rev Plant Physiol. 1960;11:191–238.

Cingolani P, Platts A, Wang LL, Coon M, Nguyen T, et al. A program for annotating and predicting the effects of single nucleotide polymorphisms, SnpEff: SNPs in the genome of Drosophila melanogaster strain w1118; iso-2; iso-3. Fly. 2012;6:80–92.

Colton-Gagnon K, Ali-Benali MA, Mayer BF, Dionne R, Bertrand A, Do Carmo S, et al. Comparative analysis of the cold acclimation and freezing tolerance capacities of seven diploid Brachypodium distachyon accessions. Ann Bot. 2014;113:681–93. doi:10. 1093/aob/mct283.

Corbesier L, Vincent C, Jang S, Fornara F, Fan Q, Searle I, et al. FT protein movement contributes to long-distance signaling in floral induction of arabidopsis. Science. 2007;316:1030–3. doi:10. 1126/science.1141752.

Dennis ES, Peacock WJ. Vernalization in cereals. J Biol. 2009;8:57. doi:10.1186/jbiol156.

DePristo MA, Banks E, Poplin R, Garimella KV, Maguire JR, Hartl C, et al. A framework for variation discovery and genotyping using next-generation DNA sequencing data. Nat Genet. 2011;43:491–8.

Distelfeld A, Li C, Dubcovsky J. Regulation of flowering in temperate cereals. Curr Opin Plant Biol. 2009;12(2):178–84. doi:10.1016/j.pbi.2008.12.010.

Dubcovsky J, Chen C, Yan L. Molecular characterization of the allelic variation at the VRN-H2 vernalization locus in barley. Mol Breed. 2005;15:395–407.

Dubcovsky J, Loukoianov A, Fu D, Valarik M, Sanchez A, Yan L. Effect of photoperiod on the regulation of wheat vernalization genes VRN1 and VRN2. Plant Mol Biol. 2006;60:469–80. doi:10.1007/s11103-005-4814-2.

Faure S, Higgins J, Turner A, Laurie DA. The FLOWERING LOCUS T-like gene family in barley (Hordeum vulgare). Genetics. 2007;176:599–609. doi:10.1534/genetics.106.069500.

Fu D, Szucs P, Yan L, Helguera M, Skinner JS, von Zitzewitz J, et al. Large deletions within the first intron in VRN-1 are associated with spring growth habit in barley and wheat. Mol Genet Genomics. 2005;273:54–65.

Garner WW, Allard HA. Effect of the relative length of day and night and other factors of the environment on growth and reproduction in plants. J Agric Res. 1920;18:553–606.

Giardine B, Riemer C, Hardison R, Burhans R, Elnitski L, Shah P, et al. Galaxy: a platform for interactive large-scale genome analysis. Genome Res. 2010;15(10):1451–5.

Goecks J, Nekrutenko A, Taylor J, The Galaxy Team. Galaxy: a comprehensive approach for supporting accessible, reproducible, and transparent computational research in the life sciences. Genome Biol. 2010;11:R86.

Gordon SP, Priest H, Marais Des DL, Schackwitz W, Figueroa M, Martin J, et al. Genome diversity in Brachypodium distachyon: deep sequencing of highly diverse inbred lines. Plant J. 2014;79:361–74. doi:10.1111/tpj.12569.

Grass Phylogeny Working Group II. New grass phylogeny resolves deep evolutionary relationships and discovers C4 origins. New Phytol. 2011;193:304–12. doi:10.1111/j.1469-8137.2011.03972.x.

Greenup A, Peacock WJ, Dennis ES, Trevaskis B. The molecular biology of seasonal flowering-responses in Arabidopsis and the cereals. Ann Bot. 2009;103:1165–72. doi:10.1093/aob/mcp063.

Griffiths S, Dunford R, Coupland G, Laurie D. The evolution of CONSTANS-like gene families in barley, rice, and Arabidopsis. Plant Physiol. 2003;131:1855–67. doi:10.1104/pp.102.016188.

Hemming MN, Peacock WJ, Dennis ES, Trevaskis B. Low-temperature and daylength cues are integrated to regulate FLOWERING LOCUS T in barley. Plant Physiol. 2008;147:355–66. doi:10.1104/pp. 108.116418.

Higgins JA, Bailey PC, Laurie DA. Comparative genomics of flowering time pathways using Brachypodium distachyon as a model for the temperate grasses. PLoS One. 2010;5, e10065. doi:10.1371/journal.pone.0010065.t001.

The International Brachypodium Initiative. Genome sequencing and analysis of the model grass Brachypodium distachyon. Nature. 2010;463:763–8. doi:10.1038/nature08747.

Jaeger KE, Wigge PA. FT protein acts as a long-range signal in Arabidopsis. Curr Biol. 2007;17:1050–4.

Karsai I, Szucs P, Mészáros K, Filichkina T, Hayes PM, Skinner JS. The Vrn-H2 locus is a major determinant of flowering time in a facultative x winter growth habit barley (Hordeum vulgare L.) mapping population. Theor Appl Genet. 2005;110:1458–66.

Kitagawa S, Shimada S, Murai K. Effect of Ppd-1 on the expression of flowering-time genes in vegetative and reproductive growth stages of wheat. Genes Genet Syst. 2012;87:161–8.

Knott JE. Effect of a localized photoperiod on spinach. Proc Am Soc Hort Sci. 1934;31:152–4.

Li C, Dubcovsky J. Wheat FT protein regulates VRN1 transcription through interactions with FDL2. Plant J. 2008;55:543–54. doi:10.1111/j.1365-313X.2008.03526.x.

Li H, Durbin R. Fast and accurate long-read alignment with Burrows-Wheeler transform. Bioinformatics. 2010;26:589–95.

Loukoianov A, Yan L, Blechl A, Sanchez A, Dubcovsky J. Regulation of VRN1 vernalization genes in normal and transgenic polyploid wheat. Plant Physiol. 2005;138:2364–73.

Lv B, Nitcher R, Han X, Wang S, Ni F, Li K, et al. Characterization of FLOWERING LOCUS T1 (FT1) gene in Brachypodium and wheat. PLoS One. 2014;9:e94171. doi:10.1371/journal.pone.0094171.t003.

Minevich G, Park DS, Blankenberg D, Poole RJ, Hobert O. CloudMap: a cloud-based pipeline for analysis of mutant genome sequences. Genetics. 2012;192:1249–69. doi:10.1534/genetics.112.144204.

Monte E, Alonso JM, Ecker JR, Zhang Y, Li X, Young J, et al. Isolation and characterization of phyC mutants in Arabidopsis reveals complex crosstalk between phytochrome signaling pathways. Plant Cell. 2003;15:1962–1980.

Oliver SN, Finnegan EJ, Dennis ES, Peacock WJ, Trevaskis B. Vernalization-induced flowering in cereals is associated with changes in histone methylation at the VERNALIZATION1 gene. Proc Natl Acad Sci U S A. 2009;106:8386–91. doi:10.1073/pnas.0903566106.

Pankin A, Campoli C, Dong X, Kilian B, Sharma R, et al. Mapping-by-sequencing identifies HvPHYTOCHROME C as a candidate gene for the early maturity 5 locus modulating the circadian clock and photoperiodic flowering in barley. Genetics. 2014;198:383–96.

Preston JC, Kellogg EA. Reconstructing the evolutionary history of paralogous APETALA1/FRUITFULL-like genes in grasses (Poaceae). Genetics. 2006;174:421–37. doi:10.1534/genetics.106.057125.

Preston JC, Sandve SR. Adaptation to seasonality and the winter freeze. Front Plant Sci. 2013;4:167. doi:10.3389/fpls.2013.00167.

Purvis ON, Gregory FG. Studies in vernalisation of cereals. Ann Bot. 1937;1:1–26.

Putterill J, Robson F, Lee K, Simon R, Coupland G. The CONSTANS gene of Arabidopsis promotes flowering and encodes a protein showing similarities to zinc finger transcription factors. Cell. 1995;80:847–57.

Ream TS, Woods DP, Amasino RM. The molecular basis of vernalization in different plant groups. Cold Spring Harb Symp Quant Biol. 2012;77:105–15. doi:10.1101/sqb.2013.77.014449.

Ream TS, Woods DP, Schwartz CJ, Sanabria CP, Mahoy JA, Walters EM, et al. Interaction of photoperiod and vernalization determines flowering time of Brachypodium distachyon. Plant Physiol. 2014;164:694–709. doi:10.1104/pp. 113.232678.

Sasani S, Hemming MN, Oliver SN, Greenup A, Tavakkol-Afshari R, Mahfoozi S, et al. The influence of vernalization and daylength on expression of flowering-time genes in the shoot apex and leaves of barley (Hordeum vulgare). J Exp Bot. 2009;60:2169–78. doi:10.1093/jxb/erp098.

Schwartz CJ, Doyle MR, Manzaneda AJ, Rey PJ, Mitchell-Olds T, et al. Natural variation of flowering time and vernalization responsiveness in Brachypodium distachyon. Bioenerg Res. 2010;3:38–46. doi:10.1007/s12155-009-9069-3.

Shimada S, Ogawa T, Kitagawa S, Suzuki T, Ikari C, Shitsukawa N, et al. A genetic network of flowering-time genes in wheat leaves, in which an APETALA1/FRUITFULL-like gene, VRN1, is upstream of FLOWERING LOCUS T. Plant J. 2009;58:668–81. doi:10.1111/j.1365-313X.2009.03806.x.

Soltis DE, Bell CD, Kim S, Soltis PS. Origin and early evolution of angiosperms. Ann N Y Acad Sci. 2008;1133:3–25.

Song YH, Shim JS, Kinmonth-Schultz HA, Imaizumi T. Photoperiodic flowering: time measurement mechanisms in leaves. Annu Rev Plant Biol. 2014;66:141222172008008. doi:10.1146/annurev-arplant-043014-115555.

Speelman EN, Van Kempton MM, Barke J, Brinkhuis H, Reichart GJ, et al. The Eocene Arctic Azolla bloom: environmental conditions, productivity and carbon drawdown. Geobiology. 2009;7:155–70.

Takano M, Inagaki N, Xie X, Yuzurihara N, Hihara F, Ishizuka T, et al. Distinct and cooperative functions of phytochromes A, B, and C in the control of deetiolation and flowering in rice. Plant Cell. 2005;17:3311–25.

Trevaskis B, Bagnall DJ, Ellis MH, Peacock WJ, Dennis ES, et al. MADS box genes control vernalization-induced flowering in cereals. Proc Natl Acad Sci U S A. 2003;100:13099–104. doi:10.1073/pnas.1635053100.

Turner A, Beales J, Faure S, Dunford R, Laurie D, et al. The pseudo-response regulator Ppd-H1 provides adaptation to photoperiod in barley. Science. 2005;310:1031–4. doi:10.1126/science.1117619.

Tyler L, Fangel JU, Fagerstrom AD, Steinwand MA, Raab TK, Willats WG, et al. Selection and phenotypic characterization of a core collection of Brachypodium distachyon inbredlines. BMC Plant Biol. 2014;14:1–15. doi:10.1186/1471-2229-14-25.

von Zitzewitz J, Szucs P, Dubcovsky J, Yan L, Francia E, Pecchioni N, et al. Molecular and structural characterization of barley vernalization genes. Plant Mol Biol. 2005;59:449–67.

Wigge PA, Kim MC, Jaeger KE, Busch W, Schmid M, Lohmann JU, et al. Integration of spatial and temporal information during floral induction in Arabidopsis. Science. 2005;309:1056–9.

Woods DP, Ream TS, Amasino RM. Memory of the vernalized state in plants including the model grass Brachypodium distachyon. Front Plant Sci. 2014a;5:99. doi:10.3389/fpls.2014.00099.

Woods DP, Ream TS, Minevich G, Hobert O, Amasino RM. PHYTOCHROME C is an essential light receptor for photoperiodic flowering in the temperate grass, Brachypodium distachyon. Genetics. 2014b;198:397–408. doi:10.1534/genetics.114.166785.

Wu L, Liu D, Wu J, Zhang R, Qin Z, Liu D, et al. Regulation of FLOWERING LOCUS T by a microRNA in Brachypodium distachyon. Plant Cell. 2013;25:4363–77. doi:10.1105/tpc.113.118620.

Yan L, Loukoianov A, Tranquilli G, Helguera M, Fahima T, Dubcovsky J, et al. Positional cloning of the wheat vernalization gene VRN1. Proc Natl Acad Sci U S A. 2003;100:6263–8. doi:10.1073/pnas.0937399100.

Yan L, Loukoianov A, Blechl A, Tranquilli G, Ramakrishna W, SanMiguel P, et al. The wheat VRN2 gene is a flowering repressor down-regulated by vernalization. Science. 2004;303:1640–4. doi:10.1126/science.1094305.

Yan L, Fu D, Li C, Blechl A, Tranquilli G, Bonafede M, et al. The wheat and barley vernalization gene VRN3 is an orthologue of FT. Proc Natl Acad Sci U S A. 2006;103:19581–6. doi:10.1073/pnas.0607142103.

Brachypodium as a Model for Grass and Cereal Diseases

Luis A.J. Mur, Valentina Mazzamurro, and Nicola Pecchioni

Abstract Brachypodium is now a well-established model species for the grass family Poaceae, and a number of groups have developed Brachypodium pathosystems, focusing on diseases of cereal and turf and bioenergy grass species. In this review we consider some interactions of pathogens with Brachypodium including those involving rust, Fusarium Head Blight, *Stagonospora* and rice blast pathogens. A particularly well-characterised interaction with Panicum mosaic virus will also be described. Characterisation of these interactions is revealing nonhost resistance mechanisms to disease and the roles of key defence hormones. A particularly fruitful area has been the use of Brachypodium genome to aid in the mapping of key disease resistance traits in larger genome Triticeae crops. It is expected that this latter, we continue to be an important feature of pathological studies involving Brachypodium but as sequencing technologies allow the large genomes of cereals to be completed, we predict that the use of Brachypodium as a platform for functional studies (e.g. involving gene knockouts) to become more important.

Keywords Pathogens • Disease • Rusts • Fusrium head blight • Rice blast • Stagonospora • Panicum mosaic virus • Resistance genes • Comparative genomics

Due to the continual evolution of new virulent strains of pathogens, crop losses due to disease remain significant and in certain instances can be devastating (Strange and Scott 2005). Losses in crop yields are an increasingly serious problem in many areas of the world, such as EU, where the urgency of environmental protection is driving the reduction of crop chemical protection (Oerke et al. 2012). New plant

L.A.J. Mur (✉)
Institute of Environmental and Rural Science, Aberystwyth University,
Edward Llwyd Building, Aberystwyth SY23 3DA, UK
e-mail: lum@aber.ac.uk

V. Mazzamurro
Department of Life Sciences, University of Modena and Reggio Emilia,
42122 Reggio Emilia, Italy

N. Pecchioni
CRA-CER Cereal Research Centre, CRA - Council for Agricultural Research
and Economic Analysis, S.S. 673 km 25,200, 71122 Foggia, Italy

J. Vogel (ed.), *Genetics and Genomics of Brachypodium*,
Plant Genetics and Genomics: Crops Models, DOI 10.1007/7397_2015_7
Published online: 02 October 2015

breeding paradigms are needed, that can exploit crop durable field resistance to pathogens (Tester and Langridge 2010). In order to discern methods of field resistance in crops scientists must deepen our understanding of plant-pathogen interactions. Model plant species—and primarily Arabidopsis—have played a central role in improving our knowledge about the responses to disease. Thus, for example, Arabidopsis has been used to develop our understanding of resistance (*R*) gene function through which plant recognise particular pathogens (Jones 2001; Belkhadir et al. 2004). Arabidopsis has also helped to discover a general pathogen detection mechanism based on the recognition of microbial-associated molecular patterns (Macho and Zipfel 2014). Most tellingly, Arabidopsis has played an invaluable role in defining the contribution of an array of signalling hormones to defense (Mur et al. 2013; Zhu 2014).

The task is now to translate such knowledge from well-established dicotyledonous model species into cereal crops. However, the rich findings derived when—for example—using Arabidopsi*s* may be difficult to translate into cereals except for a few instances such as considerations of resistance gene protein structure and function (Harris et al. 2014). There are several possible reasons for this. Perhaps the most obvious is that pathogens relevant to cereals will not infect Arabidopsis. This stated, some important classes of cereal pathogens, for example, powdery mildews, encompass species which can infect Arabidopsis (Panstruga and Schulze-Lefert 2002). Even for pathogens such as rust fungi, the dicot model plant can be used to define the basis of so-called "non-host" resistance, which is deployed against a pathogen which is not adapted to exploit any genotype of that particular plant species (Mellersh and Heath 2003; Shafiei et al. 2007). A more salient point is Arabidopsis may lack key morphological features, such seed heads, which prevent it being useful as a model for cereal pathogens such as the head blight pathogen, *Fusarium graminearum*. It may also be that there are fundamental differences in the regulation of the cereals and Arabidopsis defense response (Kogel and Langen 2005). Thus, there is clearly need for a robust model for cereal/plant-pathogen interactions. There are some very good examples of such a model: rice (Chen and Ronald 2011) and barley (Zellerhoff et al. 2010; Paulitz and Steffenson 2010). Given that these models are also very important crops, a question arises as to the value of developing the non-crop/non-forage grass species *Brachypodium distachyon* (hereafter referred to as Brachypodium) as a model. However, as stated many times, as Brachypodium is physically small and easy to grow, as well as having the genomic and functional genomic infrastructure, it lends itself to analyses in a wide range of laboratories of widely differing capabilities (Garvin et al. 2008; Mur et al. 2011). Moreover, the phylogenetic position of *B. distachyon*, as well as the availability of closely related both polyploid (*Brachypodium hybridum*) and perennial (e.g. *Brachypodium sylvaticum*) species make it a more widely exploitable model for grasses than rice or barley (Catalan et al. 2014). More pertinently to this chapter, a very large number of plant-pathogen interactions have been described for this model species which facilitate comparison between responses, and that increase the value of Brachypodium as a model (Table 1). Two other issues are discussed in this chapter: the chance of deeper understanding of non-host

Table 1 Model host-pathogen interactions involving *Brachypodium distachyon*

Fusarium graminearum	Fusarium head blight	Peraldi et al. (2011)
Fusarium culmorum	"	"
Oculimacula spp.	Eyespot	Peraldi et al. (2014)
Ramularia collo-cygni	Ramularia leaf spot	"
Puccinia brachypodii	False brome rust	Barbieri et al. (2012)
Bipolaris sorokiniana	Spot blotch and Common root rot	Pogorelko et al. (2013)
Magnaporthe grisea (*now named oryzae*)	Rice blast	Routledge et al. (2004)
Magnaporthe poae	Summer patch	Sandoya and Buanafina (2014)
Colletotrichum cereale	Anthracnose	"
Gaeumannomyces graminis	Take-all	"
Ophiosphaerella agrostis	Dead spot	"
Ophiosphaerella korrae	Necrotic ring spot	"
Pythium aphanidermatum	Blight	"
Rhizoctonia	Sheath blight	"
Scleroctinia homeocarpa	Dollar spot	"
Ustilago bromivora	Smut	Barbieri et al. (2011)
Cochliobolus heterostrophus	Corn leaf blight	Falter and Voigt (2014)
Fusarium sacchari	Sugarcane wilt	"
Pithomyces chartarum	Leaf spot	"
Pyrenophora teres	Net blotch of barley	
Stagonospora macropycnidia	Leaf spot of forage grasses	"
Stagonospora nodorum	Blotch	
Stagonospora paspali	Leaf stripe disease	
Stagonospora tainanensis	Leaf blight of sugarcane	
Barley stripe mosaic virus		Cui et al. (2012)
Panicum mosaic virus	St. Augustine decline	Mandadi and Scholthof (2012)

resistance, and the possibility to broaden the array of interactions of the model grass to well-characterised viral interactions.

Brachypodium as Host for Pathogens of Temperate and Tropical Grasses/Cereals

The evolution, diversification and domestication of cereals have led to the generation of cereal crop which are broadly adapted to either temperate or tropical zones. In the former category are to be found such as bread wheat (*Triticum aestivum*), barley (*Hordeum vulgare*) and oats (*Avena sativa*) whilst the latter includes rice

(*Oryza sativa*), maize (*Zea mays*) and sorghum (*Sorghum bicolor*). This selection pressure, largely imposed by man, has been mirrored by pathogens adapting to cause disease in particular hosts. Thus, few cereal pathogens can successfully infect the large number of hosts and, in particular, span the tropical/temperate cereal divide. Brachypodium is now emerging as a "hub" host with can support a large number of pathogenic interactions and, if not always pathogen species which can infect cereal hosts through the evolutionary divide, at least functional or phylogenetic near equivalents. This important feature augments its value for translational genomics. Thus, if genetic components conferring resistance to a pathogen were defined in Brachypodium, their equivalent could be targeted in cereals genomes. To develop this point it is worth considering some key examples based on the agronomic importance of disease.

The fungal pathogens belonging to the genus *Puccinia* are collectively known as rusts due disease symptoms are associated with a copious production of brown to yellow spores (known as urediniospore) erupting through the leaf lamina. Such spores germinate on the hosts by producing a germ tube that grows until it finds a plant stoma on which it forms an appressorium. An infection peg and a substomatal vesicle are then produced in the mesophyll space, and a primary infection hypha emerges from this vesicle. At this stage haustorial mother cells are generated and produce a specialized fungal structure in the nutrient uptake called haustoria, that is able to penetrated the plant cell wall and establish a colony from which forms a "pustule" (uredinia) (Bettgenhaeuser et al. 2014). Besides urediniospore colour, rust diseases may be subdivided based on their infection patterns; thus on wheat, include species known as stripe rust (*P. striiformis*) or stem rust (*Puccinia graminis*). Rusts have been the focus of research by plant pathologist for over 100 years as significant annual sources of cereal crop loss. For example, annual losses in wheat production due to wheat stripe rust (*Puccinia striiformis* f. sp. *tritici*) can range between 10 and 70 % depending upon the susceptibility of cultivar, earliness of the initial infection and the rate of disease development (Chen 2005). Recently, rusts have gained even more notoriety due to the emergence of wheat stem rust *Puccinia graminis* f. sp. *tritici* line known as Ug99. This emerged in Uganda in 1999 and is failed to be recognised by the *R* gene *Sr31* that is widely used in many wheat genotypes and are therefore susceptible to Ug99. New Ug99 variants have been identified which are virulent on wheat cultivars with other important *R* genes such as *Sr24* and *Sr36* (Singh et al. 2011). The challenge for Brachypodium therefore is to contribute to efforts that will increase resistance to rusts (Draper et al. 2001; Barbieri et al. 2011, 2012; Ayliffe et al. 2013; Figueroa et al. 2013).

One strategy is to assess if cereal adapted pathogens could infect certain accessions of Brachypodium and then focus on both the sources and the modes of resistance in these accessions. Initial screens examined stripe rust strains which were adapted to different cereal hosts. These found one barley adapted species (*P. striiformis* f. sp. *hordei*) where urediniospore production was noted out of necrotic leaf areas (Draper et al. 2001). This approach was also followed by others with similarly relatively poor outcomes. Thus, when Barbieri et al. (2011) screened 40 accessions belonging of five Brachypodium species; *B. phoenicoides*,

B. sylvaticum, *B. rupestre*, *B. pinnatum* and *B. distachyon*, with wheat and barley adapted *P. striiformis* species but only one accession of *B. distachyon* gave rise to good disease symptoms. This was a "hexaploid *B. distachyon*" line (PI 227011) which is a cytotype that has subsequently been reclassified as a separate allotetraploid species: *B. hybridum* (Catalan et al. 2012). Of the inbred *B. distachyon* lines only the genome sequenced Bd21 gave rise to large necrotic lesions with no urediniospore production. In another study, 140 Brachypodium accessions were screened for the responses to wheat adapted *Puccinia graminis* f. sp. *tritici*, *P. triticina*, and *P. striiformis*. Of the *B. distachyon* inbred lines tested only Bd2-3, Bd3-1, and Bd21-3 exhibited sporulation and therefore could be considered to be susceptible to *P. graminis* f. sp. *tritici* growth, with Bd21 the one line that displayed large necrotic lesions (Ayliffe et al. 2013). These authors noted a general inverse relationship between *P. graminis* f. sp. *tritici* and *P. striiformis* growth, with accessions allowing the most development of the former showing the least growth of the latter species, and vice versa. However, the most common response was the deposition of defensive callose and phenolics; the latter leading to autofluorescence under ultra-violet light. Both are features of the so-called "non-host" response where the pathogen fails to establish disease on any genotypes of given plant species. Based on these two studies, such non-host resistance would seem to be the most common responses by Brachypodium to cereal-adapted rust. Nevertheless, the potential far longer durability of such a form of resistance in the field, respect to the "host" resistance, makes it an attractive target of pathology-focused Brachypodium research (Figueroa et al. 2013; Shafiei et al. 2007; Ayliffe et al. 2008). Given this aim, it was an important advance that non-host resistance to *P. graminis* f. sp. *tritici* in accession BRTr13K was mapped to a single allele in crosses with Bd21 which should ease its genetic and biological elucidation (Ayliffe et al. 2013).

A second, alternative strategy has been to use grass-adapted rusts to define the defense mechanisms which may be analogous to those in cereals. Indeed, it has proven to be much easier to define fully susceptibility in Brachypodium to grass-adapted rusts. Thus, Ayliffe et al. (2013) found that *P. graminis* rusts having Poeae hosts showed higher ability to infect *B. distachyon* than those having Triticeae hosts. Figueroa et al. (2013) used rye (*Lolium* sp.) and Timothy grass (*Phleum pratense*) adapted rusts (respectively *P. graminis*. f. sp. *lolii* and. f. sp. *phlei-partensis*); they described Brachypodium infection which were able to complete their life-cycles to produce uredinia and urediniospores, although forming amongst necrotic patches of leaf tissue. In the studies by Barbieri et al. (2011; 2012) *Brachypodium sylvaticum*-adapted rust *Puccinia brachypodii* (false brome rust) often readily infected inbred Brachypodium lines but crucially, some exhibited different level of resistance. This allowed three quantitative loci for *P. brachypodii* resistance to be defined in a segregating Bd3-1 × Bd1-1 population. Such are undoubtedly the first of many studies and the exact definition of the sources of resistance can be expected in the near future.

Fusarium Head Blight (FHB) is another important disease of cereals including wheat and barley caused by *Fusarium graminearum* or *F. culmorum*. As its name suggests, FHB disease is primarily located with developing seed heads. It develops from spores landing on the flowering spikelets where they germinate on the exterior surfaces of florets and glumes and will colonise anthers, stigmas, and lodicules. The devastating aspect of this infection arises through the initiation of cellular necrosis, mainly through the production of trichothecene mycotoxins; most commonly deoxynivalenol (DON) and zearalenone (McMullen et al. 1997; Trail 2009; Scherm et al. 2013). Thus, losses due to FHB occur through reduced grain production and harvested grains being contaminated with mycotoxins which can cause irritation of the gastrointestinal tract cause nausea, vomiting, and diarrhoea, as well as headaches and dizziness (Goswami and Kistler 2004). Brachypodium has emerged as a good host for *Fusarium graminearum* and *F. culmorum* and gives rise to necrotic symptoms which appeared identical to those forming in cereal crops (Peraldi et al. 2011). A more detailed molecular study has confirmed the similarity of the FHB interactions with wheat and Brachypodium; in both cases symptoms being dependent on DON and displaying similar transcriptomic and metabolomic responses (Pasquet et al. 2014). Such observation allowed to developing genomic and post-genomic resources of Brachypodium to be used to investigate FHB. These helped define a mechanism through which DON could be detoxified to reduce disease symptoms. Recently, Van Nguyen et al. (2013) described how the ATF/CREB Transcription Factor (Atf1) is essential for full virulence and DON production in *Fusarium graminearum*, by overexpression and deletion mutants of *Fgatf1*. Interestingly, the constitutive expression of *Fgatf1* in the wild type fungus led to hypervirulence on *Brachypodium distachyon*, wheat and maize.

Resistance to FHB has been associated to several quantitative trait loci (QTLs) that confer partial resistance and a relatively well-characterised example is *Fhb1* (Buerstmayr et al. 2009). This QTL acts by slowing down the spread of fungus from its initial infection site and in a major advance, *Fhb1* encoding wheat cultivars were found to be able to conjugate DON to the non-toxic DON-3-glucoside (D3G) derivative (Lemmens et al. 2005). This activity has been unequivocally linked to the activity of uridine diphosphate-glycosyltransferase (UGT) (UGT13248) (Poppenberger et al. 2003; Shin et al. 2012; Schweiger et al. 2013). The Brachypodium genome has been predicted to contain 159 genes encoding potentially functional UGT of which six exhibited homology to UGT13248. All six Brachypodium UGT12348 homologues were induced by DON, with two with the highest degree of homology were able to convert DON into D3G. Further, Brachypodium along with wheat and barley was used to define a UDP-glycosidase (UGT14077) which acts on and mask zearalenone (Kovalsky Paris et al. 2014). Given such observations it is entirely possible that further mycotoxin UDP-glycosidases could be defined first in Brachypodium, perhaps based on their induction to a particular strain of FHB fungus and their activity could be validated in this species. Interestingly, Blümke et al. (2015) examined the effects of pretreating Brachypodium with DON and observed that this increased resistance to FHB. This involved DON initiating the deployment of cell wall based plants defenses to prevent colonisation of spikelet tissue. Such effects have been linked to a defense priming phenomenon (Conrath 2011) and Blümke et al. (2015)

noted that both DON treatment and priming initiate the activation of mitogen activated protein kinases (MAPkinase) and other kinases. Hence, kinases could be a common regulatory mode to both effects.

Whilst rust and FHB are well documented in Brachypodium, we can expect other important cereal diseases to be characterised in the coming years. These include fungal pathogens *Colletotrichum cereale* (the causal agent of anthracnose), the often overlooked root-infecting pathogens *Rhizoctonia solani* (causing bare patch and root rot of cereals) and *Gaeumannomyces graminis* (Take-all)—all of which are reported to be able to infect Brachypodium (Sandoya and Buanafina 2014). Take-all in particular is a major disease and the UK is where annual losses are in the range of 10 % (HGCA 2014). A Brachypodium-based pathosystem has been defined for Ramularia leaf spot (RLS) which was very similar to that established with its cereal hosts (Peraldi et al. 2014). RLS is caused by the ascomycete fungus *Ramularia collo-cygni* where an as yet ill-defined trigger causes a switch from it being a benign endophyte to a necrotrophic pathogen (Walters et al. 2008). Resistance to RLS has yet to be clearly defined, although over-expression of a stress-responsive NAC1 transcription factor in transgenic lines has shown to be effective (McGrann et al. 2014), so Brachypodium offers the possibility of defining an appropriate mechanism of resistance.

Moving to consider pathogens of tropical cereals, Brachypodium is a host for *Magnaporthe oryzae*. Originally defined as *M. grisea* this has come to be defined as a cryptic species complex with individual species mostly defined through their host specificities. Thus, there are at least two biological species and with members isolated from *Digitaria* grasses remaining as *M. grisea* but those originating from rice renamed *M. oryzae*. *M. oryzae* is causal agent of rice blast disease, a consistently devastating disease where epidemics occurring between 2001 and 2005 in China lead to the loss of 5.7 million hectares of rice (Wilson and Talbot 2009). The infection mechanisms of *M. oryzae* have been extensively characterised (Talbot 2003; Caracuel-Rios and Talbot 2007) as have sources of host resistance (Miah et al. 2013; Azizi et al. 2014; Takatsuji 2014). Further the genetic and experimental tractability of the *M. orzyae*—rice pathogen system has seen it emerge as general model for fungal-host interactions for example suggesting how genomic variability in the pathogen can be influenced by adaptation to its host (Ebbole 2007). It also can serve as a surrogate for less tractable pathosystems; for example, roots interactions with such as *Gaeumannomyces graminis* (Sesma and Osbourn 2004). The interaction of Brachypodium with *M. orzyae* (then named *M. grisea*) was described in the initial paper that first suggested Brachypodium as a model (Draper et al. 2001). Subsequently, a cytological study established the close similarity between rice-blast disease development and resistance phenotypes in Brachypodium accessions; in one case dependent on a single *R* gene (Routledge et al. 2004). Although *Magnaporthe*-based pathosystems are the subject of intensive study, Brachypodium quite quickly made a contribution to the field by demonstrating similar biochemical changes occurring within barley, rice and Brachypodium during blast symptom development. Of the many changes that were noted perhaps the most interesting was the conversion of photoassimilate into mannitol and glycerol for carbon sequestration and osmolyte production to drive fungal hyphal growth (Parker et al. 2009). One of the more important advances within the field of plant pathology over the last decade or so has been the understanding how pathogen effectors are delivered into the host as a means to mobilise nutrients and/ or suppress defenses (unless recognised by *R* gene products) (Chaudhari et al. 2014). Such effectors also feature in the *M. orgzae* virulence (Valent and Khang 2010) and Brachypodium could contribute to the definition of targets for susceptibility to pathogens and host nutrient mobilisation.

Setting Brachypodium to Work for Plant Pathologists

The most immediate role for Brachypodium to plant pathologists (as to other plant scientists) is by exploiting its small relatively low repetitive genome, through which it can play a major role in gene mapping and cloning via comparative genomic approaches. Thus, within plant pathology, the Brachypodium genome contributed to the development of markers which allowed the fine mapping of the stripe rust resistance gene *R* called *Yr26* in wheat but also led to the identification of two putative resistance gene analogues in Brachypodium (Zhang et al. 2013). In the same way, the Brachypodium genome contributed to the localization and provided markers for fine mapping of *Yrq1*, a major QTL for slow-rusting resistance to stripe rust in wheat (Cao et al. 2012). Another group seeking to map resistance to powdery mildew (caused by *Blumeria graminis* f. sp. *hordei*) amongst Spanish barley land races used the Brachypodium genome amongst other to derive suitable markers (Silvar et al. 2012, 2013). These genetic markers were used to assign resistance to QTLs barley chromosomes 7HS, 7HL and 6HL and intriguingly some were linked to gene clusters of protein kinase and encode nucleotide-binding site leucine-rich repeat (NBS-LRR) protein class of R gene products (Silvar et al. 2013). As this report and that of Yang et al. (2013) suggested the conservation of *R* gene families between Brachypodium, wheat, barley, rice and maize genomes could help to develop and test hypotheses focusing the significance of gene clustering, their way of action, and their evolutionary path. Yang et al (2013) found rapidly evolving *R* genes in maize, sorghum, Brachypodium, and rice, all encoding NBS–LRR proteins, that when inserted into rice genome confer resistance to one or more strains of rice blast disease. Since neither conventional evolutionary conservation, nor evolutionary convergence supplied a satisfactory explanation of the results, the authors suggested a mechanism termed "constrained divergence," in which *R* genes and pathogen effectors can follow only limited evolutionary pathways to increase fitness.

Stagonospora nodorum is a necrotrophic fungal pathogen causing Stagonospora nodorum blotch (SNB) disease, with significant yield losses in many wheat growing areas. Brachypodium has contributed the definition of an important source of resistance in wheat. Thus, the Brachypodium genome aided in the development of markers for the fine mapping of and identification of *Snn3-D1* which mediates the recognition of the pathogen effector SnTox3 from *S. nodorum* (Zhang et al. 2011). The fine mapping of the resistance genes *Pch1* in wheat, and *rym11* in barley was also accomplished with the help of the Brachypodium genome (Burt and Nicholson 2011; Lupken et al. 2013). Thanks to this strategy, Lupken et al. (2013) were able to reduce the genetic interval encompassing *rym11* from initial about 10 cM to 0.074 % recombination fraction, as to identify a cosegregating marker for the BaYMV/BaMMV barley mosaic virus resistance gene isolation. *Pch1* was introduced into elite wheat varieties form *Aegilops ventricosa* to confer resistance against Eyespot disease caused by the necrotrophic fungus *Tapesia yallundae* (also known as *Pseudocercosporella herpotrichoides*). However, as it in the case with many such alien introgressions there are yield penalties due to linkage drag (Burt and Nicholson 2011) used the Brachypodium genome to develop conserved orthologous sequence (COS) markers that could allow the segregation of *Pch1* away from loci conferring a yield penalty.

Such strategies have also been used in cereal virology where both susceptibility and resistance to Barley stripe mosaic virus (BSMV) has been described in Brachypodium amongst inbred lines. A clear differential interaction with North Dakota (ND18) strain and Brachypodium accessions Bd21 (susceptible) and Bd3-1 (resistant) and Bd21 x Bd3-1 recombinant inbred lines allowed mapping a single dominant *R* gene—Barley mosaic virus resistance 1 (*Bsr1*) (Cui et al. 2012). In parallel study, Lee et al. (2012) found that the Bd3-1 *Bsr1* resistance is elicited by the barley stripe mosaic virus triple gene block 1 (TGB1) movement protein. The analysis of RNA variants of the virulent Norwich (NW) vs. the ND18 strain identified two amino acid changes in NGB1 that had a major effect on resistance in Bd3-1.

Moving beyond such targeted *R* gene mapping efforts, other approaches are mining the Brachypodium genome for novel *R* genes. Pathogens are under selection pressure to evolve virulent derivatives which are not recognised by R genes. As a result resistant varieties that are the product of years of plant breeding efforts may have only a few years of usefulness in the field before rendered near useless through the derivation of new virulent pathogen strains (Parlevliet 2002). Thus, new *R* genes are required and the small genome of Brachypodium has eased the identification of 126 NBS-LRR class resistance genes in the Bd21 genome (Tan and Wu 2012). These could be aligned with orthologs in elite cereal varieties, possibly after assessment for activity in Brachypodium knockout lines which could be screened with a wide range of the pathogens known to infect this species (Table 1). In an analogous approach (Yang et al., 2013) screened the genomes of maize, Sorghum and Brachypodium genotypes to target NBS-LRR genes which exhibit homology to rice blast resistance genes. Crucially when these non-rice NBS-LRR genes were transformed into rice 15 out of 60 were able to initiate resistance against *M. oryzae*.

Beyond pathogen recognition, Brachypodium is helping to understand how defense signalling can contribute to resistance. Work in Arabidopsis has suggested that host responses to a wide range of pathogens used either by a defense module co-ordinated through salicylic acid or another co-ordinated by jasmonates/ethylene (Pieterse et al. 2012). Classically, these two defense modules are thought to be antagonistic, but synergistic interactions between them can occur (Mur et al. 2006). In one of the first metabolomic studies focusing on plant-pathogen interactions, analysis of the interaction of Brachypodium with *M. oryzae* demonstrated considerable phospholipid processing which included the derivation of jasmonic acid (Allwood et al. 2006). This work coincided with reports that jasmonates were important for rice blast resistance in rice (Mei et al. 2006). Recently Goddard et al. (2014) investigated the impact of mutation of *brassinosteroid (BR)-insensitive 1 (BRI1)* gene on disease resistance of Brachypodium and barley. Inactivation of *BRI1*, which encodes the main receptor of BR in both species, resulted in an increased disease resistance against necrotrophic and hemibiotrophic pathogens, while it has no effect on biotrophic pathogens, or on those with a prolonged asymptomatic phase. Notably, disruption of BRI1 has similar effects on disease resistance in both species, indicating conserved mechanisms of defense in the model and in barley crop.

A parallel with defense signaling in Arabidopsis was revealed through analysis of the interaction of Panicum mosaic virus (PMV) and its satellite virus (SPMV), the causal agents of the grass disease St. Augustine Decline, with Brachypodium. PMV/SPMV is a single-stranded, positive-sense RNA whose host range is known to extend to foxtail millet (*Setaria italic*a), pearl millet (*Pennisetum glaucum*), and proso millet (*Panicum miliaceum*), although infections are undoubtedly supported by other grasses (Buzen et al. 1984). The SPMV in St Augustine Decline acts to increase the accumulation of PMV by increasing the amount of PMV movement protein and therefore exacerbates symptom development (Scholthof 1999). Such complex interactions are difficult to define except in a biologically and genetically tractable model system such as Brachypodium. It was therefore a significant advance that Brachypodium has been shown to support infections of PMV producing symptoms that were equivalent to those seen in foxtail millet; and these symptoms were exacerbated by co-infection of PMV with SPMV (Mandadi and Scholthof 2012).

The quality of the genomic information offered by Brachypodium greatly facilitated a comparison of the transcriptomic responses of PMV/SPMV to those of well-characterized viral interactions in Arabidopsis (Elena et al. 2011). In this PMV-Brachypodium interaction an especially interesting response was the PMV/SPMV induced expression of Bradi3g52690 (a SDE-like helicase) which may be involved in the generation of short-interfering RNAs which could target and silence the viral genome. More general insights into the Brachypodium defense response was shown by induction of the *Pathogenesis related protein* genes *PR-1*, *PR-2*, *PR-3*, *PR-4*, and *PR-5* which are well-establish markers of SA accumulation and, notably, there were also increases of the SA biosynthesis gene *ISOCHORISMATE SYNTHASE1* (*ICS1*) (An and Mou 2011). In line with the SA-JA antagonism model this increase in the SA defense module correlated with a reduction in the expression of many (but interestingly not all) JA biosynthesis genes, namely *LIPOXYGENASE2* (*LOX2*) and *ALLENE OXIDE SYNTHASE* (*AOS*) enzymes; as of JA-responsive genes, for example *VEGETATIVE STORAGE PROTEIN1* (*VSP1*). It may appear counter-intuitive that PMV induces a defense module, but this may indicate that the JA/ET defense module is more important in defense against this virus and/or that SA could actually facilitate its infection process. This has been noted in certain necrotrophic interactions (Abd El Rahman et al. 2012). However, it was interesting that PMV/SPMV co-infection reduced the level of SA-associated gene expression by around 50 % suggesting that one function of the satellite virus could be as a defense suppressor. Mandadi et al. (2014) further expanded their etiological and molecular studies with seven infecting viruses (14 plant-virus combinations) to establish Brachypodium and *Setaria viridis* as prominent C3 and C4 plant models for plant-virus interactions. Their results identified PHYTOALEXIN DEFICIENT 4, as being consistently suppressed during most virus infections in both species, as a key upstream modulator of SA signalling in antiviral response, and evidence for conserved crosstalk between SA, JA and ethylene pathways.

Necrotrophic FHB-Brachypodium interaction also furnished evidence of deployment of the converse JA/ET defense module. For example, during FHB disease development, the putative JA biosynthetic gene Bradi5g11590 (a possible LOX); Bradi1g4588 (likely to be a 12-oxophytodienoate reductase) as well as Bradi3g23180 (a JASMONATE-INSENSITIVE 3, JAI3 homolog) which may be involved in JA signalling—were induced by DON^+ fungal strains. The metabolomic component of this study suggested that DON^+ fungal strains lead to the increased accumulation of tryptophan-derived metabolites, tryptamine, serotonin, coumaroyl-serotonin and feruloyl-serotonin (Pasquet et al. 2014). This was of especially interest given that serotonin has some anti-microbial activity (Ishihara et al. 2008) and may also be a feature in already the mentioned DON-primed resistance against FHB (Blümke et al. 2015). There is currently no evidence that the JA/ET module influenced the production of serotonin.

Where Next for Brachypodium Pathology?

Whilst progress on developing Brachypodium as a model has been impressive, more needs to be done. Other interactions with important cereal pathogens need to be developed perhaps most significantly with powdery mildew and Septoria (*Zymoseptoria* [synonym *Mycosphaerella*]) which are major sources of crop loss. This should be possible given initial reports of multiple isolates strains of *Mycosphaerella graminicola* which are able to establish disease in this host (O'Driscoll et al. 2011) and of powdery mildews infecting Brachypodium in Spain (Lange 1996). Given this, it may be expected that eventually most major cereal diseases will have their equivalent in Brachypodium.

The importance of this last point is made clear when it is considered within context of the burgeoning functional genomic resources that have been described elsewhere in this book. Thus, with the rapid advances in genomic sequencing the genetic value of the small Brachypodium genome becomes diluted as it becomes one amongst an increasing number of genomes which can be used target key traits through comparative genomics approaches. Instead, the small, diploid genome will ease functional genomic approaches to be used to elucidate the function of genes targeted in Brachypodium or orthologues to genes in other species. Based on such resources and a wide-range of model pathosystems it is to be expected that genes whole roles are interaction specific or more generally active in defense or disease susceptibility will be elucidated; and crucially, the information utilised in cereal breeding programmes.

References

Abd El Rahman T, El Oirdi M, Gonzalez-Lamothe R, Bouarab K. Necrotrophic pathogens use the salicylic acid signaling pathway to promote disease development in tomato. Mol Plant-Microbe Interact. 2012;25:1584–93.

Allwood JW, Ellis DI, Heald JK, Goodacre R, Mur LA. Metabolomic approaches reveal that phosphatidic and phosphatidyl glycerol phospholipids are major discriminatory non-polar metabolites in responses by *Brachypodium distachyon* to challenge by *Magnaporthe grisea*. Plant J. 2006;46:351–68.

An C, Mou Z. Salicylic acid and its function in plant immunity. J Integr Plant Biol. 2011;53:412–28.

Ayliffe M, Singh D, Park R, Moscou M, Pryor T. Infection of *Brachypodium distachyon* with selected grass rust pathogens. Mol Plant Microbe Interact. 2013;26:946–57.

Ayliffe M, Singh R, Lagudah E. Durable resistance to wheat stem rust needed. Curr Opin Plant Biol. 2008;11:187–92.

Azizi P, Rafii, MY, Abdullah SN, Nejat N, Maziah M, Hanafi MM, Latif MA, Sahebi M. Toward understanding of rice innate immunity against Magnaporthe oryzae. Crit Rev Biotechnol. 2014:1–10.

Barbieri M, Marcel TC, Niks RE. Host status of false brome grass to the leaf rust fungus *Puccinia brachypodii* and the Stripe Rust fungus *P. striiformis*. Plant Dis. 2011;95:1339–45.

Barbieri M, Marcel TC, Niks RE, Francia E, Pasquariello M, Mazzamurro V, et al. QTLs for resistance to the false brome rust *Puccinia brachypodii* in the model grass *Brachypodium distachyon* L. Genome. 2012;55:152–63.

Belkhadir Y, Subramaniam R, Dangl JL. Plant disease resistance protein signaling: NBS-LRR proteins and their partners. Curr Opin Plant Biol. 2004;7:391–9.

Bettgenhaeuser J, Gilbert B, Ayliffe M, Moscou MJ. Nonhost resistance to rust pathogens - a continuation of continua. Front Plant Sci. 2014;5:664.

Blümke A, Sode B, Ellinger D, Voigt CA. Reduced susceptibility to Fusarium head blight in Brachypodium distachyon through priming with the Fusarium mycotoxin deoxynivalenol. Mol Plant Pathol. 2015;16:472–83.

Buerstmayr H, Ban T, Anderson JA. QTL mapping and marker-assisted selection for Fusarium head blight resistance in wheat: a review. Plant Breed. 2009;128:1–26.

Burt C, Nicholson P. Exploiting co-linearity among grass species to map the *Aegilops ventricosa*-derived *Pch1* eyespot resistance in wheat and establish its relationship to *Pch2*. Theor Appl Genet. 2011;123:1387–400.

Buzen FG, Niblett CL, Hooper GR, Hubbard J, Newman MA. Further characterization of Panicum Mosaic-Virus and its associated satellite Virus. Phytopathology. 1984;74:313–8.

Cao X, Zhou J, Gong X, Zhao G, Jia J, Ql X. Identification and validation of a major quantitative trait locus for slow-rusting resistance to stripe rust in wheat. J Integr Plant Biol. 2012;54:330–44.

Caracuel-Rios Z, Talbot NJ. Cellular differentiation and host invasion by the rice blast fungus *Magnaporthe grisea*. Curr Opin Microbiol. 2007;10:339–45.

Catalan P, Muller J, Hasterok R, Jenkins G, Mur LA, Langdon T, et al. Evolution and taxonomic split of the model grass *Brachypodium distachyon*. Ann Bot. 2012;109:385–405.

Catalan P, Chalhoub B, Chochois V, Garvin DF, Hasterok R, Manzaneda AJ, et al. Update on the genomics and basic biology of Brachypodium: International Brachypodium Initiative (IBI). Trends Plant Sci. 2014;19:414–8.

Chaudhari P, Ahmed B, Joly DL, Germain H. Effector biology during biotrophic invasion of plant cells. Virulence. 2014;5:703–9.

Chen XM. Epidemiology and control of stripe rust [Puccinia striiformis f. sp tritici] on wheat. Can J Plant Pathol. 2005;27:314–37.

Chen XW, Ronald PC. Innate immunity in rice. Trends Plant Sci. 2011;16:451–9.

Conrath U. Molecular aspects of defence priming. Trends Plant Sci. 2011;16:524–31.

Cui Y, Lee MY, Huo N, Bragg J, Yan L, Yuan C, et al. Fine mapping of the Bsr1 barley stripe mosaic virus resistance gene in the model grass Brachypodium distachyon. PLoS One. 2012;7: e38333.

Draper J, Mur LA, Jenkins G, Ghosh-Biswas GC, Bablak P, Hasterok R, et al. *Brachypodium distachyon*. A new model system for functional genomics in grasses. Plant Physiol. 2001;127:1539–55.

Ebbole DJ. *Magnaporthe* as a model for understanding host-pathogen interactions. Annu Rev Phytopathol. 2007;45:437–56.

Elena SF, Carrera J, Rodrigo G. A systems biology approach to the evolution of plant-virus interactions. Curr Opin Plant Biol. 2011;14:372–7.

Falter C, Voigt CA. Comparative cellular analysis of pathogenic fungi with a disease incidence in *Brachypodium distachyon* and *Miscanthus* x *giganteus*. Bioenerg Res. 2014;7:958–73.

Figueroa M, Alderman S, Garvin DF, Pfender WF. Infection of *Brachypodium distachyon* by formae speciales of *Puccinia graminis*: early infection events and host-pathogen incompatibility. PLoS One. 2013;8:e56857.

Garvin DF, Gu YQ, Hasterok R, Hazen SP, Jenkins G, Mockler TC, et al. Development of genetic and genomic research resources for Brachypodium distachyon, a new model system for grass crop research. Crop Sci. 2008;48:S69–84.

Goddard R, Peraldi A, Ridout C, Nicholson P. Enhanced disease resistance caused by *BRI1* mutation is conserved between *Brachypodium distachyon* and Barley (*Hordeum vulgare*). Mol Plant Microbe Interact. 2014;27:1095–106.

Goswami RS, Kistler HC. Heading for disaster: *Fusarium graminearum* on cereal crops. Mol Plant Pathol. 2004;5:515–25.

Harris MO, Friesen TL, Xu SS, Chen MS, Giron D, Stuart JJ. Pivoting from Arabidopsis to wheat to understand how agricultural plants integrate responses to biotic stress. J Exp Bot. 2014. doi:10.1093/jxb/eru465.

Home-Grown, Cereals & Authority (HGCA). Wheat disease management guide. UK: HGCA Publications; 2014.

Ishihara A, Hashimoto Y, Tanaka C, Dubouzet JG, Nakao T, Matsuda F, et al. The tryptophan pathway is involved in the defense responses of rice against pathogenic infection via serotonin production. Plant J. 2008;54:481–95.

Jones JD. Putting knowledge of plant disease resistance genes to work. Curr Opin Plant Biol. 2001;4:281–7.

Kogel KH, Langen G. Induced disease resistance and gene expression in cereals. Cell Microbiol. 2005;7:1555–64.

Kovalsky Paris MP, Schweiger W, Hametner C, Stückler R, Muehlbauer GJ, Varga E, et al. Zearalenone-16-O-glucoside: a new masked mycotoxin. J Agric Food Chem. 2014;62:1181–9.

Lange C. Braun, U. 1995. The Powdery Mildews (Erysiphales) of Europe. Nord J Bot. 1996;16:144.

Lee MY, Yan LJ, Gorter FA, Kim BYT, Cui Y, Hu Y, et al. Brachypodium distachyon line Bd3-1 resistance is elicited by the barley stripe mosaic virus triple gene block 1 movement protein. J Gen Virol. 2012;93:2729–39.

Lemmens M, Scholz U, Berthiller F, Dall'asta C, Koutnik A, Schuhmacher R, et al. The ability to detoxify the mycotoxin deoxynivalenol colocalizes with a major quantitative trait locus for Fusarium head blight resistance in wheat. Mol Plant Microbe Interact. 2005;18:1318–24.

Lupken T, Stein N, Perovic D, Habekuss A, Kramer I, Hahnel U, et al. Genomics-based high-resolution mapping of the BaMMV/BaYMV resistance gene *rym11* in barley (Hordeum vulgare L.). Theor Appl Genet. 2013;126:1201–12.

Macho AP, Zipfel C. Plant PRRs and the activation of innate immune signaling. Mol Cell. 2014;54:263–72.

Mandadi KK, Pyle JD, Scholthof KB. Comparative analysis of antiviral responses in *Brachypodium distachyon* and *Setaria viridis* reveals conserved and unique outcomes among C3 and C4 plant defenses. Mol Plant Microbe Interact. 2014;27:1277–90.

Mandadi KK, Scholthof KB. Characterization of a viral synergism in the monocot *Brachypodium distachyon* reveals distinctly altered host molecular processes associated with disease. Plant Physiol. 2012;160:1432–52.

Mcgrann GR, Steed A, Burt C, Goddard R, Lachaux C, Bansal A, et al. Contribution of the drought tolerance-related stress-responsive NAC1 transcription factor to resistance of barley to Ramularia leaf spot. Mol Plant Pathol. 2014;16(2):201–9.

Mcmullen M, Jones R, Gallenberg D. Scab of wheat and barley: a re-emerging disease of devastating impact. Plant Dis. 1997;81:1340–8.

Mei C, Qi M, Sheng G, Yang Y. Inducible overexpression of a rice allene oxide synthase gene increases the endogenous jasmonic acid level, PR gene expression, and host resistance to fungal infection. Mol Plant Microbe Interact. 2006;19:1127–37.

Mellersh DG, Heath MC. An investigation into the involvement of defense signaling pathways in components of the nonhost resistance of *Arabidopsis thaliana* to rust fungi also reveals a model system for studying rust fungal compatibility. Mol Plant Microbe Interact. 2003;16:398–404.

Miah G, Rafii MY, Ismail MR, Puteh AB, Rahim HA, Asfaliza R, et al. Blast resistance in rice: a review of conventional breeding to molecular approaches. Mol Biol Rep. 2013;40:2369–88.

Mur LA, Kenton P, Atzorn R, Miersch O, Wasternack C. The outcomes of concentration-specific interactions between salicylate and jasmonate signaling include synergy, antagonism, and oxidative stress leading to cell death. Plant Physiol. 2006;140:249–62.

Mur LAJ, Allainguillaume J, Catalan P, Hasterok R, Jenkins G, Lesniewska K, et al. Exploiting the Brachypodium Tool Box in cereal and grass research. New Phytologist. 2011;191:334–47.

Mur LAJ, Prats E, Pierre S, Hall MA, Hebelstrup KH. Integrating nitric oxide into salicylic acid and jasmonic acid/ethylene plant defense pathways. Front in Plant Sci. 2013;4:215.

O'Driscoll A, Doohan F, Mullins E. The Mycosphaerella graminicola-Brachypodium distachyon interaction: a new model pathosystem to study Septoria tritici blotch disease of wheat. 2011 [Online]. Versailles, France. https://colloque4.inra.fr/1st_european_brachypodium_workshop/List-of-abstracts.

Oerke EC, Dehne HW, Schönbeck F, Weber A. Crop production and crop protection: estimated losses in major food and cash crops. Amsterdam: Elsevier; 2012. p. 829. ISBN 0444597948, 9780444597946.

Panstruga R, Schulze-Lefert P. Live and let live: insights into powdery mildew disease and resistance. Mol Plant Pathol. 2002;3:495–502.

Parker D, Beckmann M, Zubair H, Enot DP, Caracuel-Rios Z, Overy DP, et al. Metabolomic analysis reveals a common pattern of metabolic re-programming during invasion of three host plant species by *Magnaporthe grisea*. Plant J. 2009;59:723–37.

Parlevliet JE. Durability of resistance against fungal, bacterial and viral pathogens; present situation. Euphytica. 2002;124:147–56.

Pasquet JC, Chaouch S, Macadre C, Balzergue S, Huguet S, Martin-Magniette ML, et al. Differential gene expression and metabolomic analyses of Brachypodium distachyon infected by deoxynivalenol producing and non-producing strains of *Fusarium graminearum*. BMC Genomics. 2014;15:629.

Paulitz TC, Steffenson BJ. Biotic stress in barley: disease problems and solutions. Barley. Chichester: Wiley-Blackwell; 2010.

Peraldi A, Beccari G, Steed A, Nicholson P. *Brachypodium distachyon:* a new pathosystem to study Fusarium head blight and other Fusarium diseases of wheat. BMC Plant Biol. 2011;11:100.

Peraldi A, Griffe LL, Burt C, Mcgrann GRD, Nicholson P. *Brachypodium distachyon* exhibits compatible interactions with *Oculimacula* spp. and *Ramularia collo-cygni*, providing the first pathosystem model to study eyespot and ramularia leaf spot diseases. Plant Pathol. 2014;63:554–62.

Pieterse CMJ, VAN DER Does D, Zamioudis C, Leon-Reyes A, VAN Wees SCM. Hormonal modulation of plant immunity. Annu Rev Cell Dev Biol. 2012;28(28):489–521.

Pogorelko G, Lionetti V, Fursova O, Sundaram RM, Qi M, Whitham SA, et al. Arabidopsis and Brachypodium distachyon transgenic plants expressing Aspergillus nidulans acetylesterases have decreased degree of polysaccharide acetylation and increased resistance to pathogens. Plant Physiol. 2013;162:9–23.

Poppenberger B, Berthiller F, Lucyshyn D, Sieberer T, Schuhmacher R, Krska R, et al. Detoxification of the Fusarium mycotoxin deoxynivalenol by a UDP-glucosyltransferase from *Arabidopsis thaliana*. J Biol Chem. 2003;278:47905–14.

Routledge AP, Shelley G, Smith JV, Talbot NJ, Draper J, Mur LA. *Magnaporthe grisea* interactions with the model grass *Brachypodium distachyon* closely resemble those with rice (Oryza sativa). Mol Plant Pathol. 2004;5:253–65.

Sandoya GV, Buanafina MMD. Differential responses of *Brachypodium distachyon* genotypes to insect and fungal pathogens. Physiol Mol Plant Pathol. 2014;85:53–64.

Scherm B, Balmas V, Spanu F, Pani G, Delogu G, Pasquali M, et al. *Fusarium culmorum*: causal agent of foot and root rot and head blight on wheat. Mol Plant Pathol. 2013;14:323–41.

Scholthof KBG. A synergism induced by satellite panicum mosaic virus. Mol Plant Microbe Interact. 1999;12:163–6.

Schweiger W, Pasquet JC, Nussbaumer T, Paris MP, Wiesenberger G, Macadre C, et al. Functional characterization of two clusters of *Brachypodium distachyon* UDP-glycosyltransferases encoding putative deoxynivalenol detoxification genes. Mol Plant Microbe Interact. 2013;26:781–92.

Sesma A, Osbourn AE. The rice leaf blast pathogen undergoes developmental processes typical of root-infecting fungi. Nature. 2004;431:582–6.

Shafiei R, Hang C, Kang JG, Loake GJ. Identification of loci controlling non-host disease resistance in Arabidopsis against the leaf rust pathogen *Puccinia triticina*. Mol Plant Pathol. 2007;8:773–84.

Shin S, Torres-Acosta JA, Heinen SJ, McCormick S, Lemmens M, Paris MPK, et al. Transgenic *Arabidopsis thaliana* expressing a barley UDP-glucosyltransferase exhibit resistance to the mycotoxin deoxynivalenol. J Exp Bot. 2012;63:4731–40.

Silvar C, Perovic D, Nussbaumer T, Spannagl M, Usadel B, Casas A, et al. Towards positional isolation of three quantitative trait loci conferring resistance to powdery mildew in two Spanish barley landraces. PLoS One. 2013;8:e67336.

Silvar C, Perovic D, Scholz U, Casas AM, Igartua E, Ordon F. Fine mapping and comparative genomics integration of two quantitative trait loci controlling resistance to powdery mildew in a Spanish barley landrace. Theor Appl Genet. 2012;124:49–62.

Singh RP, Hodson DP, Huerta-Espino J, Jin Y, Bhavani S, Njau P, et al. The Emergence of Ug99 races of the Stem Rust fungus is a threat to world wheat production. Annu Rev Phytopathol. 2011;49(49):465–81.

Strange RN, Scott PR. Plant disease: a threat to global food security. Annu Rev Phytopathol. 2005;43:83–116.

Takatsuji H. Development of disease-resistant rice using regulatory components of induced disease resistance. Front Plant Sci. 2014;5:630.

Talbot NJ. On the trail of a cereal killer: exploring the biology of *Magnaporthe grisea*. Annu Rev Microbiol. 2003;57:177–202.

Tan S, Wu S. Genome wide analysis of nucleotide-binding site disease resistance genes in *Brachypodium distachyon*. Comp Funct Genomics. 2012;2012:418208.

Tester M, Langridge P. Breeding technologies to increase crop production in a changing world. Science. 2010;327:818–22.

Trail F. For blighted waves of grain: *Fusarium graminearum* in the postgenomics era. Plant Physiol. 2009;149:103–10.

Valent B, Khang CH. Recent advances in rice blast effector research. Curr Opin Plant Biol. 2010;13:434–41.

VAN Nguyen T, Kroger C, Bonnighausen J, Schafer W, Bormann J. The ATF/CREB Transcription factor atf1 is essential for full virulence, deoxynivalenol production, and stress tolerance in the cereal pathogen *Fusarium graminearum*. Mol Plant Microbe Interact. 2013;26:1378–94.

Walters DR, Havis ND, Oxley SJP. *Ramularia collo-cygni*: the biology of an emerging pathogen of barley. FEMS Microbiol Lett. 2008;279:1–7.

Wilson RA, Talbot NJ. Under pressure: investigating the biology of plant infection by *Magnaporthe oryzae*. Nat Rev Microbiol. 2009;7:185–95.

Yang S, Li J, Zhang X, Zhang Q, Huang J, Chen JQ, et al. Rapidly evolving R genes in diverse grass species confer resistance to rice blast disease. Proc Natl Acad Sci U S A. 2013;110:18572–7.

Zellerhoff N, Himmelbach A, Dong W, Bieri S, Schaffrath U, Schweizer P. Nonhost resistance of barley to different fungal pathogens is associated with largely distinct, quantitative transcriptional responses. Plant Physiol. 2010;152:2053–66.

Zhang Z, Friesen TL, Xu SS, Shi G, Liu Z, Rasmussen JB, et al. Two putatively homoeologous wheat genes mediate recognition of SnTox3 to confer effector-triggered susceptibility to *Stagonospora nodorum*. Plant J. 2011;65:27–38.

Zhang X, Han D, Zeng Q, Duan Y, Yuan F, Shi J, et al. Fine mapping of wheat stripe rust resistance gene Yr26 based on collinearity of wheat with Brachypodium distachyon and rice. PLoS One. 2013;8:e57885.

Zhu Z. Molecular basis for jasmonate and ethylene signal interactions in Arabidopsis. J Exp Bot. 2014;65:5743–8.

Brachypodium and the Abiotic Environment

David L. Des Marais and Thomas E. Juenger

Abstract *Brachypodium distachyon* and its congeners are found in diverse environments throughout the temperate regions of the world. *Brachypodium* species also show considerable variation in life history strategy, with species representing short-season annual habits and other species persisting for multiple years. This variation in ecological setting and life history suggests the existence of considerable genetic diversity in adaptation to the abiotic environment, both in constitutive tolerance to local conditions and in the capacity of single genotypes to acclimate to changing or unpredictable conditions. We review the limited but growing empirical literature on the physiology, development, and molecular biology of the interaction of *Brachypodium* with the abiotic environment. We highlight how these findings inform studies of ecologically and agriculturally-important plant species, and identify areas of future research that will extend the utility of *Brachypodium* as a model genetic system for understanding plant-environment interactions.

Keywords *Brachypodium distachyon* • Drought • Abiotic stress • Cold • Adaptation • Acclimation • Plasticity • GxE

Introduction

Plants are confronted by a range of challenging environmental conditions that may vary on the scale of hours, days, months, years or even generations. Plant species have evolved diverse strategies to cope with this variable abiotic environment. These strategies include the evolution of both constitutive as well as inducible—plastic—traits that provide adaptive abiotic stress tolerance. *Brachypodium*

D.L. Des Marais (✉)
Department of Organismic and Evolutionary Biology and The Arnold Arboretum, Harvard University, Cambridge, MA, USA
e-mail: desmarais@fas.harvard.edu

T.E. Juenger (✉)
Department of Integrative Biology and Institute for Cellular and Molecular Biology, The University of Texas at Austin, Austin, TX, USA
e-mail: tjuenger@austin.utexas.edu

J. Vogel (ed.), *Genetics and Genomics of Brachypodium*,
Plant Genetics and Genomics: Crops Models, DOI 10.1007/7397_2015_13
Published online: 02 October 2015

distachyon is found in diverse environments across its native range surrounding the Mediterranean and Middle East. Natural populations of *B. distachyon* (*sensu* Schippmann 1991) occur from the mid-Atlantic islands east to Afghanistan, from north Africa into central France, spanning habitats near sea level to over 1800 m. *B. distachyon* has also successfully established invasive populations throughout temperate regions of the world as a result of recent human-mediated dispersal (Khan 1984). Other species of *Brachypodium* are found in equally diverse environments, albeit largely restricted to temperate climates. This habitat diversity suggests that populations of *B. distachyon*—and its congeners—may display varying degrees of local adaptation and/or a range of phenotypic plasticity that facilitate population persistence under many abiotic and biotic conditions. Here, we review the small but growing literature documenting molecular, physiological, and developmental aspects of *Brachypodium* interactions with the abiotic environment. We also discuss which of these features of *Brachypodium* vary within and between natural populations, and which are expressed facultatively (i.e. representing phenotypic plasticity or acclimation responses). Throughout, we will highlight the utility of *B. distachyon* and its congeners as reference genetic systems for temperate C3 cereal and biofuel crops, as well as discuss prospects for *Brachypodium* as a model for elucidating the evolution and ecology of plant-abiotic interactions. We focus on *Brachypodium* responses to soil drying and to temperature extremes, due to the limited number of studies assessing other aspects of plant-environment interactions.

As discussed in other chapters of the current volume, the genus *Brachypodium* is phylogenetically sister to a large clade of C3 grasses, the Pooideae. The Pooideae includes many cereal and forage crops and is broadly distributed in temperate climates around the world. Divergence time estimates from molecular analyses indicate that *B. distachyon* diverged from other Pooideae grasses ca. 32–39 mya, and from rice ca. 40–54 mya (The International Brachypodium Initiative 2010). Due to its phylogenetic position, its C3 photosynthetic system, and its adaptation to temperate climates, *B. distachyon* has been proposed as a genetic model system for Pooid cereal and forage species. For example, the growth form, physiology, and abiotic stress responses of *Brachypodium* should much better reflect characteristics of wheat (*Triticum*), barley (*Hordeum*) and rye (*Secale*) in comparison to the C4 grass *Zea mays* or sub-tropically-adapted *Oryza sativa*. Plants which exhibit C3 photosynthesis are generally distributed at higher latitudes, where the increased efficiency under high temperature afforded by C4 photosynthesis is less favorable. Higher-latitude grasses are also subject to a greater annual range of temperatures, including temperatures at or well-below 0 °C.

The *Brachypodium* genus has several interesting attributes for the analysis of plant-environment interactions. First, the genus contains both annual (e.g. *B. distachyon*; Brkljacic et al. 2011) and perennial species (e.g. *B. sylvaticum*; Steinwand et al. 2013) that may be contrasted to assess differential strategies of biomass acquisition and allocation between life history types. Moreover, perennial species vary in the persistence of their stems (Khan 1984) suggesting genetic diversity in biomass allocation strategies. Finally, species (and, in some cases, populations within species) representing a range of ploidies allow for evaluation of classic questions in evolutionary ecology dealing with niche occupancy of diploid vs. higher-ploidy genetic systems.

Life History Variation Within and Among *Brachypodium* Species

Distressingly little has been published regarding the ecological setting and life history strategies of *Brachypodium* species. This gap in our understanding of basic ecology is unfortunate because variation in life history strategies can have a strong influence on plant adaptation to abiotic stress. In Fig. 1, we present annual patterns of precipitation and temperature for the site-of-origin for four commonly-studied *B. distachyon* accessions; the cool, wet winters and warm, dry summers typical of the Mediterranean climate are apparent in all four sites. These climate diagrams

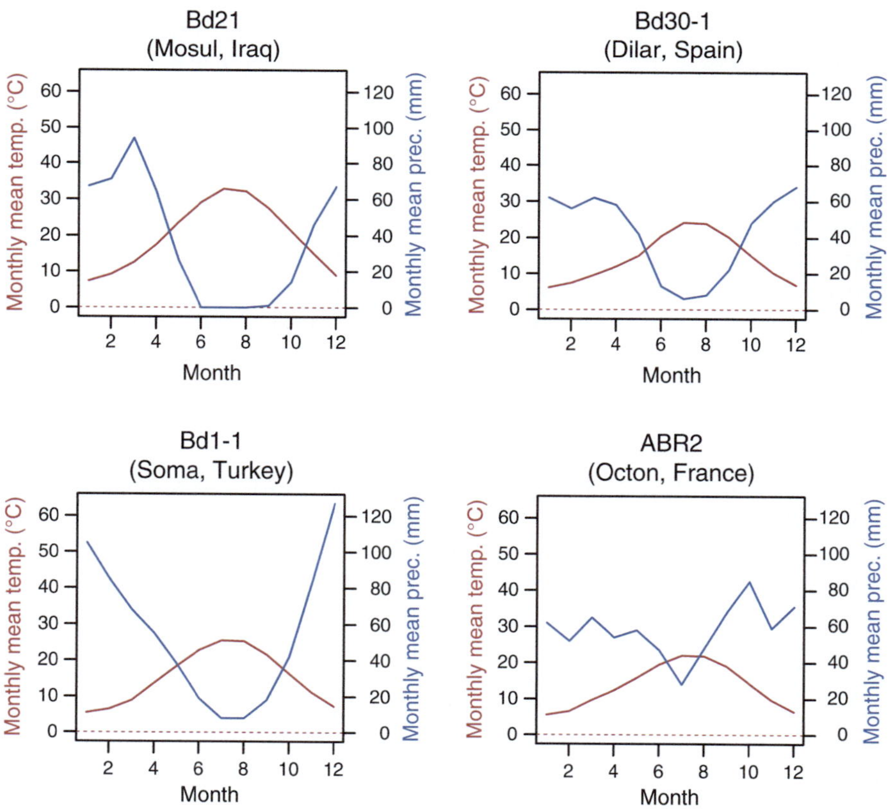

Fig. 1 Climate diagrams for sites of origin of four commonly-studied natural accession of *Brachypodium distachyon*. Climate data are from Hijmans et al. (2005)

also highlight the differences in water availability among populations during the portion of the year in which temperatures are likely conducive to growth. In southern Europe, *B. distachyon* is observed to germinate in the spring, setting seed in the early summer and senescing in the late summer and autumn (P. Catalan and A. Manzaneda pers. comm.). This represents an annual life history strategy, specifically a "spring-annual" strategy. The reference accession of *B. distachyon*, Bd21 from northern Iraq, and several other accessions descended from wild-collected plants are spring-annuals. These accessions flower readily under cultivation without extended cold treatment, or vernalization. Here, the timing of early spring establishment, rapid growth, and characteristics to escape hot and dry summer conditions—or to avoid their effects on cellular function—are paramount to ecological success and persistence.

In contrast to these rapid-flowering accessions, lab and greenhouse studies with *B. distachyon* reveal that many other accessions are relatively late flowering—"winter annuals"—and require cold stratification of seed or vernalization of adult plants to flower. Moreover, vernalization responses are often associated with flowering time responsiveness to photoperiod conditions (Schwartz et al. 2010; Ream et al. 2014). For example, the Bd18-1 accession originating from central Turkey requires >4 weeks of cold to transition from vegetative growth to flowering in lab studies (Schwartz et al. 2010). Seeds of many winter annual species germinate in the cool wet fall followed by establishment, slow and continuous growth through much of winter, and rapid growth and flowering in spring and early summer. In locations with exceptionally cold winter climates, plants may overwinter as dormant crowns or plants. It is likely that winter annual plants will be well established and larger in spring, with more stored resources, as compared to spring annual plants because of the head start this life history strategy affords them. The larger size may allow them to complete reproduction earlier in the spring before summer droughts, or to better avoid dehydration associated with drying environments because of expanded root systems and access to a deeper profile of soil moisture. Despite compelling evidence from lab studies for "overwintering" behavior in *B. distachyon,* we are not aware of any published reports of overwintering populations from field observations.

Life history strategies often show strong correlations with other developmental traits and physiological strategies for coping with abiotic stress. In some species spring-annual plants have higher growth rates, lower water use efficiency, and lower cold tolerance than do winter-annual plants. For example, accessions of *Arabidopsis thaliana* exhibit strong genetic tradeoffs between water-use efficiency and flowering time (McKay et al. 2003). This may reflect an adaptive axis associated with the evolution of drought-escaping spring annuals and drought tolerant winter annuals (Lovell et al. 2013; Kenny et al. 2014). Pleiotropic effects at a number of flowering time candidates (e.g., FRI and FLC) have been implicated as driving these tradeoffs (Mendez-Vigo et al. 2011; Lovell et al. 2013). These physiological correlates of life history have not been studied directly in *B. distachyon* but are not uncommon in related C3 grain crops. For example, quantitative trait loci (QTL) for carbon isotope discrimination (an integrated measure of water-use efficiency in C3 plants) were associated with wheat heading date and plant height suggesting a possible pleiotropic relationship (Rebetzke et al. 2008). A better understanding of the genetic basis of these physiological tradeoffs, if they exist in *Brachypodium*, may give insights into constraints to drought adaptation in natural populations and facilitate efforts to improve crop tolerance to abiotic stresses (Juenger 2013).

While the annual *B. distachyon* has received the majority of attention from the research community, most *Brachypodium* species exhibit a perennial life history strategy. Perenniality is thought to evolve primarily as a compromise between allocation tradeoffs, survivorship risks, and subsequent fitness benefits (Bradshaw and Putwain 1972; Partridge and Harvey 1988). For example, plants can allocate limited resources to storage, maintenance, defense, growth and reproduction. Plants that allocate more resources to maintenance or storage, for example in belowground structures or meristems, are more likely to persist through extended periods of cold and drought and exhibit perennial-like life histories. These plants forgo current reproduction for increased survivorship but may accrue fitness benefits from reproduction in future growing seasons. In contrast, annual plants partition resources to maximize current reproduction at the expense of lower survivorship and future reproduction. As such, annual and perennial plants are also expected to have very different strategies for coping with the environment and abiotic stresses. The most obvious of these relates to annual variation in temperature: perennial species must cope with winter cold, whereas annuals likely only experience cold as seed if they germinate prior to the final winter frost, or if they persist until the first frost. Moreover, perennial plants are much more likely to shed leaf material and persist in a dormant state through extended periods of drought or cold. Unlike many short-lived annual species, perennial grasses utilize reserve meristems as perennating structures that likely facilitate regrowth following damage from abiotic stress. Differences among the perennial species of *Brachypodium* may be interesting in this regard (Schippmann 1991; Catalan and Olmstead 2000). *B. arbuscula* and *B. retusum* produce branched stems that may persist for many years. In contrast, the other perennial species (e.g. *B. sylvaticum*, *B. pinnatum*, and others) produce unbranched tillers that last for only 1–2 years. The perennials also differ in their investment in rhizomatous tissues. Most perennial species are capable of producing long rhizomes, whereas *B. mexicanum* produces much shorter rhizomes. Finally, perennial plants may exhibit considerable legacy effects across growing seasons associated with abiotic stress. This phenomenon is commonly observed in grassland communities, where poor precipitation in an earlier season results in meristem limitation in subsequent seasons and resulting lags in productivity. Comparative studies of annual and perennial *Brachypodium* may begin to provide insight into the mechanisms allowing plant persistence especially under stressful environmental conditions, and provide possible avenues for the development of hardy abiotic stress tolerance perennial grain and biofuel crops (Glover et al. 2010).

Our extended discussion of the interplay of life history and ecophysiology highlights the fact our poor understanding of the ecology of naturally occurring *Brachypodium* populations represents a significant impediment to interpreting plant function in the context of the field environment. This information is critical as it provides context for the abiotic stress and selective regimes experienced by plants in nature and because it may help to interpret the adaptive significance of trait variation observed in realistic stress tolerance experiments. This gap can hopefully be addressed with future field demographic studies, the collection of new plant material across the natural geographical range of *B. distachyon,* and through careful curation of existing herbarium specimens.

Plant-Water Relations

Herbaceous plants have evolved a number of strategies or syndromes for surviving in the often harsh and unpredictable abiotic environment (Ludlow 1989). Under short-term non-lethal stress, many plants essentially cease photosynthesis and therefore growth. More moderate abiotic stress, including stresses that arise gradually, may allow for plants to acclimate to some degree. Considering first stress induced by soil drying, eco-physiologists recognize three very general non-exclusive syndromes of response. The first syndrome, dehydration avoidance, encompasses a suite of traits which plants deploy to prevent drying environments from resulting in dehydration of tissues and cells. Some features of dehydration avoidance observed in herbaceous plants include increased investment and growth in roots, increased water use efficiency, changes in cellular osmotic potential, and changes in cell wall characteristics. A second general syndrome is dehydration tolerance. These strategies are often deployed when reduced cellular water content becomes unavoidable, and include physiological responses such as the synthesis of proteins which stabilize cellular components to prevent damage or facilitate recovering from stress following water deficit. There are many parallels between dehydration tolerance and acclimation to severe cold, discussed below. Finally, herbaceous plants often rely on drought escape, which may include either genetically programmed early flowering (e.g. before an annually predictable end-of-season drying or terminal drought) or the induction of flowering in response to soil drying. Recent work points to a genetic basis of this latter aspect of drought escape (Han et al. 2013; Schmalenbach et al. 2014), though evidence for its frequency in natural populations is largely anecdotal. We will consider general aspects of plant response to other stressors in the following sections.

Ecological Context

Most of our limited understanding of *Brachypodium* performance in the field arises from work done in *B. distachyon*. Complicating matters, a recent comprehensive taxonomic reassessment of *B. distachyon* demonstrated that diverse natural collections previously described as ploidy variants of *B. distachyon* actually represent three separate taxonomic entities: *B. distachyon*, *B. stacei*, and their allotetraploid hybrid, *B. hybridum* (Catalan et al. 2012). (See the chapter by Catalan in this volume for more on these considerations.) For our purposes, this taxonomic realignment means that those studies on *B. distachyon* published prior to 2012 which did not explicitly study known genotypes developed by the community cannot provide a clear picture of the specific ecology of *B. distachyon*. With that caveat, several studies point to the generally hardy nature of *B. distachyon sensu lato* under challenging environmental conditions. A study in a California grassland, representing the non-native range of *Brachypodium*, found that *B. hybridum* outcompeted resident native annual grasses during several years of drought (Armstrong and Huenneke 1993; note that these authors refer to the plants as "*B. distachyon*," though the plants were most likely *B. hybridum* as *B. distachyon sensu stricto* is unknown from California). This advantage increased over successive years of the drought, which the authors attributed to the larger seed size of *B. hybridum* facilitating rapid root growth early in the season. Rapid root establishment represents a drought avoidance strategy, allowing plants to outcompete neighbors for limited and dwindling soil moisture.

Natural populations of *B. distachyon* may show variation in drought escape strategy, as well. Aronson and colleagues performed a manipulative experiment contrasting the constitutive flowering times and drought-induced flowering response of desert (short season) vs Mediterranean-climate (longer season) Israeli populations of *B. distachyon* (Aronson et al. 1992). These authors found that, in a common garden, plants from the desert population flowered more rapidly than did Mediterranean plants regardless of the imposed watering regime. This result suggests that desert plants were genetically programmed to progress through their life cycle quickly, perhaps as an adaptation to flower before a predictable terminal drought. Strikingly, the authors also found differences in flowering response to soil water content. Mediterranean populations bolted and produced their first anthers more rapidly when soil moisture was limited, as compared to well-watered conditions. In contrast, desert plants show little drought response in the timing of bolting but significantly accelerated their progression through subsequent reproductive transitions (anthesis, seed set, and senescence). *B. distachyon sensu stricto* is not described from Israel, and so it is very likely that these populations represent *B. hybridum* and/or *B. stacei*. Future work may reveal that these genetic differences in flowering strategies may reflect divergence between the two species, or the hybrid nature of *B. hybridum*.

The role of hybridization in *B. distachyon* water relations was more directly assessed by Manzaneda and colleagues investigating Spanish populations of *B. distachyon* grown in the greenhouse (Manzaneda et al. 2012). By assessing integrated lifetime water use efficiency (WUE) of diverse natural genotypes, the authors found that allotetraploid *B. distachyon* (which we now know represents *B. hybridum*; see above) exhibited higher WUE and was generally found in more arid locations in Spain as compared to diploid *B. distachyon s.s.* Recent work assessing predicted ecological niches (using Ecological Niche Modeling) of the three species has suggested some ecological differentiation among the species

(López-Alvarez et al. 2015). Reinforcing the findings of Manzaneda and colleagues, ENM predicts that *B. distachyon* is found at higher elevations, experiencing cooler temperatures with more moisture than the comparatively lower-elevation, warmer, drier-adapted *B. stacei*. Allotetraploid *B. hybridum* is found to be largely intermediate in its climate preference compared to its two parental species. These patterns are particularly apparent on the Iberia peninsula, where all three species are found.

Additional ecological context for *B. distachyon* has been provided by two studies linking variation in DNA markers with composite indices of climate variation. These climate Genome Wide Association Studies (GWAS) offer a powerful way to infer putative functional relationships between natural variation and complex environmental parameters. Working with populations of Spanish *B. distachyon*, *B. hybridum*, and *B. stacei*, Hammami et al. identified fairly strong genetic clustering by species, and some correspondence between species and major axes explaining climate variation (Hammami et al. 2014). Dell'Acqua and colleagues extended this line of analysis by using next-generation sequencing on a smaller number of *B. distachyon* populations from Turkey, and performing both GWAS and genome scans for evidence of recent directional selection (Dell'Acqua et al. 2014). The higher genomic resolution afforded by this approach allowed the researchers to identify specific genes that were closely linked to climate-associated SNPs. The small sample size and deliberately conservative statistical analysis used in this analysis restricted power to some extent but the authors were able to identify a handful of genes putatively involved in local adaptation to climate. These included a MYB transcription factor, several other signaling proteins, and an enzyme hypothesized to play a role in wax synthesis (an important component of cuticles, and thus water retention in dry environments).

Developmental and Physiological Patterns of Response to Soil Drying

We currently lack a comprehensive understanding of how *B. distachyon*, and *Brachypodium* more generally, responds to soil drying. The extensive drought studies with wheat may illuminate key traits likely involved in drought adaptation in related C3 grasses like *Brachypodium*. Lack of progress on this key issue arises from an absence of manipulative experiments in the field environment, and from the lack of a consensus on how best to apply drying treatments in growth chambers or greenhouses. Moving forward, it is essential that investigators consider the confounding effects of light level, day length, relative humidity, growing media (including nutrient availability), and temperature on their experimental manipulation of water availability. Withholding water for 10 days from plants grown in a growth chamber set to conditions favorable for *Arabidopsis thaliana* will give dramatically different results than withholding water from plants grown in full sunlight and spring temperatures representative of, for example, Mosul, Iraq (the collection locale for the reference Bd21 accession). The severity of stress, in terms of the rate of onset and the target soil water potential, will also have a considerable influence on experimental results, as will the life stage at which a stress is applied.

Natural accessions of *B. distachyon* show a range of responses to soil drying. Luo and colleagues withheld water from 57 soil-grown accessions until the wilting point, and assessed a range of developmental and physiological characters (Luo et al. 2011). Most accessions reached the wilting point in 6–7 days under the temperature and light conditions used. Generally, the dry-treatment plants had reduced leaf water content and reduced chlorophyll fluorescence (indicating reduced components of photosynthetic efficiency) and produced fewer leaves and tillers as compared to control plants. These observed effects of soil drying are typical of many small herbaceous plants (Chaves 2002). The authors characterized the accessions on a tolerance scale of their devising, finding that the BdTR2b, Bd3-1, and Bd18-1 were the most susceptible to soil drying and that B1-1, Tek-9 and Tek-10 were the least susceptible. Accessions along this continuum were most obviously differentiated by the extent to which leaf water content and chlorophyll fluorescence were affected by the soil drying. Interpreting results in relation to dehydration tolerance from this type of uncontrolled terminal drying experiment is difficult owing to interrelated functional effects of transpiration, photosynthetic rate, turgor pressure, and root:shoot biomass ratios. For example, the plants that wilted more quickly may simply have been larger plants that more rapidly exhausted soil moisture in a small pot; a supervised controlled dry-down experiment can control for these issues to some extent. We will return to the important issues of experimental design and ecophysiological interpretation at the end of this chapter.

The complex ways in which soil drying can affect plant tissues was carefully dissected by Verelst et al. (2013). Working with the reference accession of *B. distachyon*, Bd21, and using detailed understanding of leaf development from Arabidopsis and cereal crops, the authors distinguished among regions of the leaf in which cells are proliferating (e.g. undergoing mitosis), expanding, or mature. Leaves from plants grown under either moderate or severe soil drying were significantly shorter than control plants. Strikingly, the number of cells in stressed leaves was statistically indistinguishable from control leaves. Careful analysis of leaf development revealed that the reduction in leaf length without change in cell number was due to unabated cell division in the proliferation zone but greatly reduced cell expansion in the region of the leaf immediately distal to the meristematic tissue. The authors note that this result contrasts with data from diverse plant species—including *A. thaliana*, wheat, barley, maize and rice—showing both reduced cell division and reduced expansion under soil drying conditions. We will return to this study below when we consider molecular aspects of *Brachypodium* responses to soil drying.

As described earlier, roots play a key role in plant adaptation and acclimation to soil water availability. Chochois et al. recently described soil-drying response of 16 natural accessions of *B. distachyon*, and found substantial between-accession differences in both root morphology and biomass (Chochois et al. 2015). These authors found that most measured root traits have very high heritability, and emphasize the important implications of a strong genetic basis for improving agriculturally-relevant traits. Under soil drying, most accessions increased investment in the primary seminal root (the root which arises from the seed during germination and forms the deep-penetrating tissue in this species), at the expense of roots arising from the coleoptile and leaf nodes. The extent to which this occurred, and the relative biomass allocation to different root types, varied considerably among accessions.

Molecular Responses of *Brachypodium* to Soil Drying

Researchers have utilized three inter-related avenues of research to explore the molecular aspects of *Brachypodium* response to soil drying. The first body of work builds on decades of work in reference plant species such as Arabidopsis, rice, and maize, and has identified apparent homologs of proteins demonstrated in these divergent species to play an important role in signaling responses to imposed stress. Perhaps unsurprisingly, the reference *B. distachyon* contains dozens of members of gene families known from other species to play an important role in environmental signaling. These families include the AREB/ABF, DREB, and WRKY families of environmentally-responsive transcription factors, as well as gene families more directly tied to regulation of growth (Mochida et al. 2011; Tripathi et al. 2012). Such *in silico* analyses are an excellent starting point to test hypotheses about conservation of gene function and the role of gene family member turnover in the evolution of environmental response, and these analyses also provide provisionary annotations essential for global analyses of gene expression.

A second avenue of research into the molecular responses of *Brachypodium* is to estimate directly global transcriptional responses to imposed soil drying (or a proxy). Building on the widely-cited and influential AtGenExpress study in *A. thaliana* (Kilian et al. 2007), which used a common Affymetrix microarray platform to assess transcriptional responses to a series of standardized environmental stressors in the lab, Priest et al. designed a microarray based on the Bd21 version 1.0 genome assembly (Priest et al. 2014). The Kilian dataset for *A. thaliana* has been hugely influential in subsequent studies of environmentally-responsive gene expression and was used extensively to annotate gene functions in the *A. thaliana* genome. (By extension, these annotations have been propagated across many additional plant genomes). Priest and colleagues used their microarray (BradiAR1b520742) to measure transcriptional response of Bd21 plants to a battery of stressors across a time series, including benchtop desiccation as a proxy for vegetative dehydration response. The resulting dataset is massive and several interesting patterns emerged. Desiccation stress results in the upregulation of ~1000 genes, with fairly little change in the magnitude of response over the sampled period of 1–24 h. By contrast, ~1000 genes are downregulated early during desiccation and the number of downregulated genes grows substantially over time until the transcription of nearly 5000 genes are affected by 24 h. This latter result may be a simple effect of the plants shutting down most cellular processes as the leaves lose turgor from this very severe imposed stress. Indeed, function enrichment analyses of these downregulated genes indicated that many genes involved in cell cycle, DNA replication, and cell growth were turned off in their drying treatment. This transcriptional signature of cellular processes contrasts strongly with the results of Verelst et al. (2013), discussed above, finding unimpeded cell proliferation under moderate soil drying.

Priest et al. further looked for broad patterns in transcriptional network structure by clustering genes with similar transcriptional responses (e.g. upregulated over time during desiccation, but downregulated under elevated temperature or salinity) and then performing enrichment analysis on gene ontology and promoter sequence motifs. Such networks of co-regulated genes are excellent hypothesis generators that can narrow subsequent searches for key regulatory components such as transcription factors and sequence motifs associated with the regulation of target genes. The ABRE sequence motif, variants of which are frequently found to be associated with drought stress-responsive genes in many plant species (Shinozaki and Yamaguchi-Shinozaki 2007), was identified in this type of analysis. Interestingly, two groups of genes were downregulated under desiccation yet showed no overall enrichment for known cellular functions, despite the observation that most members of each group were associated with at least one annotated function. However, collectively, these groups were strongly enriched for several short sequence motifs in their promoter regions, suggesting a mechanism for their co-regulation even though the cellular functions of this coordinated regulation are not yet known.

Measuring the transcriptional response of diverse genotypes to environmental perturbation is another powerful tool for identifying the molecular genetic mechanisms of response and of cross-environment genetic correlations (Des Marais et al. 2013). A recent study of transcriptional responses to gradual soil drying across six natural accessions of *B. distachyon* revealed substantial among-accession diversity in response, or Genotype by Environment interaction (GxE) (Gordon et al. 2014). This analysis, which used mRNASeq to measure transcript abundance of ~15,000 genes, found 870 genes showed consistent treatment response across the six accessions while 300 genes responded to the treatment in a accession-specific manner (e.g. upregulated in one accession but downregulated in others). Multifactor experiments such as this are powerful because they are able to identify both genes that respond in a similar manner across diverse natural accessions, which possibly represent the conserved adaptive strategy of a species, and because they identify genes that respond differently in accessions. These latter genes, showing expression GxE, are of interest because they may reflect genes which are transcriptionally non-responsive under stressful conditions in some natural populations (e.g. because expressing the gene incurs a selective cost) but which are responsive to that stress in other populations (e.g. because NOT expressing them incurs a greater cost). As such, natural genetic diversity in transcriptional response can serve as the source material for the evolution of inducible or plastic response phenotypes and may underlie local adaptation to climate (Lasky et al. 2014; Marchand et al. 2014). Studying expression GxE can also be a powerful means to make functional inferences about the molecular control of ecophysiological and developmental traits (Des Marais et al. 2012; Rengel et al. 2012).

The preceding studies focused exclusively on transcription levels of protein-coding genes, but increasing evidence suggests that small RNA species respond to soil drying as well. As a companion to the study of leaf developmental responses to drying, discussed earlier (Verelst et al. 2013), Bertolini and colleagues used next-generation sequencing technologies to characterize the complement of miRNA species in *B. distachyon* (Bertolini et al. 2013). These authors isolated both miRNA species previously described in other plant species as well as several miRNAs which appear to be unique to *B. distachyon*. Thirty-two miRNAs had distinct expression profiles when contrasting proliferating and expanding cells, and four miRNAs were significantly differentially expressed by soil drying in proliferating cells. Many of the miRNAs identified by Bertolini et al. using short-read sequencing were also identified by Budak and Akpinar (2011) using a microarray constructed of phylogenetically-conserved miRNA species.

Finally, conventional forward and reverse genetic approaches can be used to explore *Brachypodium* molecular responses to soil drying. We are aware of only a single study to date that has identified and validated a single gene effect for response to soil drying. Sun et al. assessed the functional effects of a WRKY transcription factor, BdWRKY36, and concluded that it was a positive regulator of drought response (Sun et al. 2014). BdWRKY36 appears to be constitutively expressed in all tissues, and is upregulated when leaves are exposed to benchtop desiccation, as well as at low temperature (4 °C) and high temperature (40 °C). The authors overexpressed BdWRKY36 in tobacco and found that, under soil drying as compared to control condition, this construct conferred increased leaf relative water content and chlorophyll abundance, decreased ion leakage and accumulation of reactive oxygen species (ROS), and increased expression of several well-known genes encoding ROS-scavenging proteins. The development of *B. distachyon* mutant populations through T-DNA insertion (Bragg et al. 2012) and TILLING (Dalmais et al. 2013) and will facilitate future efforts to discover genes involved in drought responses using forward genetic screens.

Brachypodium Response to Cold

Many populations of annual plants like *B. distachyon* experience severe cold, such as during season-ending frost, or during unusually cold temperatures in the early Spring. *Brachypodium* perennial species certainly experience severe cold, though we are unaware of any research addressing cold tolerance or acclimation response in these species. Annual plants which germinate in the Autumn use extended periods of cold as one of several vernalization cues to transition from vegetative to reproductive growth. The body of research dealing with *B. distachyon* flowering is reviewed elsewhere in this volume. We focus here on research into the acclimation of *B. distachyon* to low temperatures.

Plants challenged with low temperatures often show two inter-related responses. First, they must survive the initial period of cold. Second, this initial cold can also stimulate an acclimation response that allows the plant to survive subsequent, often more severe, cold. Colton-Gagnon and colleagues assessed both of these aspects of cold response in seven natural accessions of *B. distachyon* (Colton-Gagnon et al. 2014). Four of these accessions were previously known to exhibit a "spring-annual" life history: these plants were competent to transition from vegetative to reproductive growth without a low temperature cue. The remaining three accessions exhibit typical "winter-annual" strategy: they require an extended cold cue to stimulate the transition to flowering. (Recall that the life history of these accessions has not been evaluated empirically in their native habitats.) These authors found that the seven accessions had largely similar non-acclimated cold tolerance, with 50 % mortality (LT50) at around −8 °C. Acclimation via 28 days at 4 °C reduced LT50 to around −12 °C, again with no statistical difference in survival between "spring-annual" accessions (which might be inferred to experience cold temperatures infrequently) and "winter-annual" accessions (which presumably would experience cold temperatures). Cold acclimation in all accessions involved synthesis of fructans, free cellular Proline, and water soluble sugars. Accumulation of these molecules is frequently observed in response to low temperature and is generally believed to play a role in protecting cellular components from freezing damage. Fructan synthesis under cold and drought conditions is a key feature of Pooid grasses and may serve to provide a ready source of cellular energy or to stabilize cell membranes during osmotic stress (Hincha et al. 2000). Colton-Gagnon et al. caution that free cellular Proline concentration alone is a poor predictor of *Brachypodium* cold tolerance as the accessions studied showed wide variation in Proline abundance but little variation in cold tolerance. These authors conclude by arguing that the constitutive cold tolerance of *B. distachyon* "spring-annual" accessions does not match the general definition of temperate cereal species spring accessions, and argue instead that these accessions should be classified as having a "facultative" life history strategy.

Li et al. paired bioinformatic analysis of the Bd21 genome with comparative physiology and provide a nice evolutionary perspective on the cold tolerance of *B. distachyon* in relation to other members of the Pooideae (Li et al. 2012). They focus on genes encoding three groups of proteins believed to be core determinants of cold tolerance in Pooid grasses. The Ice Recrystallization Inhibition Proteins (IRIP) minimize cell damage by limiting ice crystal growth in the apoplast (Sidebottom et al. 2000). Based on their gene tree analysis, Li et al. conclude that IRIP proteins evolved before the divergence of *Brachypodium* from other Pooid grasses, but that independent radiations of these gene families occurred in the two lineages. C-repeat Binding Factor (CBF) transcriptional regulators are widely distributed in plants and regulate plant-wide response to low temperatures (Fowler et al. 2007). Interestingly, Li et al. find that an entire sub-family of CBFs—CBF4—is missing from *B. distachyon*, suggesting that this group of transcription factors evolved subsequent to the divergence of *Brachypodium*. A third group of proteins believed to play a central role in Pooid cold tolerance are the Fructan Synthases (FST). FST genes are involved in the conversion of sucrose into fructose, which is polymerized to form fructans (see above). Li et al. show that *B. distachyon* has FST-like proteins, but that these proteins lack key domains associated with fructan synthase activity in core Pooids. Li et al. show that *B. distachyon* does accumulate fructans during cold acclimation, reinforcing the findings of Colton-Gagnon and colleagues, but that they accumulate fructans to a lesser extent than what is observed in core Pooid species.

The microarray study by Priest et al. (2014), described above in the context of transcriptional responses to soil drying, also assayed *B. distachyon* response to moderate cold treatment. Compared to the strong transcriptomic response observed under drying (~5000 genes downregulated after 24 h of desiccation), the response to 24 h of 4 °C treatment was quite moderate and dominated by the transcriptional induction of just a few hundred genes. This moderate response likely reflects the constitutive tolerance of *B. distachyon* to mild cold observed by Colton-Gagnon et al. (2014), described above. Priest and colleagues identified 63 genes which were upregulated by cold treatment over time but showed very little transcriptional response to their other treatments (desiccation, high temperature, and salt). These 63 genes were enriched for functions associated with transcriptional regulation, RNA metabolic processes, and regulation more generally; these genes might be interesting candidates to target for artificial selection as they may have limited pleiotropic effects in other stressful environments.

Brachypodium Responses to Elevated Temperatures

Relatively little is known about the molecular responses of *B. distachyon* or its congeners to stress due to elevated temperature. The proximate cause of elevated heat stress is the reduced efficiency—and biophysical instability at extreme temperatures—of cellular structures and processes. At the molecular level, this phenomenon has been studied in the greatest detail in plants as it relates to the photosynthetic apparatus (Allakhverdiev et al. 2008). Plants cope with elevated temperature stress on short time scales via biosynthesis of protective molecules (e.g. heat shock proteins) and via increased transpiration as a means of evaporative cooling. The efficacy of this latter response may be compromised if the temperature stress is accompanied by soil drying, which is often observed in the field environment.

Boden and colleagues assessed the response of Bd21 to slightly elevated temperature (27 °C), focusing particularly on the differential effects of this treatment when applied at the vegetative or reproductive stage (Boden et al. 2013). *A. thaliana* plants grown at 27 °C transition to flowering without the normally requisite long day cue but this pattern is not seen in *B. distachyon*. Moreover, *B. distachyon* plants grown at 27 °C do not transition to flowering earlier than plants grown at 22 °C. These two observations suggest that 27 °C is not perceived as a stressful condition by the Bd21 accession, though the authors also find that the seeds of plants grown at 27 °C fill for a shorter duration and therefore weigh less at maturity than as compared to seeds of plants grown at 22 °C. Prior research in *A. thaliana* demonstrated a temperature-dependent effect of the occupancy of H2A.Z histones near the Transcriptional Start Sites (TSS) of genes, which inhibits transcription of those genes (Kumar and Wigge 2010). Increased temperature reduces H2A.Z occupancy, and thereby facilitates genome-wide transcriptional changes. The authors identified an apparent Bd21 ortholog of gene encoding an *A. thaliana* H2A.Z histone, *BdHTA9*, finding that its occupancy at the TSS of several candidate genes is not affected by elevated temperature in seedlings (i.e. vegetative tissue) but is affected in developing grains. This temperature effect in grains is not due to differences in transcription of *BdHTA9* itself. The authors surmise that H2A.Z histones may therefore account for some of the difference in transcriptional activity, and thereby development, when contrasting vegetative and reproductive tissues under elevated temperature. Their work also reiterates the important point that environmental stress is perceived differently by different parts of the plant.

The microarray study by Priest et al. (2014), also assayed *B. distachyon* response to high temperature treatment (24 h at 42 °C, a temperature which Bd21 plants may well have occasionally experienced in their native environment in northern Iraq; see Fig. 1). Here, the observed response was primarily transcript downregulation during the course of treatment. Just a few hundred genes were upregulated—dominated by heat shock proteins, genes associated with protein folding, and the dubiously helpful GO term "response to heat"—while over 1000 genes were downregulated after 12 h at 42 °C. Many of the downregulated genes reflected a slowing of plant growth at this high temperature: genes associated with the cell cycle, DNA replication and cell-wall biogenesis. In this latter respect, the high stress treatment imposed by Priest et al. had functionally similar effects to their desiccation treatment and may indicate that plants were conserving resources with the aim to resume growth following cessation of stress.

Interactions Between Environmental Stressors

Two outstanding questions in plant breeding and evolutionary genetics are how plants optimize fitness in multiple environments and how plants respond to combinations of stressful conditions such as heat and low soil water content, or abiotic and biotic stressors. A straightforward way to conceptualize this first issue is to ask whether a genetic locus affecting performance in one environment (e.g. drought) has pleiotropic effects in a second environment (e.g. high temperature). We recently reviewed the empirical literature on pleiotropy across environments, focusing on QTL studies in wild and crop species (Des Marais et al. 2013). We found that the preponderance of pleiotropic loci had much larger additive effects in one environment than the other, or had a significant additive effect in only one environment. This pattern of trait expression is of interest because it suggests that fitness trade-offs between environments may be minimal for these naturally-segregating alleles, and is in contrast to the yield trade-offs often observed in studies of loss-of-function mutations or transgenic efforts to improve crop performance (Hall and Richards 2013).

At the molecular level, we can investigate mechanisms of cross-environment correlations by assessing whether different stressors invoke similar cellular responses. This has been explored in *B. distachyon* by the Priest et al. microarray study, discussed earlier (Priest et al. 2014). These authors found considerable overlap in the genes whose expression was induced under desiccation and under salt stress. Although these genes did not show any significant functional enrichments the authors did identify a number of enriched sequence motifs in their promoters, suggesting a common regulatory control for response to these two classic osmotic stressors.

A final, understudied, aspect of plant-abiotic environment interactions is the molecular, developmental, and physiological responses of plants to multiple stressors encountered at the same time (Rizhsky et al. 2004; Makumburage and Stapleton 2011; Prasch and Sonnewald 2013). While the field environment very often presents plants with complex challenges, such as high temperature in conjunction with low soil water content, or abiotic stress in conjunction with biotic stress, most molecular and physiological studies focus on just one imposed stress treatment at a time. *B. distachyon* is an excellent model system in which to address the mechanisms of combined stress. Its small stature facilitates the large, highly replicated, multi-factor experimental manipulations necessary for identifying the genetic basis of response to stressors encountered independently and in combination. We are not aware of any published studies looking at combined stressors in *Brachypodium*.

Moving Forward: *Brachypodium* as a Reference Genus for Plant-Abiotic Interactions

There is a growing need to improve crop productivity and to better understand plant adaptation in natural populations in the face of abiotic stress. In particular, changes in precipitation and temperature regimes driven by climate change are a potential threat to future crop productivity and natural ecosystems worldwide (Wheeler and von Braun 2013; Pacifici et al. 2015). *Brachypodium* is an emerging model system with extensive genetic, genomic, and natural variation resources that can be brought to bear on these issues (see other chapters in this volume). In particular, it is likely that natural populations of *Brachypodium* have evolved novel mechanisms of stress avoidance and tolerance that could be leveraged for improvement of several grain crops. A better understanding of natural variation in abiotic stress in *Brachypodium*, especially at the molecular level, provides a unique opportunity for gaining insights into local adaptation and as well as novel avenues for improving crops. Here we outline some important summary points and future opportunities in studies of abiotic stress with *Brachypodium*:

1. Researchers should strive to impose realistic abiotic stresses to facilitate the discovery of adaptive responses and the interpretation of physiological and molecular mechanisms. Experimental design should carefully consider the timing and duration of stress treatment, the stage of plant development when imposed, and the possible interactions and indirect effects of manipulations (e.g., nutrient limitation or increased leaf temperature associated with soil drying). Matching stress assays with existing and well-established protocols for grain crops may facilitate the transfer of insights to important agricultural systems (Fischer 2011).
2. Plants experience the environment as a complex milieu of spatially and temporally varying conditions. Plants are remarkable in their ability to acclimate and physiologically respond to this complex, changing, and heterogeneous environment. Nevertheless, the vast majority of abiotic stress studies center on the manipulation of single environment factors across few levels of variation. We anticipate that combinations of stresses may induce signaling and physiological responses that are unique, that provide cross-protection, and that are especially relevant to crop improvement (Mittler and Blumwald 2010). We believe studies targeting multiple abiotic stresses in combination will be especially insightful for developing models of physiological functions and elucidating key molecular pathways.
3. We know surprisingly little about *Brachypodium* from natural populations and in a field setting. Additional information from field natural history observations, field experiments, and the continued collection of new germplasm will be key to the success of *Brachypodium* as an emerging model grass. In particular, life history and phenological data are critical needs as they determine the matching of environmental conditions with developmental transitions, the timing of stressful environments, and anticipated strategies for ecological adaptation.
4. Plant responses to abiotic stress often involve tradeoffs—traits that increase fitness or yield under stress come at a cost to performance in benign conditions (Des Marais et al. 2013). These pleiotropic relationships are a major limitation to traditional breeding programs and for adaptation in the wild. A central challenge in plant biology is to identify genetic constraints and through an understanding of their mechanism manipulate them for crop improvement. The rapid development of quantitative genetic and genomic approaches provide immense opportunity for teasing apart these mechanisms and for manipulating the costs and benefits of key abiotic stress tolerance traits.

Acknowledgements Thanks to Jesse Lasky for generating Fig. 1. T.E.J. is supported by the DOE Office of Science, Office of Biological and Environmental Research (BER), grant no. DE-SC0008451 and NSF funding through the Plant Genome Research Program (IOS-0922457).

References

Allakhverdiev SI, Kreslavski VD, Klimov VV, Los DA, Carpentier R, Mohanty P. Heat stress: an overview of molecular responses in photosynthesis. Photosynth Res. 2008;98(1–3):541–50.
Armstrong JK, Huenneke LF. Spatial and temporal variation in species composition in California grasslands: the interaction of drought and substratum. The vegetation of ultramafic (Serpentine) soils: Proceedings of the First International Conference on Serpentine Ecology. Andover, Hampshire: Intercept Ltd; 1993.
Aronson J, Kigel J, Shmida A, Klein J. Adaptive phenology of desert and Mediterranean populations of annual plants grown with and without water stress. Oecologia. 1992;89:17–26.
Bertolini E, Verelst W, Horner DS, Gianfranceschi L, Piccolo V, Inze D, et al. Addressing the role of microRNAs in reprogramming leaf growth during drought stress in Brachypodium distachyon. Mol Plant. 2013;6(2):423–43.
Boden SA, Kavanova M, Finnegan EJ, Wigge PA. Thermal stress effects on grain yield in Brachypodium distachyon occur via H2A.Z-nucleosomes. Genome Biol. 2013;14(6):R65.
Bradshaw AD, Putwain PD. Life history variation in *Poa annua*. Evolution. 1972;31:233–46.
Bragg JN, Wu J, Gordon SP, Guttman ME, Thilmony R, Lazo GR, et al. Generation and characterization of the Western Regional Research Center Brachypodium T-DNA insertional mutant collection. PLoS One. 2012;7(9):e41916.
Brkljacic J, Grotewold E, Scholl R, Mockler T, Garvin DF, Vain P, et al. Brachypodium as a model for the grasses: today and the future. Plant Physiol. 2011;157(1):3–13.
Budak H, Akpinar A. Dehydration stress-responsive miRNA in Brachypodium distachyon: evident by genome-wide screening of microRNAs expression. OMICS. 2011;15(11):791–9.
Catalan P, Muller J, Hasterok R, Jenkins G, Mur LA, Langdon T, et al. Evolution and taxonomic split of the model grass Brachypodium distachyon. Ann Bot. 2012;109(2):385–405.
Catalan P, Olmstead RG. Phylogenetic reconstruction of the genus *Brachypodium* P. Beauv (Poaceae) from combined sequences of chloroplast *ndh*F and nuclear ITS. Plant Syst Evol. 2000;220:1–19.
Chaves MM. How plants cope with water stress in the field? Photosynthesis and growth. Ann Bot. 2002;89(7):907–16.
Chochois V, Vogel J, Rebetzke GJ, Watt M. Variation in adult plant phenotypes and partitioning among seed and stem-borne roots across *Brachypodium distachyon* accessions to exploit in breeding cereals for well-watered and drought environments. Plant Physiol. 2015;168 (3):953–67.
Colton-Gagnon K, Ali-Benali MA, Mayer BF, Dionne R, Bertrand A, Do Carmo S, et al. Comparative analysis of the cold acclimation and freezing tolerance capacities of seven diploid Brachypodium distachyon accessions. Ann Bot. 2014;113(4):681–93.
Dalmais M, Antelme S, Ho-Yue-Kuang S, Wang Y, Darracq O, d'Yvoire MB, et al. A TILLING platform for functional genomics in *Brachypodium distachyon*. PLoS One. 2013;8(6):e65503.
Dell'Acqua M, Zuccolo A, Tuna M, Gianfranceschi L, Pe ME. Targeting environmental adaptation in the monocot model *Brachypodium distachyon*: a multi-faceted approach. BMC Genomics. 2014;15:801.
Des Marais DL, Hernandez KH, Juenger TE. Genotype-by-environment interaction and plasticity: exploring genomic responses of plants to the abiotic environment. Annu Rev Ecol Evol Syst. 2013;44:5–29.
Des Marais DL, McKay JK, Richards JH, Sen S, Wayne T, Juenger TE. Physiological genomics of response to soil drying in diverse Arabidopsis accessions. Plant Cell. 2012;24(3):893–914.
Fischer RA. Wheat physiology: a review of recent developments. Crop Pasture Sci. 2011;62:95–114.
Fowler S, Cook D, Thomashow MF. The CBF cold-response pathway. In: Jenks MA, Hasegawa PM, editors. Plant abiotic stress. Chichester: Wiley-Blackwell; 2007. p. 71–99.
Glover JD, Reganold JP, Bell LW, Borevitz J, Brummer EC, Buckler ES, et al. Agriculture. Increased food and ecosystem security via perennial grains. Science. 2010;328(5986):1638–9.

Gordon S, Priest H, Des Marais DL, Schackwitz W, Figueroa M, Martin J, et al. Genome diversity in *Brachypodium distachyon*: Deep sequencing of highly diverse inbred lines. Plant J. 2014;79:361–74.

Hall AJ, Richards RA. Prognosis for genetic improvement of yield potential and water-limited yield of major grain crops. Field Crop Res. 2013;143:18–33.

Hammami R, Jouve N, Soler C, Frieiro E, González JM. Genetic diversity of SSR and ISSR markers in wild populations of *Brachypodium distachyon* and its close relatives *B. stacei* and *B. hybridum* (Poaceae). Plant Syst Evol. 2014;300(9):2029–40.

Han Y, Zhang X, Wang Y, Ming F. The suppression of WRKY44 by GIGANTEA-miR172 pathway is involved in drought response of Arabidopsis thaliana. PLoS One. 2013;8(11): e73541.

Hijmans RJ, Cameron SE, Parra JL, Jones PG, Jarvis A. Very high resolution interpolated climate surfaces for global land areas. Int J Climatol. 2005;25(15):1965–78.

Hincha DK, Hellwege EM, Heyer AG, Crowe JH. Plant fructans stabilize phosphatidylcholine liposomes during freeze-drying. Eur J Biochem. 2000;267:535–40.

Initiative TIB. Genome sequencing and analysis of the model grass Brachypodium distachyon. Nature. 2010;463:763–8.

Juenger TE. Natural variation and genetic constraints on drought tolerance. Curr Opin Plant Biol. 2013;16:274–81.

Kenny A, McKay J, Richards JH, Juenger T. Direct and indirect section on flowering time, water-use efficiency (deltaC13), and their plasticty to drought in Arabidopsis thaliana. Ecol Evol. 2014;43:4505–21.

Khan MA. Biosystematic studies in *Brachypodium* (Poaceae). Leicester: University of Leicester; 1984.

Kilian J, Whitehead D, Horak J, Wanke D, Weinl S, Batistic O, et al. The AtGenExpress global stress expression data set: protocols, evaluation and model data analysis of UV-B light, drought and cold stress responses. Plant J. 2007;50(2):347–63.

Kumar SV, Wigge PA. H2A.Z-containing nucleosomes mediate the thermosensory response in Arabidopsis. Cell. 2010;140:136–47.

Lasky JR, Des Marais DL, Lowry DB, Povolotskaya I, McKay J, Richards JH, et al. Natural variation in abiotic stress responsive gene expression and local adaptation to climate in *Arabidopsis thaliana*. Mol Biol Evol. 2014;31:2283–96.

Li C, Rudi H, Stockinger EJ, Cheng H, Cao M, Fox SE, et al. Comparative analyses reveal potential uses of *Brachypodium distachyon* as a model for cold stress responses in temperate grasses. BMC Plant Biol. 2012;12:65.

López-Alvarez D, Manzaneda AJ, Rey PJ, Giraldo P, Benavente E, Allainguillaume J, et al. Environmental niche variation and evolutionary diversification of the *Brachypodium distachyon* grass complex species in their native circum-Mediterranean range. Am J Bot. 2015;102(7):1073–88.

Lovell JT, Juenger TE, Michaels SD, Lasky JR, Platt A, Richards JH, et al. Pleiotropy of FRIGIDA enhances the potential for multivariate adaptation. Proc Biol Sci. 2013;280 (1763):20131043.

Ludlow MM. Strategies of response to water stress. In: Kreeb KH, Richter H, Hinckley TM, editors. Structural and functional responses to environmental stresses. The Hague: SPB Academic; 1989. p. 269–81.

Luo N, Liu J, Yu X, Jiang Y. Natural variation of drought response in *Brachypodium distachyon*. Physiol Plant. 2011;141:19–29.

Makumburage GB, Stapleton AE. Phenotype uniformity in combined-stress environments has a different genetic architecture than in single-stress treatments. Front Plant Sci. 2011;2:12.

Manzaneda AJ, Rey PJ, Bastida JM, Weiss-Lehman C, Raskin E, Mitchell-Olds T. Environmental aridity is associated with cytotype segregation and polyploidy occurrence in Brachypodium distachyon (Poaceae). New Phytol. 2012;193:797–805.

Marchand G, Huynh-Thu VA, Kane NC, Arribat S, Vares D, Rengel D, et al. Bridging physiological and evolutionary time-scales in a gene regulatory network. New Phytol. 2014;203 (2):685–96.

McKay JK, Richards JH, Mitchell-Olds T. Genetics of drought adaptation in *Arabidopsis thaliana*: I. Pleiotropy contributes to genetic correlations among ecological traits. Mol Ecol. 2003;12:1137–51.

Mendez-Vigo B, Pico FX, Martinez-Zapater JM, Alonso-Blanco C. Altitudinal and climatic adaptation is mediated by flowering traits and *FRI*, *FLC*, and *PHYC* genes in Arabidopsis. Plant Physiol. 2011;157:1942–55.

Mittler R, Blumwald E. Genetic engineering for modern agriculture: challenges and perspectives. Annu Rev Plant Biol. 2010;61:443–62.

Mochida K, Yoshida T, Sakurai T, Yamaguchi-Shinozaki K, Shinozaki K, Tran LS. In silico analysis of transcription factor repertoires and prediction of stress-responsive transcription factors from six major gramineae plants. DNA Res. 2011;18(5):321–32.

Pacifici M, Foden WB, Visconti P, Watson JEM, Butchart SHM, Kovacs KM, et al. Assessing species vulnerability to climate change. Nat Clim Change. 2015;5(3):215–24.

Partridge RB, Harvey PH. The ecological context of life history evolution. Science. 1988;241:1449–55.

Prasch CM, Sonnewald U. Simultaneous application of heat, drought, and virus to Arabidopsis plants reveals significant shifts in signaling networks. Plant Physiol. 2013;162(4):1849–66.

Priest HD, Fox SE, Rowley ER, Murray JR, Michael TP, Mockler TC. Analysis of global gene expression in *Brachypodium distachyon* reveals extensive network plasticity in response to abiotic stress. PLoS One. 2014;9(1):e87499.

Ream TS, Woods DP, Schwartz CJ, Sanabria CP, Mahoy JA, Walters EM, et al. Interaction of photoperiod and vernalization determines flowering time of Brachypodium distachyon. Plant Physiol. 2014;164(2):694–709.

Rebetzke GJ, Condon AG, Farquhar GD, Appels R, Richards RA. Quantitative trait loci for carbon isotope discrimination are repeatable across environments and wheat mapping populations. Theor Appl Genet. 2008;118:123–37.

Rengel D, Arribat S, Maury P, Martin-Magniette ML, Hourlier T, Laporte M, et al. A gene-phenotype network based on genetic variability for drought responses reveals key physiological processes in controlled and natural environments. PLoS One. 2012;7(10):e45249.

Rizhsky L, Liang H, Shuman J, Shulaev V, Davletova S, Mittler R. When defense pathways collide. The response of Arabidopsis to a combination of drought and heat stress. Plant Physiol. 2004;134:1683–96.

Schippmann U. Revision der europaischen Arten der Gattung *Brachypodium* Palisot de Beauvois (Poaceae). Boissiera. 1991;45:1–250.

Schmalenbach I, Zhang L, Reymond M, Jimenez-Gomez JM. The relationship between flowering time and growth responses to drought in the Arabidopsis Landsberg erecta x Antwerp-1 population. Front Plant Sci. 2014;5:609.

Schwartz CJ, Doyle MR, Manzaneda AJ, Rey PJ, Mitchell-Olds T, Amasino RM. Natural variation of flowering time and vernalization responsiveness in *Brachypodium distachyon*. Bioenergy Res. 2010;3:38–46.

Shinozaki K, Yamaguchi-Shinozaki K. Gene networks involved in drought stress response and tolerance. J Exp Bot. 2007;58(2):221–7.

Sidebottom C, Buckley S, Pudney P, Twigg S, Jarman C, Holt C, et al. Phytochemistry: heat-stable antifreeze protein from grass. Nature. 2000;406:256.

Steinwand MA, Young HA, Bragg JN, Tobias CM, Vogel JP. Brachypodium sylvaticum, a model for perennial grasses: transformation and inbred line development. PLoS One. 2013;8(9): e75180.

Sun J, Hu W, Zhou R, Wang L, Wang X, Wang Q, et al. The Brachypodium distachyon BdWRKY36 gene confers tolerance to drought stress in transgenic tobacco plants. Plant Cell Rep. 2014;34(1):23–35.

Tripathi P, Rabara RC, Langum TJ, Boken AK, Rushton DL, Boomsma DD, et al. The WRKY transcription factor family in *Brachypodium distachyon*. BMC Genomics. 2012;13:270.

Verelst W, Bertolini E, De Bodt S, Vandepoele K, Demeulenaere M, Pe ME, et al. Molecular and physiological analysis of growth-limiting drought stress in *Brachypodium distachyon* leaves. Mol Plant. 2013;6(2):311–22.

Wheeler T, von Braun J. Climate change impacts on global food security. Science. 2013;341 (6145):508–13.

The Genus *Brachypodium* as a Model for Perenniality and Polyploidy

Sean P. Gordon, Lifeng Liu, and John P. Vogel

Abstract The genus *Brachypodium* contains annual and perennial species with both diploid and polyploid genomes. Like the annual species *B. distachyon*, some of the perennial and polyploid species have traits compatible with use as a model system (e.g. small genomes, rapid generation time, self-fertile and easy to grow). Thus, there is an opportunity to leverage the resources and knowledge developed for *B. distachyon* to use other *Brachypodium* species as models for perenniality and the regulation and evolution of polyploid genomes. There are two factors driving an increased interest in perenniality. First, several perennial grasses are being developed as biomass crops for the sustainable production of biofuel and it would be useful to have a perennial model system to rapidly test biotechnological crop improvement strategies for undesirable impacts on perenniality and winter hardiness. In addition, a deeper understanding of the molecular mechanisms underlying perenniality could be used to design strategies for improving energy crops, for example, by changing resource allocation during growth or by altering the onset of dormancy. The second factor driving increased interest in perenniality is the potential environmental benefits of developing perennial grain crops. *B. sylvaticum* is a perennial with attributes suitable for use as a perennial model system. A high efficiency transformation system has been developed and a genome sequencing project is underway. Since many important crops, including emerging biomass crops, are polyploid, there is a pressing need to understand the rules governing the evolution and regulation of polyploid genomes. Unfortunately, it is difficult to study polyploid crop genomes because of their size and the difficulty of manipulating those plants in the laboratory. By contrast, *B. hybridum* has a small polyploid genome and is easy to work with in the laboratory. In addition, analysis of the *B. hybridum* genome, will be greatly aided by the genome sequences of the two extant diploid species (*B. distachyon* and *B. stacei*) that apparently gave rise to

S.P. Gordon • J.P. Vogel (✉)
DOE, Joint Genome Institute, Walnut Creek, CA, USA
e-mail: jpvogel@lbl.gov

L. Liu
DOE, Joint Genome Institute, Walnut Creek, CA, USA

USDA-Agriculture Research Service, Western Regional Research Center, Albany, CA, USA

J. Vogel (ed.), *Genetics and Genomics of Brachypodium*,
Plant Genetics and Genomics: Crops Models, DOI 10.1007/7397_2015_19
Published online: 10 October 2015

B. hybridum. Availability of high quality reference genomes for these three species will be a powerful resource for the study of polyploidy.

Keywords Biofuel • *Brachypodium sylvaticum* • *Brachypodium stacei* • *Brachypodium hybridum* • Perennial grain • Transformation • Gene containment • Homeolog

Introduction

While *Brachypodium distachyon* is an excellent general model for the grasses two traits that make it a tractable genetic model, its annual nature and diploid genome, also make it unsuitable to study perenniality and polyploidy directly. An annual lifecycle is desirable in a model because it allows more generations to be analyzed in a shorter time. Since the vast majority of grain crops are annual, an annual lifecycle is an advantage in this context as well. However, an annual model by itself cannot be used to identify the molecular mechanisms that result in a perennial lifecycle or to identify transgenes or other factors that negatively impact perenniality and winter hardiness. Similarly, the diploid nature of the *B. distachyon* genome, while absolutely essential for its use as a general genetic model system, make it unsuitable to tease apart the rules governing polyploid genome regulation and evolution. Fortunately, closely related perennial and polyploid *Brachypodium* species can be used to study perenniality and polyploidy.

The Need for a Perennial Model

Perennial crops have several advantages over annual crops. Perennial crops are more efficient at capturing sunlight over the growing season because they reach canopy closure much earlier in the growing season. Thus, they can achieve much higher yields (Dohleman and Long 2009). They also have deeper roots that can harvest more water and nutrients decreasing the need for irrigation and fertilizer. Since they are not planted every year, planting costs and energy requirements are much lower. In addition, once established they outcompete weeds and reduce erosion. These advantages have made perennial grasses like switchgrass and *Miscanthus* leading candidates for dedicated biomass crops (Carroll and Somerville 2009; Dohleman and Long 2009). Since these emerging crops are not yet domesticated, there is enormous opportunity for improving traits like yield, stress tolerance, nutrient use efficiency and conversion efficiency. Unfortunately, their complex genetics, polyploid genomes, large size, long generation times and the difficulty measuring relevant traits make breeding these grasses through conventional approaches slow and difficult. Thus, approaches that can accelerate crop improvement are needed for these emerging crops (DOE 2006). Two such approaches, biotechnological methods of crop improvement and knowledge-based breeding strategies would be greatly enabled by a perennial model system.

Currently, virtually all efforts to use biotechnology to improve biomass crops are based on data from annual species like *Arabidopsis thaliana*, rice, or *B. distachyon*. Thus, it is unknown whether these approaches will have an impact on winter hardiness or perenniality until transgenic biomass crops are created and tested. This is problematic because many transgenic manipulations may negatively impact perenniality or hardiness and it is very expensive to create and test transgenic biomass crops. Approaches that alter hormone levels to enhance stress tolerance may be especially prone to undesirable effects (Rivero et al. 2007; Peleg et al. 2011). Thus, it would be extremely useful to have a model perennial grass to rapidly test transgenic strategies for unintended effects on hardiness and/or perenniality before investing in creating transgenic biomass crops.

A deeper knowledge of the mechanisms and molecular machinery underlying perenniality could be used to optimize biomass crops through biotechnology or targeted molecular breeding. For example, it may be desirable to delay the onset of dormancy, delay flowering, alter the number of meristems saved for the following year, or tweak overall source-sink allocations to favor higher yield of stems. The knowledge necessary for these approaches can most easily be obtained from a model perennial system.

Another driver for developing a model perennial grass is the increasing interest in developing perennial grain crops (Cox et al. 2006; Bell et al. 2010; Glover and Reganold 2010; Glover et al. 2010; https://landinstitute.org/). The advantages mentioned above that have made perennial crops the leading candidates for dedicated biomass crops would also be advantages for grain crops (Carroll and Somerville 2009; Dohleman and Long 2009; Glover et al. 2010). Given the significant potential environmental and economic advantages of perennials grains one reasonably asks why are all major grain crops annual? It has often been assumed that since perennials invest more in roots and overwintering structures than annuals that they must have lower yield of reproductive structures e.g. grain. This assumption is wrong for two reasons. First, while perennial crops definitely invest more resources in their roots and crowns than annual crops, they are so much more productive that they can invest more in perennial structures and still achieve much higher yield than annual crops. Since perennials begin growth long before annuals can be planted in the spring and grow much more rapidly due to their large established root system they reach canopy closure early in the season and harvest much more sunlight over the entire growing season (Dohleman and Long 2009). Indeed, a perennial grass could have a much lower harvest index (the ratio of grain produced to the total biomass) and still have a much higher yield per acre than an annual. However, once established, perennial crops do not necessarily have a lower harvest index than annuals. A review of the energy flow into reproductive structures for several high yielding perennial crops (Van Tassel et al. 2010) revealed that many perennial crops allocate as much photosynthate to reproduction as do annual crops. The best performing perennial crop was apple which devoted an amazing 65 % of fixed carbon into reproductive structures (Palmer 1988). This is much higher than the average (48 %) and nearly as high as the maximum (68 %) recorded for maize, one of the most productive annual grains (Ciampitti and Vyn 2012). Thus, there is no inherent biological reason why annual grains should be superior. Why then are all the major grains annual? The answer may be simply that annual grains were more suited to production and selection by Neolithic people at the dawn of agriculture (Van Tassel et al. 2010). This seems intuitive because early farmers that planted the best seeds from annual grasses year after year would have seen more rapid improvements than farmers who either harvested the same patch of perennial grasses each year or planted seeds from perennial grasses that would have to compete with already established perennial grasses. The situation is different for woody perennial crops like fruit trees or grapes. For these crops, early farmers could easily identify a single superior clone or sport that would then be propagated and spread. Under these scenarios it is easy to see how annual grains would be favored and woody perennial fruits could be domesticated. Fortunately, modern plant breeding is no longer constrained by the limitations faced by Neolithic people. Some success in developing new perennial grains by breeding perennial grasses or by crossing annual grains with perennial relatives has been achieved (Cox et al. 2006; Bell et al. 2010; Glover and Reganold 2010; Glover et al. 2010) (https://landinstitute.org/). However, significant hurdles remain. It will be very difficult for a truly new grain to gain widespread public acceptance and push aside maize, rice and wheat. Thus, breeding perennial relatives with existing grains like wheat to create a perennial grain equivalent to existing grains is very attractive. Unfortunately, it has proven difficult due to biological limitations. For example, while perennial wheatgrass can be crossed with wheat it appears that an entire chromosome from perennial wheatgrass is required to maintain perenniality (Hayes et al. 2012). This makes it difficult to reach the same grain quality of modern wheat in a perennial form. If we had a complete understanding of the genes and regulatory networks required for perenniality it would be possible to either engineer current grains to be perennial or design more advanced breeding strategies based on maker assisted selection.

Brachypodium sylvaticum as a Model Perennial Grass

B. sylvaticum is a perennial grass with characteristics (e.g. short generation time, small stature, easily grown, small genome, self-fertile) required for a model system (Steinwand et al. 2013). The perenniality of *B. sylvaticum* was amply demonstrated by a study in Finland where researchers followed the survival of all the individual plants on a plot of land over 20 years (Hæggström and Skytén 1996). Amazingly, the four individual plants of *B. sylvaticum* present at the beginning of the study persisted for the entire 20 years of monitoring. The close relationship of *B. sylvaticum* and *B. distachyon* will allow comparative analyses of annual and perennial lifestyles. In addition, their close relationship will allow *B. sylvaticum*

Fig. 1 *Brachypodium sylvaticum* freezing tolerance. (**a**) Control plants not exposed to cold. (**b**) Plants without vernalization moved from warm growing conditions into −10 °C for 8 h. Plants were photographed 2 weeks after freezing treatment. (**c**) Plants vernalized at 4 °C for 4 weeks prior to incubation at −10 °C for 8 h. Plants were photographed 2 weeks after freezing treatment. Scale bar = 2 cm

researchers to leverage the extensive suite of tools and resources created for *B. distachyon*.

In temperate climates perennials must survive freezing conditions during the winter. Typically, freezing tolerance gradually increases in response to cooling temperatures and shorter daylength in the fall. Significantly, *B. sylvaticum* exhibits significant freezing tolerance that is enhanced by cold acclimation (Fig. 1). In addition, significant natural variation in freezing tolerance is present in *B. sylvaticum* inbred lines (Fig. 1). Thus, *B. sylvaticum* can be used to study cold acclimation and freezing tolerance.

Efficient *Agrobacterium*-mediated transformation is an absolute requirement for a modern model plant. Fortunately, a highly efficient method to transform *B. sylvaticum* has been developed (Steinwand et al. 2013). Remarkably, the transformation rate in the first published report was extremely high (average efficiency 67 %) for one accession, Ain-1. Indeed, while the transformation takes longer, the efficiency is actually higher than *B. distachyon* transformation. With an efficient transformation method in hand, the most significant hurdle to establishing *B. sylvaticum* as a model perennial grass was cleared.

Another requirement for a model plant is a diverse collection of freely available germplasm. Fortunately, the USDA National Plant Germplasm System maintains stocks of many plants, including *B. sylvaticum* (http://www.ars-grin.gov/npgs/). Researchers analyzed 32 accessions labeled *B. sylvaticum* from a wide geographic area to confirm genome size and chromosome number prior to creating inbred lines. Seventeen lines had c-values (~1 pg/nucleus) consistent with prior reports and indicative of a genome size of approximately 340 Mb (Steinwand et al. 2013). The chromosome number of two representatives from the 1 pg/nucleus c-value group was determined to be consistent with previous reports ($2n = 18$) (Long 1989; Shi et al. 1993; Khan and Stace 1999). Fifteen lines were then inbred for 2–5 generations and made available to the research community. Subsequently, additional inbreeding (up to nine generations of single seed descent) was conducted for a subset of lines (unpublished). Interestingly, four lines had much larger genomes (c-values ~1.5 or ~2) and higher chromosome counts ($2n = 27$ or 36) indicating that they were either a different species and/or polyploid (Steinwand et al. 2013).

Unlike *B. distachyon*, *B. sylvaticum* flowers open and the anthers produce a large amount of pollen (Steinwand et al. 2013). This suggests that, while the *B. sylvaticum* accessions tested were self-fertile, they probably outcross frequently. This is supported by the observation that 19 % of genotypes for ten SSR markers were heterozygous. An outcrossing rate of 5.4 % under greenhouse conditions was determined by following the flow of a reporter gene from transgenic to non-transgenic plants (Steinwand et al. 2013). The significant outcrossing rate has two practical implications. First, it is necessary to inbreed lines from wild material for multiple generations to create nearly homozygous lines. Second, for many genetic studies it is necessary to prevent outcrossing by bagging flowers or spatial/temporal separation of flowering lines. The fact that *B. sylvaticum* flowers open and produce a large amount of pollen also makes it easier to make controlled crosses than *B. distachyon*. Crosses between several *B. sylvaticum* lines have been made with a reasonable frequency on the very first attempt (unpublished).

The ability to control or limit the flow of a transgene beyond the field planted, gene containment, will likely be a prerequisite for any commercial application of transgenic biomass crops. Its perennial nature, frequent outcrossing, and copious pollen production make *B. sylvaticum* an excellent system to optimize strategies for gene containment. Perenniality is an advantage in this context because sterile lines, a part of most gene containment strategies, can be maintained and studied indefinitely. This is a major advantage over annual models where sterile lines are ephemeral and must be maintained in a heterozygous condition, if possible, or recreated for every experiment. Frequent outcrossing is crucial to actually testing gene flow to determine if the system is working. Copious pollen production allows researchers to easily determine the effect of transgenes on pollen viability.

The next major requirement for a model system is a high quality genome assembly and annotation. *B. sylvaticum* has been sequenced using PacBio long read sequencing technology (unpublished). The assembly consists of a small number (<2500) of large contigs that are currently being arranged into pseudomolecules with the aid of a high density genetic map based on 288 F2 lines genotyped by reduced representation sequencing. Annotation will be aided by transcriptome sequence from a diverse set of tissues and treatments. With inbred lines available, an efficient transformation system and a soon to be available genome sequence, the future of *B. sylvaticum* as a perennial model for the grasses is very bright.

The Need for a Polyploid Model System

While polyploid species can be found on many branches of the tree of life, they are unusually common among plants. It is estimated that 47–70 % of all angiosperms are polyploid (reviewed by Ramsey and Schemske 1998). Polyploidy is even more common among grasses with about 80 % of grass species possessing polyploid genomes that are the result of many independent polyploidization events ranging from very recent to ancient (Stebbins 1985; Hunziker and Stebbins 1987; Estep et al. 2014). The diversity of genome size and ploidy levels seen throughout the grass family raises a number of interesting questions: Why are grasses so tolerant of changes in ploidy? What happens when two genomes come together over the short and long term? Why are polyploid forms sometimes more adaptable? Are there rules governing genome evolution after polyploidization? How do repeated polyploidization events alter genome evolution? The answers to these questions are economically important because many grass crops are polyploid (e.g. wheat, sugarcane, switchgrass).

In addition, the production of new hybrids and allopolyploids holds great promise for introducing useful traits into crop plants and improving yield (Mason and Batley 2015). However, with a few notable exceptions (e.g. wheat), the synthesis or even re-synthesis of polyploids in the laboratory is not an exact and predictable science. This highlights a lack of knowledge about how to overcome barriers to hybridization such as establishment of stable chromosome pairing and prevention of pairing between homoeologous chromosomes. Furthermore, the exact mechanisms that govern the maintenance and expression of homoeologous genes in polyploids are unknown, but may involve epigenetic regulation (Song and Chen 2015).

To develop a deeper understanding of polyploid genome evolution and regulation, it would be extremely useful to have a tractable polyploid grass model system. But what should such a model system look like? First, it should consist of three species: a tetraploid species and extant representatives of the two species that contributed the sub-genomes. Second, the genomes should be compact to facilitate analysis. Third, all three species should be small, easy to grow and self-fertile. Fourth, the three species should be transformable to test hypothesis.

A Trio of *Brachypodium* Species as a Tractable Model for Polyploidy

As described in chapter "Phylogeny and Evolution of the Genus *Brachypodium*", the genus *Brachypodium* includes species that vary greatly in ploidy and chromosome number. Thus, a trio of *Brachypodium* species may be suitable to use as a model for polyploidy. The recent separation of *B. distachyon* into three species (*B. distachyon, B. stacei* and *B. hybridum*) gives us three strong candidates. Initially it was thought that these three species were a polyploid series of *B. distachyon* because they looked similar and had 2n chromosome numbers of 10, 20, 30 (Robertson 1981). However, careful analysis revealed that the 2n = 10 and 2n = 20 forms were both diploid with nearly identical genome sizes and that the 2n = 30 form was probably an allotetraploid with sub-genomes similar or identical to the genomes of the two diploid forms (Hasterok et al. 2004). The situation was clarified with the publication of a paper that split *B. distachyon* into three species: *B. distachyon* 2n = 2x = 10; *B. stacei* 2n = 2x = 20; *B. hybridum* 2n = 4x = 30 (Catalán et al. 2012). A benefit of the initial taxonomic confusion is that *B. hybridum*, and to a lesser extent *B. stacei*, were included in the early work to develop resources and methods for *B. distachyon*. Thus, we know that *B. hybridum* and *B. stacei* are easy to grow under controlled conditions, self-fertile and have small genomes (*B. stacei* ~270 Mb and *B. hybridum* ~540 Mb) (Draper et al. 2001; Vogel et al. 2006). In addition, *B. hybridum* was the first *Brachypodium* species to be transformed (Draper et al. 2001). Indeed, *B. hybridum* lines stood out as the best lines for *Agrobacterium*-mediated transformation in the first published report (Vogel et al. 2006). Significantly, some of the *B. hybridum* accessions developed a highly embryogenic callus from mature seeds. We are not aware of any attempt to transform *B. stacei*, but given the success in transforming *B. distachyon*, *B. hybridum* and *B. sylvaticum* it if highly likely that it will be possible to efficiently transform *B. stacei*. Thus, *B. hybridum*, *B, distachyon* and *B. stacei* are excellent candidates to serve as a model system for polyploidy.

One reason for the success of some polyploidy crops is that polyploids are sometimes larger, more vigorous and more stress tolerant than their diploid progenitors. While this phenomenon has been widely utilized in plant breeding and seems to be due to increased cell size and a heterosis-like effect caused by bringing two genomes together, the exact reasons for this are poorly understood. Thus it is not surprising that *B. hybridum* seems to be more adaptable than its putative diploid parents. This can be seen in their native ranges where *B. hybridum* is more common and occupies more environmental niches than *B. distachyon* or *B. stacei* (see chapter "Phylogeny and Evolution of the Genus *Brachypodium*"). In addition, *B. hybridum* has become established around the world suggesting that it is more adaptable to variable conditions (Bevan et al. 2011). New collections of *B. hybridum* and to a lesser extent *B. stacei* are being brought into the laboratory for studies of natural diversity, environmental adaptation and evolution (see chapters "Phylogeny and Evolution of the Genus *Brachypodium*" and "Genomic Diversity and Climate Adaptation in *Brachypodium*"). Thus, this trio of species is already serving as a model polyploidy system.

High quality genome assemblies for all three species will be required for these species to serve as a polyploid model system. The *B. distachyon* genome has already been sequenced and the version 3 release is essentially a base perfect finished genome with ambiguity confined to the centromeric regions (IBI 2010; http://phytozome.jgi.doe.gov). To complete the trio of genomes the DOE Joint Genome institute is currently sequencing the genomes of *B. stacei* and *B. hybridum*. The initial *B. stacei* genome assembly and annotation has been completed (unpublished). As expected, the assembly consists of ten pseudomolecules and the annotation contains a similar number of genes as the current *B. distachyon* annotation (unpublished). The *B. hybridum* genome has also been sequenced and an initial assembly created. This assembly is now being ordered into pseudomolecules using a high-density genetic map created by skim sequencing an F2 mapping population. The assembly of polyploidy genomes can be confounded by similar sequences in the two subgenomes. This problem is more pronounced for closely related sub-genomes. Fortunately, with a divergence of time of ~16 million years ago *B. stacei* and *B. distachyon*, the progenitors of *B. hybridum*, had ample time to diverge at the sequence level (Chapter "Phylogeny and Evolution of the Genus *Brachypodium*"). The initial assembly is close to the expected genome size and a comparison to a high-density genetic map indicates a manageable number of false joins. Thus, a completed genome assembly should be available soon.

High-quality annotated genomes of diploid progenitors together with their derived allopolyploid will be a powerful system for deciphering the exact mechanistic details of polyploid establishment, stabilization and evolution. For example, formation of allopolyploids in the laboratory is associated with epigenetic changes including DNA methylation and histone modifications (Song and Chen 2015). rRNA gene silencing has been shown to occur in *B. hybridum* with *B. distachyon*-like rRNA genes transcriptionally active *B. stacei*-like rRNA genes repressed (Idziak et al. 2008). Thus, epigenetic changes may play a role in gene regulation in *B. hybridum*. Epigenetic changes in plant allopolyploids also include disregulation of transposable elements in the nascent allopolyploid that may contribute to initial genome destabilization (Song and Chen 2015). An in depth comparison of the *B. distachyon*, *B. stacei* and *B. hybridum* genomes will allow researchers to determine in exquisite detail the fate of transposable elements after the polyploidization event. Genome-wide examination in such detail is not currently possible using existing crop or model systems.

A critical step in the maintenance of polyploids is the establishment of normalized meiosis and chromosome segregation. In wheat the Pairing homeologous 1 (*Ph1*) locus is a major locus for preventing association of nonhomologous chromosomes, thus stabilizing meiosis (Riley and Chapman 1958). Interestingly, wheat *Ph1* was mapped to a 2.5 Mb region in part using markers derived from *B. sylvaticum* (Griffiths et al. 2006). This region contains a cluster of cyclin-dependent kinase-like genes that may influence kinase activity, important for specificity of chromosome pairing (Greer et al. 2012). Thus, a syntenically related locus to *Ph1* exists within *Brachypodium* species. In accordance with the strong syntenic relation to wheat at this locus, *B. sylvaticum* has a syntenic cyclin-dependent kinase ortholog in this region. Therefore, an immediate question is whether the locus syntenic to *Ph1* in *Brachypodium* plays a similar role in ensuring proper chromosome pairing in *Brachypodium* allopolyploids. Alternatively, completely different loci involving different mechanisms may be employed by *Brachypodium* allopolyploids possibly including some of the other loci implicated in regulating nonhomologous chromosome pairing in wheat (Greer et al. 2012).

Other areas that can be addressed with the trio of *Brachypodium* genomes are the mechanisms underlying the preferential expression of genes from one genome (genome dominance) and the related loss of original gene function through multiple mechanisms (genome fractionation). In maize, subgenomes derived from an ancient genome doubling are biased in genome fractionation and homoeologous gene expression (Woodhouse et al. 2010). In contrast, studies in wheat indicate no overall subgenome bias in homoeologous gene expression. However, bias in homoeologous gene expression was observed in specific wheat tissues (Pfeifer et al. 2014). Analysis of homoeologous gene expression in *B. hybridum* relative to its diploid progenitors in combination with full genome sequences may shed light on possible mechanisms of genome dominance and biased genome fractionation. Interestingly, it appears that *B. hybridum* has arisen more than once through independent crosses between *B. distachyon* and *B. stacei* (López-Alvarez et al. 2012, Chapter "Phylogeny and Evolution of the Genus *Brachypodium*"). These multiple origins may be useful as natural replicates to determine the reproducibility of genome dominance and fractionation.

Complete genome sequences of *B. distachyon, B. stacei* and *B. hybridum* will make it possible to study mechanistic details of polyploids. Such details are required to better understand, manipulate and predict the outcome of hybridization events in the laboratory and ultimately contribute knowledge that may allow improvement of crop species.

Outlook

The development of tools to use *B. sylvaticum* to study perenniality and a trio of *Brachypodium* species to study polyploid genome regulation and evolution is proceeding apace. This growth is fueled by an increasing interest in the development of perennial grasses as dedicated biomass crops, the promise of environmentally superior perennial grains and the desire to understand polyploid genomes in order to improve numerous polyploid crops. The biological attributes (small size, rapid generation time, small genome, self-fertility) of *B. sylvaticum*, *B. stacei*, *B. hybridum* and, of course, *B. distachyon* make them well suited for use as model organisms. In addition, *B. sylvaticum*, *B. distachyon* and *B. hybridum* can be very efficiently transformed which facilitates many experimental manipulations. The final major piece necessary for a model system, a high quality genome sequence and annotation, is already in place for *B. distachyon* and *B. stacei* and will be in place soon for *B. sylvaticum* and *B. hybridum*. Thus, the future for the genus *Brachypodium* to serve as a model for perennial and polyploid grasses is very bright.

Acknowledgement The work conducted by the U.S. Department of Energy Joint Genome Institute, a DOE Office of Science User Facility, is supported under Contract No. DE-AC02-05CH11231. Additional funding was provided by Office of Biological and Environmental Research, Office of Science, US Department of Energy, interagency agreement DE-SC0006999.

References

Bell LW, Wade LJ, Ewing MA. Perennial wheat: a review of environmental and agronomic prospects for development in Australia. Crop Pasture Sci. 2010;61(9):679–90.

Bevan MW, Garvin DF, Vogel JP. *Brachypodium distachyon* genomics for sustainable food and fuel production. Curr Opin Biotechnol. 2011;21(2):211–7.

Carroll A, Somerville C. Cellulosic biofuels. Annu Rev Plant Biol. 2009;60:165–82.

Catalán P, Müller J, Hasterok R, Jenkins G, Mur LAJ, Langdon T, et al. Evolution and taxonomic split of the model grass *Brachypodium distachyon*. Ann Bot. 2012;109(2):385–405.

Ciampitti IA, Vyn TJ. Physiological perspectives of changes over time in maize yield dependency on nitrogen uptake and associated nitrogen efficiencies: a review. Field Crop Res. 2012;133:48–67.

Cox TS, Glover JD, Van Tassel DL, Cox CM, DeHaan LR. Prospects for developing perennial grain crops. BioScience. 2006;56(8):649–59.

DOE, editor. Breaking the biological barriers to cellulosic ethanol: a joint research agenda. U.S. Department of Energy, Office of Science and Office of Energy Efficiency; 2006.

Dohleman FG, Long SP. More productive than maize in the Midwest: how does Miscanthus do it? Plant Physiol. 2009;150(4):2104–15.

Draper J, Mur LAJ, Jenkins G, Ghosh-Biswas GC, Bablak P, Hasterok R, et al. *Brachypodium distachyon*. A new model system for functional genomics in grasses. Plant Physiol. 2001;127 (4):1539–55.

Estep MC, McKain MR, Diaz DV, Zhong J, Hodge JG, Hodkinson TR, et al. Allopolyploidy, diversification, and the Miocene grassland expansion. Proc Natl Acad Sci U S A. 2014;111 (42):15149–54.

Glover JD, Reganold JP. Perennial grains food security for the future. Issues Sci Technol. 2010;26 (2):41–7.

Glover JD, Reganold JP, Bell LW, Borevitz J, Brummer EC, Buckler ES, et al. Increased food and ecosystem security via perennial grains. Science. 2010;328(5986):1638–9.

Greer E, Martín AC, Pendle A, Colas I, Jones AME, Moore G, et al. The *Ph1* locus suppresses Cdk2-type activity during premeiosis and meiosis in wheat. Plant Cell. 2012;24(1):152–62.

Griffiths S, Sharp R, Foote TN, Bertin I, Wanous M, Reader S, et al. Molecular characterization of *Ph1* as a major chromosome pairing locus in polyploid wheat. Nature. 2006;439 (7077):749–52.

Hæggström CA, Skytén R. Flowering and individual survival of a population of the grass *Brachypodium sylvaticum* in Nåtö, Åland Islands, SW Finland. Ann Bot Fennici. 1996;33 (1):1–10.

Hasterok R, Draper J, Jenkins G. Laying the cytotaxonomic foundations of a new model grass, *Brachypodium distachyon* (L.) beauv. Chromosome Res. 2004;12(4):397–403.

Hayes RC, Newell MT, DeHaan LR, Murphy KM, Crane S, Norton MR, et al. Perennial cereal crops: an initial evaluation of wheat derivatives. Field Crop Res. 2012;133:68–89.

Hunziker JH, Stebbins GL. Chromosomal evolution in the Gramineae. In: Soderstrom TR, Hilu KW, Campbell CS, Barkworth ME, editors. Grass systematics and evolution. Washington, DC: Smithsonian Institution; 1987. p. 179–87.

IBI. Genome sequencing and analysis of the model grass *Brachypodium distachyon*. Nature. 2010;463:763–8.

Idziak D, Hasterok R. Cytogenetic evidence of nucleolar dominance in allotetraploid species of *Brachypodium*. Genome. 2008;51(5):387–91. doi:10.1139/G08-017.

Khan MA, Stace CA. Breeding relationships in the genus *Brachypodium* (Poaceae: Pooideae). Nord J Bot. 1999;19(3):257–69.

Long G. Morphological and physiological variation in *Brachypodium sylvaticum*. PhD, University of Wales College of Cardiff; 1989.

López-Alvarez D, López-Herranz ML, Betekhtin A, CatalÃ¡n P. A DNA barcoding method to discriminate between the model plant *Brachypodium distachyon* and its close relatives *B. stacei* and *B. hybridum* (Poaceae). PLoS One. 2012;7(12).

Mason AS, Batley J. Creating new interspecific hybrid and polyploid crops. Trends Biotechnol. 2015;33(8):436–41.

Palmer JW. Annual dry matter production and partitioning over the first 5 years of a bed system of Crispin/M.27 apple trees at four spacings. J Appl Ecol. 1988;25(2):569–78.

Peleg Z, Reguera M, Tumimbang E, Walia H, Blumwald E. Cytokinin-mediated source/sink modifications improve drought tolerance and increase grain yield in rice under water-stress. Plant Biotechnol J. 2011;9(7):747–58.

Pfeifer M, Kugler KG, Sandve SR, Zhan B, Rudi H, Hvidsten TR, et al. Genome interplay in the grain transcriptome of hexaploid bread wheat. Science. 2014;345(6194):1250091.

Ramsey J, Schemske DW. Pathways, mechanisms, and rates of polyploid formation in flowering plants. Annu Rev Ecol Syst. 1998;29:467–501.

Riley R, Chapman V. Genetic control of the cytologically diploid behaviour of hexaploid wheat. Nature. 1958;182(4637):713–5.

Rivero RM, Kojima M, Gepstein A, Sakakibara H, Mittler R, Gepstein S, et al. Delayed leaf senescence induces extreme drought tolerance in a flowering plant. Proc Natl Acad Sci U S A. 2007;104(49):19631–6.

Robertson IH. Chromosome numbers in *Brachypodium* Beauv. (Gramineae). Genetica. 1981;56 (1):55–60.

Shi Y, Draper J, Stace C. Ribosomal DNA variation and its phylogenetic implication in the genus *Brachypodium* (Poaceae). Plant Syst Evol. 1993;188(3-4):125–38.

Song Q, Chen ZJ. Epigenetic and developmental regulation in plant polyploids. Curr Opin Plant Biol. 2015;24:101–9.

Stebbins GL. Polyploidy, hybridization, and the invasion of new habitats. Ann Missouri Bot Gard. 1985;72(4):824–32.

Steinwand MA, Young HA, Bragg JN, Tobias CM, Vogel JP. *Brachypodium sylvaticum*, a model for perennial grasses: transformation and inbred line development. PLoS One. 2013;8(9).

Van Tassel DL, DeHaan LR, Cox CS. Missing domesticated plant forms: can artificial selection fill the gap? Evol Appl. 2010;3:434–52.

Vogel JP, Garvin DF, Leong OM, Hayden DM. *Agrobacterium*-mediated transformation and inbred line development in the model grass *Brachypodium distachyon*. Plant Cell Tiss Org Cult. 2006;85:199–211.

Woodhouse MR, Schnable JC, Pedersen BS, Lyons E, Lisch D, Subramaniam S, et al. Following tetraploidy in maize, a short deletion mechanism removed genes preferentially from one of the two homeologs. PLoS Biol. 2010;8(6):e1000409.

Brachypodium as an Arabidopsis for the Grasses: Are We There Yet?

Christopher W.P. Lyons and Karen-Beth G. Scholthof

Abstract *Brachypodium distachyon* has recently been developed as a small, easily manipulated grass to further translational and applied agronomic research paradigms outside of the scope of the Arabidopsis programs. The aim of the workers behind this fledgling Brachypodium community is that this grass will lead to theoretical and practical advances within cereal, bio-energy, and forage research. Some of the successes of Brachypodium within the context of Arabidopsis are noted and structured around the research triad—basic, translational, and applied work—with a particular focus on Brachypodium as a tool for advancing plant virology and pathosystems biology work.

Keywords Brachypodium • Poaceae • Switchgrass • Plant pathology • Plant virology • Bioenergy • Model plant • Systems biology

Introduction

Although it is arguably the only broadly studied plant model, *Arabidopsis thaliana* (Arabidopsis) shares features typical of the other model systems—it is easy to maintain in the laboratory, it has a rapid regeneration time, and it is amenable to physiological and genetic manipulation. But most importantly, Arabidopsis has been accepted, even embraced, by the life sciences research community. Clearly, Arabidopsis has earned the accolades and acceptance associated with other experimental model organisms. Arabidopsis has even been hailed as the model representative of all plants (Leonelli 2007a, b). If this is true, then is there room for *Brachypodium distachyon* (Brachypodium) as a model organism? If so, what might the job be for this small grass in the coming years?

In the early 1900s in European gardens, Arabidopsis was considered a weed; in the century since then it has become the model organism for plant science research. The history of Arabidopsis has shown us the capability of using one particular model to explore research topics spanning basic, translational, and application-driven outcomes.

C.W.P. Lyons • K.-B.G. Scholthof (✉)
Department of Plant Pathology and Microbiology, Texas A&M University,
College Station, TX 77843-2132, USA
e-mail: kbgs@tamu.edu

J. Vogel (ed.), *Genetics and Genomics of Brachypodium*,
Plant Genetics and Genomics: Crops Models, DOI 10.1007/7397_2015_8
Published online: 02 October 2015

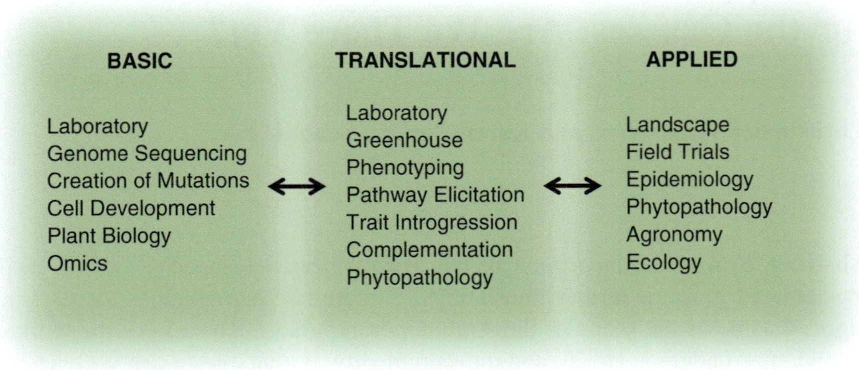

Fig. 1 The scientific triptych. The development of model organisms, including Arabidopsis, has historically taken a primarily reductionist approach. The research flow has been mostly in a unidirectional manner towards increasing complexity by application of previous knowledge. In the twentieth century, knowledge was predicated on the increased value of basic (theoretical) and translational (laboratory) research and lesser values were attributed to applied (field) research. Now, in a post-genomics era, due in part to the ready availability of omics and high-throughput screening technologies, there appears to be concurrent research and some fluidity within this model developmental process in all three areas. The workflow of resources and data, as exemplified by the use of Brachypodium, is less reductionist than that of standard model organisms. Here, one or multiple parts of the triptych are intended to be representative of the influence and support of the other(s), reflecting a holistic approach

Yet these successes should not entirely be taken as support for a broad universality of all plants, the ways they interact with their environment (either naturally, in the laboratory, or within artificial agroecosystems), or their ease of use in solving real-world problems (Fig. 1). In-part due to a perceived need to solve crop-based problems, a second plant model was proposed that would specifically be used to understand the fundamentals of grass-specific biology—to include plant pathology and agronomy. This focus would lead to the study of the close evolutionary relationship between Brachypodium and the cereals—which would be used as one rationale to promote it as a model organism.

For the establishment of Brachypodium as a viable system, a catalog of phenotypic traits and their associated genomic loci would have to be created and archived to tackle the translational and applied needs of plant breeders and scientists in allied fields (Catalan et al. 2014). Brachypodium was imbued with a job of solving crucial problems associated directly with grasses and cereals—plants humans use as sources of food, feed, forage, recreation, and bioenergy. Since the inception of the idea of specifically using Brachypodium to address these needs in the early 1990s, the expected model organism-specific tools were quickly established, using Arabidopsis and other model systems as guides (Fig. 2). Today many questions still remain: How do we determine which problems need to be worked-on; how do we foster a community that allows for the rapid coordination of addressing these problems and is willing to develop the tools needed to do such a job; and, in general, what is the feasibility of supporting that work using Brachypodium as a model organism in a post-genomics era.

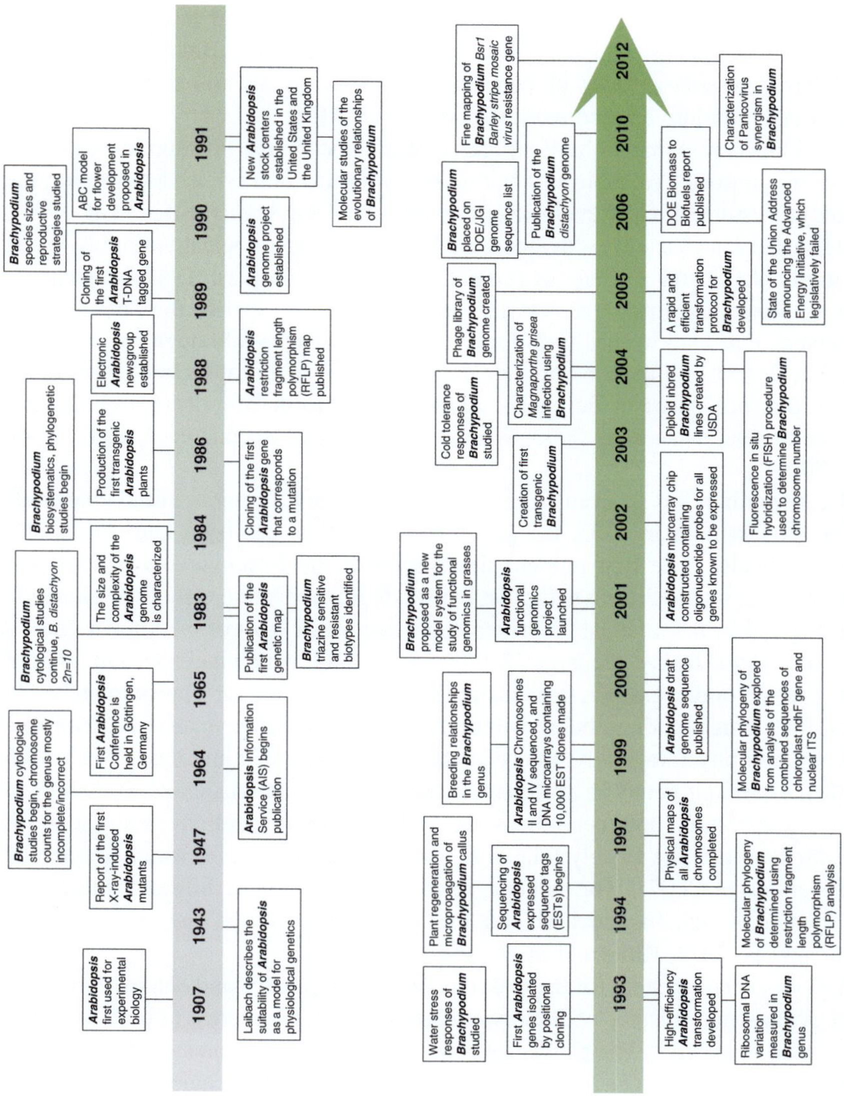

Fig. 2 Timeline of notable events in the histories of *Arabidopsis thaliana* and *Brachypodium distachyon* (Figure modified from Somerville and Koornneef (2002). *Nature Reviews Genetics*, with permission)

Classification of a Model Plant

It would seem a given that a settled systematics is key to selecting and establishing a model organism. Yet, a curious aspect of both Brachypodium and Arabidopsis is the difficulty of resolving their taxonomic classifications. One might imagine that if a plant is designated a model, then it should be firmly rooted in the tree of life. Incongruously, being able to taxonomically place both Brachypodium and Arabidopsis is still a focus of current research—from the most basic through translational and into applied work (Catalan et al. 2014).

Arabidopsis researchers have struggled to understand the "comprehensive concept of evolutionary relationships within (its) mustard family" in determining the molecular systematics and evolution of *Arabidopsis* and *Arabis* (Koch et al. 1999). While earlier classification schemes were based on morphological variation, it was not until RLFP and nuclear ribosomal ITS and chloroplast DNA sequencing analyses were performed that we start seeing a move to clarify the phylogenetic framework that lead to the *Arabidopsis thaliana* lineage—although further work would be done to compare worldwide populations of Arabidopsis to determine the genetic distance between European, Asian, India, North American, and "laboratory" ecotypes (Zwan et al. 2000; Warwick et al. 2006; Heenan et al. 2002; Koch et al. 2000).

Brachypodium has an even richer taxonomic history. From its initial characterization by Linneaus in 1753, the genus *Brachypodium* has been hard to classify phylogenetically. The first modern description of the *Brachypodium* genus was made in 1812, by the French naturalist Palisot de Beauvois in *Essai D'une Nouvelle Agrostographie*, in which he described and categorized most of the known grasses in the old world. Today, the large *Pooideae* subfamily, which includes a number of grain crops, forage grasses, and turfgrasses is comprised of approximately ten tribes, with Brachypodium being within the distinct tribe *Brachypoideae*. Morphologically and physiologically, many of the species included and described within the *Brachypodium* genus share many of the same characteristics—being short in stature, self-fertile, and perennial—the only known exceptions being that of *B. distachyon* and the closely related *B. stacei* and *B. hybridum* which are annual species (Garvin 2007). The genetic relationships between different species of *Brachypodium* remains unclear and the systematics of the genus is ongoing (Catalan et al. 2014; Khan and Stace 1999).

A phylogenetic system for the classification of the grasses and elucidation of the exact evolutionary history of *Brachypodium* is of crucial interest to both botanists and agronomists alike. The genus has been "floating"—being anchored as possibly being close to wheat (*Triticum*), fescue, brome, or barley clades—while being simultaneously described as being in tribe of its own (*Brachypodiinae*) and within *Triticeae* by others (Khan and Stace 1999; Hasterok et al. 2004; Pacurar 2009; Shi 1991; Tateoka 1962). The recent determination of the genome sequence of the reference *B. distachyon* accession, Bd21, and the sequencing of an additional seven highly divergent lines of *B. distachyon*, as well as *B. stacei* (diploid) and

B. hybridum (the amphidiploid of these parents), has enhanced our knowledge of the biology and taxonomy of *Brachypodium* species (Gordon et al. 2014; López-Alvarez et al. 2012). From this, the Brachypodium workers will have additional tools to fulfill the promise of a model system for the grasses, akin to the status achieved by Arabidopsis.

Developing Brachypodium: From Practice to Theory (and Back Again)

In 2002, in reminiscing about the status of plant biology and model organisms a decade earlier, The Multinational Arabidopsis Steering Committee recalled posing the question: "Why not a plant?" (The Multinational Arabidopsis Steering Committee 2002). Yet, the bigger question might have been "Why Arabidopsis?" when it was proposed as the first plant to have its genome sequenced. Many of the model-associated characteristics ascribed to Arabidopsis are not unique (Figs. 1 and 2), but its small stature, short generation time, and small genome were important enough at the time to justify its boosters' push for it to become the most widely used model organism in all of molecular genetics (Leonelli 2007a, b; Meyerowitz 1987). Yet as early as 1995, there were suggestions that Brachypodium "has several features of its genome and growth habit ... that may allow it to be developed as a model molecular genetic system representative of the temperate grasses" (Bablak et al. 1995). Some model-like biological characteristics of Brachypodium were a "small physical size, a rapid life cycle, and undemanding growth requirements that make it amenable to high-throughput screening routines" (Draper et al. 2001), much like Arabidopsis. Although all these features were shared between the two plants, what additional resources and outcomes would be needed to justify Brachypodium's development as a model on-par with that of the highly used and prized Arabidopsis?

The ability to create artificial mutations through mutagenesis or to discover natural mutations conferring phenotypic variation in an organism is crucial in the development of a model. These mutants and the ability to discover or make them, we would argue, are the mainstay of genetic and translational functional genomic studies. For example, a monumental first step in the history of Arabidopsis research was published in 1969 by one of the founders of developmental plant biology, George Rédei, with the successful creation of 65 nutritional mutants that, at the time, were reported as disrupting at least four different loci responsible for the thiamine biosynthesis pathway (Li and Rédei 1969). By selectively recovering seeds exposed to either X-rays or EMS, Rédei showed that Arabidopsis could be a manageable and malleable laboratory system for the generation of distinct variants which could be used in further biochemical and physiological characterization assays (Rédei and Li 1969).

For Brachypodium, the emergence of genetics came from a different approach. In 1983, natural *s*-triazine-resistant mutants of *B. distachyon* growing along roadsides in Israel (Gressel et al. 1983) were used to ask: "Can wild species become problem weeds because of herbicide resistance?" (Gressel and Kleifeld 1994). This question reflected ongoing concerns that wild species could become invasive, crowding-out or competing with crops in agricultural fields. Thus, for the first time, Brachypodium was taken from the field to the lab, grown in a greenhouse, and tested for herbicide resistance—a *bona fide* example of a meshing of theoretical, translational, and applied research to test a specific hypothesis that had agronomic and ecological implications (Fig. 1). The authors summarized that Brachypodium could be used as an exemplar to determine the potential spread of herbicide resistance or antibiotic gene transfer from genetically modified crops to weeds (Gressel et al. 1983; Gressel and Kleifeld 1994; Matzrafi et al. 2014).

More recent studies make use of Brachypodium mutants developed in the lab (Bragg et al. 2012). Two examples of this include the use of T-DNA transformed Brachypodium accessions to trace the phytohormone brassinosteroid pathway which regulates early plant growth and development (Xu et al. 2015) and carotenoid cleavage dioxygenase genes that are responsible for catalysis of carotenoids to apocarotenoids (Priya and Siva 2014). While basic research like this has been completed and the results using Brachypodium are as fundamental as some of the work previously performed using Arabidopsis as the plant of study (Nishimura and Dangl 2010; Vallabhaneni et al. 2010), little work from either model system has demonstrated success in being translated directly from the lab and applied to field crops.

A clear goal of using Arabidopsis as a basic model was to identify a wide range of plant-specific gene functions, thus establishing Arabidopsis as model for rapid, systematic identification of genes for crop improvement (Initiative 2000). With Arabidopsis, plant science finally had "the foundations for more comprehensive comparison of conserved processes in all eukaryotes" for genomic comparison with the then newly sequenced—Drosophila and *C. elegans*—animal model systems. As Arabidopsis use bloomed in the laboratory (Lyons and Scholthof 2015), an idea was building that Brachypodium could be developed to meet perceived gaps in translational and applied research areas not reached, or impossible to reach due to evolutionary differences, by the Arabidopsis community—specifically the biology, pathology, and physiology of cereal and bioenergy grasses.

In 1993, *B. sylvaticum* was specifically mentioned, along with rice (which was also being developed as a model system at this time), as possibly being useful for comparative genetics (Moore et al. 1993). Through the combined work of a number of dedicated researchers having previously seen the success and drive behind the sequencing of the Arabidopsis genome, it was in 2010—10 years after Arabidopsis—that the genome analysis of Brachypodium was completed (International Brachypodium Initiative 2010)—providing the final tools needed for those groups to work towards greater basic, translational, and applied research goals. Will model plants retain their status in the research community? Perhaps not, as noted by Garvin, "given the rapid progress in genome sequencing of crops, one question that emerges is whether future investment in new model species to serve as surrogates for crops is necessary" (Garvin 2007). Nevertheless, the expected model organism resources are in place for Brachypodium: it is a laboratory-friendly plant that has a short lifecycle of 6–10 weeks, is inexpensive to maintain, requires little space, and is generally not fussy; an omics toolkit with knockout lines and transgenic protocols; and a community that is sharing materials, meeting at international conferences, and actively promoting their organism of choice. In the sections below, we argue to continue this active promotion of Brachypodium as a model with select examples from recent literature.

Translational Work in Era of Omics: Integrated Phytopathological Systems

Plant pathology provides a useful basis to look at and utilize Brachypodium across disciplinary subdivisions of basic, translational (or transitional), and applied research (Fig. 1). For plant pathologists, such neat lines are blurred, with most of our work occurring in transitional areas. While there is always the need to pursue fundamental knowledge, we remain cognizant of the short and long-term needs of growers and other stakeholders. Yet, genomics and systems biology research approaches have broadened the scope of science, allowing for model organisms to be used as tools for plant pathology.

Brachypodium is susceptible to a number of fungal, bacterial, and viral pathogens (Mandadi et al. 2014; Mandadi and Scholthof 2012, 2013; Sandoya and de O Buanafina 2014; Parker et al. 2008; Ayliffe et al. 2013) and has been used to understand plant responses to challenge by those pathogens, some of which cannot be studied using Arabidopsis or other dicotyledonous plant hosts. An example of the characterization of one of these pathogenic responses is the focus of a recent paper aiming to study a wounding and grazing-associated gene response in Brachypodium—a response with links to conserved defense signaling in eudicots (Mur et al. 2004). *Bdpin1*, a serine proteinase inhibitor, was shown to be rapidly induced by both *Magnaporthe grisea* infection and physical wounding as part of larger local and systemic responses. By incorporating basic molecular biology techniques including DNA and RNA blots, cosmid library screenings, and GUS reporter assays (constructs expressed transiently via bombardment in Brachypodium, rice, wheat, and maize and stably through the creation of transgenic tobacco), Mur et al. (2004) provide an example of using all of these models to show the feasibility of expressing a gene in agronomically important cereals for phenotypic and molecular characterization. This work indicates that similar local wound responses occur in both monocot and dicot models upon challenge and mechanical damage, but there are dissimilar systemic responses. These results support other continuing work on *cis*-elements which may have diverged between the two angiosperm classes (Wang et al. 2009). Similar studies have explored the characterization of Brachypodium lipid profiles (Islam et al. 2012); alternative splicing landscapes (Mandadi and Scholthof 2015); wound-, endosperm-, and pollen-specific promoters (Coussens et al. 2012); and WRKY transcription factors identified by bioinformatics approaches (Mandadi and Scholthof 2012; Tripathi et al. 2012). These outcomes suggest that Brachypodium, following from Arabidopsis, can be used for the deliberate identification and characterization of agronomically-important genes and traits from a tractable genetic system while specifically addressing studies of pathogen interactions and disease.

Brachypodium Virology: The Research Triptych

Virologists are keenly interested in the identification, molecular characterization, and deployment of resistance genes for virus control in the field; re-engineering plant viruses as tools for gene delivery in plants; and dissecting the molecular interplay between the virus and the host in the cellular milieu. Here, we focus on several interactions between Brachypodium and plant viruses—our area of specialization—and the interplay of basic, translational, and field research. For this, we will discuss *Panicum mosaic virus* (PMV) (Mandadi et al. 2014; Mandadi and Scholthof 2012, 2013, 2015), *Barley stripe mosaic virus* (BSMV) (Cui et al. 2012; Lee et al. 2012), and *Bamboo mosaic virus* (BaMV) (Liou et al. 2014). Our intent here is to demonstrate that it is possible to work deeply within one aspect of the triptych (Fig. 2), while retaining the intent to use Brachypodium as a model organism to integrate and move between these arbitrary boundaries to identify and solve real problems within the broad rubric of systems biology.

Viruses demonstrate the utility of Brachypodium as a model system to explore host and pathogen biology and their interactions. As an example, in 2002, BSMV was constructed as a virus vector as a proof-of-principle tool for gene silencing, using a host-derived phytoene desaturase (PDS) gene for challenge on barley, maize, and wheat. This work was followed-up in 2011, with the construction of a 35*S*-promoter based BSMV construct suitable for agro-infiltration, again using PDS, which showed silencing (revealed by leaf bleaching) of Brachypodium and *N. benthamiana* (Yuan et al. 2011), as well as ginger (family *Zingiberaceae*, a monocotyledonous plant outside the *Poaceae*) (Renner et al. 2009). More recently, *Bamboo mosaic virus* (BaMV; genus *Potexvirus*) and its satellite RNA (satRNA) have both been engineered to efficiency silence Brachypodium (and *N. benthamiana*) using PDS and other genes involved in chlorophyll production or maintenance (Liou et al. 2014). For this, an agro-infiltration system was used, establishing an infection in *Chenopodium quinoa*, followed by passage to Brachypodium leaves.

The use of BSMV as a tool to identify genetic viral resistance in Brachypodium is another example of the interplay between the field and the laboratory (Cui et al. 2012). This particular study made use of well-established strategies used by plant breeders, including screening susceptible and resistant Brachypodium recombinant inbred lines (RIL), identifying SNPs, and scoring insertion and deletion sequences (INDELs). High-throughput genomic sequencing defined the *Bsr1* gene locus within a 23 kbp region of Chromosome 3 (Cui et al. 2012). This region contained five open reading frames including an *R*-gene motif and four genes of unknown function. Interestingly, *Bsr1*-conferred dominant antiviral-resistance is temperature sensitive, with a BSMV-resistance breaking phenotype occurring at >27 °C (Cui et al. 2012). This opens up many new venues for both basic, translational, and applied research to define exactly which plant gene(s) are key to resistance and which virus nucleic acid sequences or protein(s) within the triple gene block (movement and suppressor proteins) are the virus-associated molecular patterns (VAMPs) (Mandadi and Scholthof 2013) or effectors eliciting the resistance response. By identifying the minor genes for resistance in barley and major genes in oats to determine the origins and evolution of resistance, it may be possible to deploy the Brachypodium *Bsr1* in the field in barley towards reducing upwards of 50 % yield losses in a key cereal. Further research on the biology of BSMV may reveal if resistance breaking in the field is associated with flowering—which may provide a window for the virus to enter the nascent seed—or if the *Bsr1* gene protects the embryo from infection, as seed-transmission is a component of the epidemiology of BSMV. It also is worthwhile to compare and contrast these findings with those of Arabidopsis for characterization and identification of the diversity of host resistance genes (Ouibrahim and Caranta 2013).

In addition to BSMV and BaMV (and its satRNA), Brachypodium is also a host for PMV and its satellite virus (SPMV), *Brome mosaic virus*, Sorghum yellow banding virus, Maize mild mottle virus, *Foxtail mosaic virus*, and *Wheat streak mosaic virus* (Mandadi et al. 2014; Mandadi and Scholthof 2012, 2013, 2015). We have shown that there are generally conserved innate immune responses shared between Brachypodium, Arabidopsis, and *N. benthamiana* during plant virus infections (Mandadi and Scholthof 2013). PMV and SPMV also affect host mRNA splicing, which we hypothesize is achieved by localization of the capsid proteins to the nucleus (PMV CP) and nucleolus (SPMV CP) (Mandadi and Scholthof 2015). The alternative splicing landscapes altered during a Brachypodium-Panicovirus interaction have revealed evidence of conserved splicing patterns among monocots and dicots (Mandadi and Scholthof 2015). This work closely follows from evidence in Arabidopsis that cellular stress affects splicing of a spliceosome protein, SCL33 (Reddy and Shad 2011; Thomas et al. 2012), which we found also occurs in Brachypodium in response to PMV and PMV + SPMV infections (Mandadi and Scholthof 2015). Although there are no near-term outcomes based on this fundamental work, it will be fascinating to determine the similarities and differences of host splicing mechanisms between these two plants that last shared a common ancestor ca. 150 M years ago.

An exemplar of field (applied) virus and grass research informing laboratory research over the course of several decades is given in the history of endemic PMV infections of switchgrass (*Panicum virgatum*) and St. Augustinegrass (*Stenotaphrum secundatum*) in the southern and midwestern regions of the United States. PMV infections were described in 1957, having persisted in switchgrass nurseries in Kansas for more than 5 years (Sill and Pickett 1957; Sill and Desai 1960). The disease susceptibility ratings showed a range of host:virus outcomes—from severe necrosis and no seed set to apparent immunity. The virus was infectious on eight species of *Panicum* and three species of *Setaria*, including *S. italica* (foxtail millet) by rub-inoculation. By 1962, Sill and co-workers had reported PMV "occurs naturally in [switchgrass] an important native perennial grass" (Sill and Talens 1962). In switchgrass, the most florid PMV symptoms occurred during the summer months, with daytime temperatures of 25–36 °C, suggesting the virus was uniquely adapted to the environment and PMV is an "endemic virus in the Great Plains States" (Sill and Talens 1962). Also in 1962, PMV was identified in St. Augustinegrass in several counties in south Texas. PMV moved rapidly through the Gulf Coast states, infecting millions of acres of St. Augustinegrass, an economically important turf- and forage-grass. To combat this epidemic, St. Augustinegrass lines were mutagenized with ^{60}Co, selected as "immune" or "resistant," and used by commercial growers to produce sod and sprigs for residential lawns. As there were no genetics for either the virus or the host plants and limited physicochemical characterizations of PMV and SPMV (Niblett and Paulsen 1975), defining "resistance" would prove to be problematic. In the late-1980s the Panicovirus complex was brought into the lab for reverse genetic studies (Scholthof 1999; Turina et al. 1998, 2000; Masuta et al. 1987). Yet, the lack of a model plant greatly delayed fundamental and translational work on host interactions. With the release of the Brachypodium draft genome in 2010, it became possible to model host:virus interactions at the phenotypic as well as omics levels, largely by comparison with Arabidopsis (Mandadi et al. 2014; Mandadi and Scholthof 2012, 2013, 2015).

In 2012, PMV once again became an object of field investigation. Disease surveys of switchgrass breeding nurseries in Nebraska revealed that PMV was the culprit behind an influx of virus infections and that further pathosystem work with Brachypodium had application. With U.S. interests in developing feedstocks for bioenergy—with one such priority being the development and use of switchgrass to meet this need—suddenly PMV was a potential threat to production of this perennial grass (Stewart et al. 2013, 2015). Following the field survey (applied research), we transitioned to the laboratory, using Brachypodium to understand the molecular biology of the virus and host following infection. Our ongoing research aims to determine if there are conserved as well as unique host antiviral immune responses in Brachypodium (Mandadi et al. 2014; Mandadi and Scholthof 2012, 2015) that may prove useful for rapid selection of *Poaceae* genes for resistance to several viruses in addition to PMV (Stewart et al. 2013, 2015). Thus, Brachypodium has become an agent of discovery across research platforms. And, we suggest Brachypodium is a unique model organism—being able to "do" work in the cloud, the laboratory, and in the field—work ranging from fundamental host:virus genetics to virus epidemiology (Lyons and Scholthof 2015; Garvin 2007) (Fig. 1).

Future research advances using Brachypodium will arguably be predicated on a foundation of big data, omics, and the laboratory reagents and protocols developed around Arabidopsis, maize, rice, and other crop plants (Fig. 2). As an established research tool, Brachypodium is being used to determine physiological (Vogel 2008; Buanafina 2009) and molecular responses to changing environmental conditions such as temperature and drought stress, waterlogging (Shiono and Yamada 2014), nutrient availability (Ingram et al. 2012), salinity, and herbicides (Matzrafi et al. 2014). Using high-throughput systems, such as the newly developed GrainScan for seed size and colorimetric analysis (Whan et al. 2014) and other semi-automated imaging techniques to study root system architecture (Ingram et al. 2012) and physical phenotypic characteristics that may translate well to applied needs within bioenergy, feed, and forage crop breeding and germplasm selection, Brachypodium—but more correctly, the knowledge gained from using it for study—is moving beyond the translational scene of the laboratory and into the field.

Beyond the Lab and Field: Community Building

Since 1991, a strong effort put forth by the Arabidopsis community has produced in an enviable set of resources, most valuable of which is the Arabidopsis Biological Resource Center (ABRC) hosted by Ohio State University—a concrete example of the ethics of sharing (see http://abrc.osu.edu/files/Ethics_Sharing_Resources_v1.pdf). Sharing includes donations, distribution, and maintenance of viable seed. The ABRC estimates that upwards of one-million mutant plant lines are available to investigators worldwide for a small fee. Even with the successes, the disparity is evident—the majority of donations are from USA groups, while the majority of users come from China—clearly this culture of sharing is fragile. That said, if Brachypodium is to continue having the research status of being a basic, translational, and applied grass for cereals, then the creation of a stock center—with sustained funding from a federal agency and a host university—should be a priority. Currently there is a loose-knit community of providers of seed, materials (T-DNA insertion lines), protocols, and bioinformatic resources (Brachypodium.org), etc.—the development of which was all built (fortunately) on the expertise of early Brachypodium adopters coming from the Arabidopsis community, as we have discussed previously (Lyons and Scholthof 2015).

Sharing becomes increasingly important in light of reduced availability of competitive-funding for plant science research overlaid with developing (oftentimes expensive and labor-intensive) key tools for model organism use. Although Brachypodium was developed as a tool for crop improvement vis-à-vis laboratory, translational, and field experimentation, it is unlikely that grower organizations will fund fundamental, long-term research. Industry benefaction is a potential route, but it may reduce the aspect of sharing resources. These issues might be worth further discussion, perhaps with the ABRC, as well as at meetings attended by Brachypodium users and enthusiasts. We are fortunate that the Arabidopsis workers, and to a similar extent the Maize Genetics Cooperation Stock Center (under the auspices of the USDA and the University of Illinois-Urbana), have provided a template and a menu of options towards developing and funding a Brachypodium Resource Center.

The Post-Genomics Era?

The original concept of a single plant model to represent the kingdom *Plantae*, seen in the use and success of Arabidopsis, has been challenged, at least to a limited extent, with the development of other plants towards answering specific questions and meeting convergent, well-defined goals within the larger plant biology, plant pathology, and crop science communities (Lyons and Scholthof 2015). When looking at the sheer number of Arabidopsis publications in the past decade it is clear that financial support and reagent-sharing communities have been key to its successes. Yet as plant pathologists, we find Arabidopsis limited in its ability to transition to problem-solving in the field. And, today, our most urgent concerns dictate a shift in focus to the *Poaceae* with its plethora of issues seen affecting cereals, feed, forage, bioenergy, and recreational grasses. Areas of research on wild and cultivated grasses include ongoing intractable needs to strengthen root architecture, adapt to abiotic and biotic stresses, and address nutrient uptake and deficiencies. These real problems can be recapitulated in the lab using Brachypodium. With concurrent field studies, we may be able to more rapidly advance the translational utility of this model species than if maize, rice, or sorghum were used instead.

The Brachypodium community has made amazing progress in the past decade, through the deliberate construction of a toolbox of resources and program-leading experiments that will drive the next era of research. Yet it is, we suggest, a critical scientific period in which the Brachypodium community needs to focus-on and show the utility of Brachypodium as a translational tool for crop improvement. Of course, fundamental plant science research will continue to compare and contrast Brachypodium with Arabidopsis and establish the differences between those two plants—one being a monocot and one being a dicot—likely leading to new conceptual breakthroughs. But, we are absolutely dependent on grasses for our survival—we consume grains, more so than Brussels sprouts—and from this, Brachypodium is now poised to be the "working grass hero" (Garvin 2007) to rationally develop agricultural research programs to improve grain, forage, and energy grasses in the twenty-first century.

Acknowledgments We appreciate the helpful suggestions provided by Herman Scholthof and Kranthi K. Mandadi during manuscript preparation. This research was funded in part by a Science and Technology Studies grant from the National Science Foundation (NSF 1456878) awarded to K.-B.G.S.

References

Ayliffe M, Singh D, Park R, Moscou M, Pryor T. Infection of *Brachypodium distachyon* with selected grass rust pathogens. Mol Plant Microbe Interact. 2013;26:946–57.

Bablak P, Draper J, Davey MR, Lynch PT. Plant regeneration and micropropagation of *Brachypodium distachyon*. Plant Cell Tiss Org Cult. 1995;42:97–107.

Bragg JN, Wu J, Gordon SP, Guttman ME, Thilmony R, Lazo GR, et al. Generation and characterization of the Western Regional Research Center Brachypodium T-DNA insertional mutant collection. PLoS ONE. 2012;7:e41916.

Buanafina MMO. Feruloylation in grasses: current and future perspectives. Mol Plant. 2009;2:861–72.

Catalan P, Chalhoub B, Chochois V, Garvin DF, Hasterok R, Manzaneda AJ, et al. Update on the genomics and basic biology of *Brachypodium*. International Brachypodium Initiative (IBI). Trends Plant Sci. 2014;19:414–8.

Coussens G, Aesaert S, Verelst W, Demeulenaere M, De Buck S, Njuguna E, et al. *Brachypodium distachyon* promoters as efficient building blocks for transgenic research in maize. J Exp Bot. 2012;63:4263–73.

Cui Y, Lee MY, Huo N, Bragg J, Yan L, Yuan C, et al. Fine mapping of the *Bsr1 Barley stripe mosaic virus* resistance gene in the model grass *Brachypodium distachyon*. PLoS ONE. 2012;7: e38333.

Draper J, Mur LAJ, Jenkins G, Ghosh-Biswas GC, Bablak P, Hasterok R, et al. *Brachypodium distachyon*: a new model system for functional genomics in grasses. Plant Physiol. 2001;127:1539–55.

Garvin D. *Brachypodium distachyon*: a new model system for structural and functional analysis of grass genomes. Model plants and crop improvement. Boca Raton, FL: CRC Press; 2007a. p. 109–23.

Garvin DF. Brachypodium: a new monocot model plant system emerges. J Sci Food Agric. 2007b;87:1177–9.

Gordon SP, Priest H, Des Marais DL, Schackwitz W, Figueroa M, Martin J, et al. Genome diversity in *Brachypodium distachyon*: deep sequencing of highly diverse inbred lines. Plant J. 2014;79:361–74.

Gressel J, Kleifeld Y. Can wild species become problem weeds because of herbicide resistance? *Brachypodium distachyon*: a case study. Crop Prot. 1994;13:563–6.

Gressel J, Regev Y, Malkin S, Kleifeld Y. Characterization of an *s*-triazine-resistant biotype of *Brachypodium distachyon*. Weed Sci. 1983;31:450–6.

Hasterok R, Draper J, Jenkins G. Laying the cytotaxonomic foundations of a new model grass, *Brachypodium distachyon* (L.) Beauv. Chromosome Res. 2004;12:397–403.

Heenan PB, Mitchell AD, Koch M. Molecular systematics of the New Zealand *Pachycladon* (*Brassicaceae*) complex: generic circumscription and relationships to *Arabidopsis* sens. lat. and *Arabis* sens. lat. New Zeal J Bot. 2002;40:543–62.

Ingram PA, Zhu J, Shariff A, Davis IW, Benfey PN, Elich T. High-throughput imaging and analysis of root system architecture in *Brachypodium distachyon* under differential nutrient availability. Philos Trans R Soc B. 2012;367:1559–69.

International Brachypodium Initiative. Genome sequencing and analysis of the model grass *Brachypodium distachyon*. Nature. 2010;463:763–8.

Islam MN, Chambers JP, Ng CK-Y. Lipid profiling of the model temperate grass, *Brachypodium distachyon*. Metabolomics. 2012;8:598–613.

Khan MA, Stace CA. Breeding relationships in the genus *Brachypodium* (*Poaceae: Pooideae*). Nord J Bot. 1999;19:257–69.

Koch M, Bishop J, Mitchell-Olds T. Molecular systematics and evolution of *Arabidopsis* and *Arabis*. Plant Biol. 1999;1:529–37.

Koch MA, Haubold B, Mitchell-Olds T. Comparative evolutionary analysis of chalcone synthase and alcohol dehydrogenase loci in *Arabidopsis*, *Arabis*, and related genera (*Brassicaceae*). Mol Biol Evol. 2000;17:1483–98.

Lee M, Yan L, Gorter F, Kim B, Cui Y, Hu Y, et al. *Brachypodium distachyon* line Bd3-1 resistance is elicited by the *Barley stripe mosaic virus* triple gene block 1 movement protein. J Gen Virol. 2012;93:2729–39.

Leonelli S. Growing weed, producing knowledge: an epistemic history of *Arabidopsis thaliana*. Hist Philos Life Sci. 2007a;29:193–223.

Leonelli S. Weed for thought: using *Arabidopsis thaliana* to understand plant biology. PhD dissertation. Amsterdam, the Netherlands: Vrije Universiteit; 2007b.

Li SL, Rédei GP. Thiamine mutants of the crucifer, Arabidopsis. Biochem Genet. 1969;3:163–70.

Liou M-R, Huang Y-W, Hu C-C, Lin N-S, Hsu Y-H. A dual gene-silencing vector system for monocot and dicot plants. Plant Biotechnol J. 2014;12:330–43.

López-Alvarez D, López-Herranz ML, Betekhtin A, Catalán P. A DNA barcoding method to discriminate between the model plant *Brachypodium distachyon* and its close relatives *B. stacei* and *B. hybridum* (*Poaceae*). PLoS ONE. 2012;7:e51058.

Lyons CWP, Scholthof K-BG. Watching grass grow: the emergence of *Brachypodium distachyon* as a model for the *Poaceae*. In: Phillips D, Kingsland S, editors. Archimedes: new perspectives on the history of life sciences and agriculture, vol. 40. New York, NY: Springer; 2015. p. 479–501.

Mandadi KK, Scholthof K-BG. Characterization of a viral synergism in the monocot *Brachypodium distachyon* reveals distinctly altered host molecular processes associated with disease. Plant Physiol. 2012;160:1432–52.

Mandadi KK, Scholthof K-BG. Plant immune responses against viruses: how does a virus cause disease? Plant Cell. 2013;25:1489–505.

Mandadi KK, Scholthof K-BG. Genome-wide analysis of alternative splicing landscapes modulated during plant-virus interaction in *Brachypodium distachyon*. Plant Cell. 2015;27:1–15.

Mandadi KK, Pyle JD, Scholthof K-BG. Comparative analysis of antiviral responses in *Brachypodium distachyon* and *Setaria viridis* reveals conserved and unique outcomes among C_3 and C_4 plant defenses. Mol Plant Microbe Interact. 2014;27:1277–90.

Masuta C, Zuidema D, Hunter BG, Heaton LA, Sopher DS, Jackson AO. Analysis of the genome of satellite panicum mosaic virus. Virology. 1987;159:329–38.

Matzrafi M, Gadri Y, Frenkel E, Rubin B, Peleg Z. Evolution of herbicide resistance mechanisms in grass weeds. Plant Sci. 2014;229:43–52.

Meyerowitz EM. *Arabidopsis thaliana*. Annu Rev Genet. 1987;21:93–111.

Moore G, Gale MD, Kurata N, Flavell RB. Molecular analysis of small grain cereal genomes: current status and prospects. Nat Biotechnol. 1993;11:584–9.

Mur L, Xu R, Casson S, Stoddart W, Routledge A, Draper J. Characterization of a proteinase inhibitor from *Brachypodium distachyon* suggests the conservation of defence signalling pathways between dicotyledonous plants and grasses. Mol Plant Pathol. 2004;5:267–80.

Niblett CL, Paulsen AQ. Purification and further characterization of panicum mosaic virus. Phytopathology. 1975;65:1157–60.

Nishimura MT, Dangl JL. Arabidopsis and the plant immune system. Plant J. 2010;61:1053–66.

Ouibrahim L, Caranta C. Exploitation of natural genetic diversity to study plant–virus interactions: what can we learn from *Arabidopsis thaliana*? Mol Plant Pathol. 2013;14:844–54.

Pacurar DI. Turning a wild plant into a model–a déjà vu story. Not Bot Horti Agrobo. 2009;37:17–24.

Parker D, Beckmann M, Enot DP, Overy DP, Rios ZC, Gilbert M, et al. Rice blast infection of *Brachypodium distachyon* as a model system to study dynamic host/pathogen interactions. Nat Protoc. 2008;3:435–45.

Priya R, Siva R. Phylogenetic analysis and evolutionary studies of plant carotenoid cleavage dioxygenase gene. Gene. 2014;548:223–33.

Reddy ASN, Shad AG. Plant serine/arginine-rich proteins: roles in precursor messenger RNA splicing, plant development, and stress responses. Wiley Interdiscip Rev RNA. 2011;2:875–89.

Rédei GP, Li SL. Effects of X rays and ethyl methanesulfonate on the chlorophyll *b* locus in the soma and on the thiamine loci in the germline of Arabidopsis. Genetics. 1969;61:453–9.

Renner T, Bragg J, Driscoll HE, Cho J, Jackson AO, Specht CD. Virus-induced gene silencing in the culinary ginger (*Zingiber officinale*): an effective mechanism for down-regulating gene expression in tropical monocots. Mol Plant. 2009;2:1084–94.

Sandoya GV, de O Buanafina MM. Differential responses of *Brachypodium distachyon* genotypes to insect and fungal pathogens. Physiol Mol Plant Pathol. 2014;85:53–64.

Scholthof K-BG. A synergism induced by satellite panicum mosaic virus. Mol Plant Microbe Interact. 1999;12:163–6.

Shi Y. Molecular studies of the evolutionary relationships of Brachypodium (Poaceae). PhD dissertation. Botany, United Kingdom: University of Leicester; 1991.

Shiono K, Yamada S. Waterlogging tolerance and capacity for oxygen transport in *Brachypodium distachyon* (Bd21). Plant Root. 2014;8:5–12.

Sill Jr WH, Desai MK. Additional characteristics of panicum mosaic virus. Plant Dis Rep. 1960;44:487.

Sill WH, Pickett RC. A new virus disease of switchgrass, *Panicum virgatum* L. Plant Dis Rep. 1957;41:241–9.

Sill WH, Talens LT. New hosts and characteristics of the panicum mosaic virus. Plant Dis Rep. 1962;46:780–3.

Somerville C, Koornneef M. A fortunate choice: the history of Arabidopsis as a model plant. Nat Rev Genet. 2002;3:883–9.

Stewart CL, Yuen GY, Vogel K, Pyle JD, Scholthof K-BG. *Panicum mosaic virus*: a potential threat to biofuel switchgrass production. Phytopathology. 2013;103:S2.138.

Stewart CL, Pyle JD, Jochum CC, Vogel KP, Yuen GY, Scholthof K-BG. Multi-year pathogen survey of biofuel switchgrass breeding nurseries reveals high incidence of infections by Panicum mosaic virus and its satellite virus. Phytopathology. 2015;105:1146–54.

Tateoka T. A cytological study of some Mexican grasses. B Torrey Bot Club. 1962;89:77–82.

The Arabidopsis Genome Initiative. Analysis of the genome sequence of the flowering plant *Arabidopsis thaliana*. Nature. 2000;408:796–815.

The Multinational Arabidopsis Steering Committee. The Multinational Coordinated *Arabidopsis thaliana* Functional Genomics Project. Report. 2002. http://www.nsf.gov/pubs/2002/bio0202/functional.htm.

Thomas J, Palusa SG, Prasad KVSK, Ali GS, Surabhi G-K, Ben-Hur A, et al. Identification of an intronic splicing regulatory element involved in auto-regulation of alternative splicing of SCL33 pre-mRNA. Plant J. 2012;72:935–46.

Tripathi P, Rabara RC, Langum TJ, Boken AK, Rushton DL, Boomsma DD, et al. The WRKY transcription factor family in *Brachypodium distachyon*. BMC Genomics. 2012;13:270.

Turina M, Maruoka M, Monis J, Jackson AO, Scholthof K-BG. Nucleotide sequence and infectivity of a full-length cDNA clone of panicum mosaic virus. Virology. 1998;241:141–55.

Turina M, Desvoyes B, Scholthof KB. A gene cluster encoded by panicum mosaic virus is associated with virus movement. Virology. 2000;266:120–8.

Vallabhaneni R, Bradbury LMT, Wurtzel ET. The carotenoid dioxygenase gene family in maize, sorghum, and rice. Arch Biochem Biophys. 2010;504:104–11.

Vogel J. Unique aspects of the grass cell wall. Curr Opin Plant Biol. 2008;11:301–7.

Wang B, Sang Y, Song J, Gao X-Q, Zhang X. Expression of a rice *OsARGOS* gene in *Arabidopsis* promotes cell division and expansion and increases organ size. J Genet Genomics. 2009;36:31–40.

Warwick SI, Al-Shehbaz IA, Sauder CA. Phylogenetic position of *Arabis arenicola* and generic limits of *Aphragmus* and *Eutrema* (*Brassicaceae*) based on sequences of nuclear ribosomal DNA. Can J Bot. 2006;84:269–81.

Whan AP, Smith AB, Cavanagh CR, Ral J-PF, Shaw LM, Howitt CA, et al. GrainScan: a low cost, fast method for grain size and colour measurements. Plant Methods. 2014;10:23.

Xu Y, Zhang X, Li Q, Cheng Z, Lou H, Ge L, et al. BdBRD1, a brassinosteroid C-6 oxidase homolog in *Brachypodium distachyon* L., is required for multiple organ development. Plant Physiol Biochem. 2015;86:91–9.

Yuan C, Li C, Yan L, Jackson AO, Liu Z, Han C, et al. A high throughput *Barley stripe mosaic virus* vector for virus induced gene silencing in monocots and dicots. PLoS ONE. 2011;6:e26468.

Zwan CV, Brodie SA, Campanella JJ. The intraspecific phylogenetics of *Arabidopsis thaliana* in worldwide populations. Syst Bot. 2000;25:47–59.

Index

J. Vogel (ed.), *Genetics and Genomics of Brachypodium*,
Plant Genetics and Genomics: Crops and Models, DOI 10.1007/7397_2016

H

N

O

P

U

MIX
Papier aus verantwortungsvollen Quellen
Paper from responsible sources
FSC® C105338
FSC
www.fsc.org

If you have any concerns about our products,
you can contact us on
ProductSafety@springernature.com

In case Publisher is established outside the EU,
the EU authorized representative is:
Springer Nature Customer Service Center GmbH
Europaplatz 3, 69115 Heidelberg, Germany

Printed by Libri Plureos GmbH
in Hamburg, Germany